AN ANALYTICAL CALCULUS

VOLUME II

AN ANALYTICAL CALCULUS

FOR SCHOOL AND UNIVERSITY

BY

E. A. MAXWELL

Fellow of Queens' College, Cambridge

VOLUME II

CAMBRIDGE

AT THE UNIVERSITY PRESS

1966

CAMBRIDGE UNIVERSITY PRESS
Cambridge, New York, Melbourne, Madrid, Cape Town, Singapore, São Paulo, Delhi

Cambridge University Press
The Edinburgh Building, Cambridge CB2 8RU, UK

Published in the United States of America by Cambridge University Press, New York

www.cambridge.org
Information on this title: www.cambridge.org/9780521056977

First published 1954
Reprinted 1966
This digitally printed version 2008

A catalogue record for this publication is available from the British Library

ISBN 978-0-521-05697-7 hardback
ISBN 978-0-521-09036-0 paperback

CONTENTS

Chapter X: Curves—continued

CHAPTER XI: COMPLEX NUMBERS

CHAPTER XII: SYSTEMATIC INTEGRATION

CHAPTER XIII: INTEGRALS INVOLVING 'INFINITY'

PREFACE

Appreciation for help received was expressed in the Preface to Volume I, but I would record how much deeper my indebtedness becomes as the work progresses.

E. A. M.

QUEENS' COLLEGE, CAMBRIDGE
June, 1953

CHAPTER VII

THE LOGARITHMIC AND EXPONENTIAL FUNCTIONS

THE particular functions which we have used in the earlier chapters (Volume I) are the powers of x, the ordinary trigonometric functions, and combinations of them such as polynomials.

We now introduce an entirely new function, the *logarithm*. The need for it arises, for example, when we seek to evaluate the integral

$$\int x^n \, dx$$

for $n = -1$. The standard formula

$$\int x^n \, dx = \frac{1}{n+1} x^{n+1}$$

becomes meaningless; the integral cannot be evaluated in terms of the functions at present at our disposal.

1. The logarithm. Consider the integral

$$\int \frac{dx}{x}.$$

To make the discussion precise, we shall fix the lower limit, giving it the value unity; the effect of this is merely to remove ambiguity about the arbitrary constant. The integral is a function of its upper limit, which we denote by the letter x, replacing the variable in the integration by the letter t. The function is thus

$$f(x) \equiv \int_1^x \frac{dt}{t},$$

where (Vol. I, p. 87) $$f'(x) = \frac{1}{x}.$$

The function defined in this way is called the *logarithm* of x, usually written

$$\log x$$

or

$$\log_e x,$$

the suffix e being inserted for reasons to be given later (p. 27). Thus

$$\log x = \int_1^x \frac{dt}{t}.$$

2. First properties of the logarithm. We now prove some of the basic properties to which the logarithm owes its importance. The reader will note the very close connexion with 'logarithms to the base 10', with which he is presumably familiar.

(i) $$\log 1 = 0.$$

This follows immediately, since (Vol. I, p. 83)

$$\int_1^1 \frac{dt}{t} = 0.$$

(ii) $$\log xy = \log x + \log y.$$

For

$$\log xy = \int_1^{xy} \frac{dt}{t}$$

$$= \int_1^x \frac{dt}{t} + \int_x^{xy} \frac{dt}{t} \quad \text{(Vol. I, p. 83).}$$

Now use the substitution

$$t = xu$$

in the latter integral. We have the relation

$$dt = x\,du$$

(remembering that x is CONSTANT here, the variable of integration being t). Also the values x, xy of t correspond to the values $1, y$ of u. Hence

$$\int_x^{xy} \frac{dt}{t} = \int_1^y \frac{x\,du}{xu} = \int_1^y \frac{du}{u}$$

$$= \log y.$$

We therefore have the required relation

$$\log xy = \log x + \log y.$$

COROLLARY. $\log(x/y) = \log x - \log y$.

For
$$\log x = \log\left\{\left(\frac{x}{y}\right)y\right\}$$
$$= \log(x/y) + \log y.$$

(iii) $$\log(x^n) = n \log x.$$

In the relation
$$\log x = \int_1^x \frac{dt}{t},$$
make the substitution
$$u = t^n.$$
We have the relation
$$du = nt^{n-1}dt.$$
Also the values $1, x$ of t correspond to the values $1, x^n$ of u. Therefore, since
$$\frac{1}{t} = \frac{nt^{n-1}}{nt^n},$$
we have
$$\log x = \int_1^x \frac{nt^{n-1}dt}{nt^n}$$
$$= \int_1^{x^n} \frac{du}{nu} \text{ (applying the substitution)}$$
$$= \frac{1}{n}\log(x^n),$$
so that
$$\log(x^n) = n \log x.$$

Note. n may have any real value, and is not necessarily a positive integer.

(iv) *The value of* $\log x$ *increases indefinitely as* x *does.*

Suppose that N is any large number and m the largest integer such that $2^m < N$. Then
$$\log N = \int_1^2 \frac{dt}{t} + \int_2^4 \frac{dt}{t} + \int_4^8 \frac{dt}{t} + \ldots + \int_{2^{m-1}}^{2^m} \frac{dt}{t} + \int_{2^m}^N \frac{dt}{t}.$$

Now consider
$$\int_{2^{p-1}}^{2^p} \frac{dt}{t}.$$
Throughout the interval $(2^{p-1}, 2^p)$, the variable t is less than 2^p, so that $\frac{1}{t}$ exceeds $\frac{1}{2^p}$. Hence, by the basic definition of an integral (Vol. I, p. 81),
$$\int_{2^{p-1}}^{2^p} \frac{dt}{t} > \int_{2^{p-1}}^{2^p} \frac{dt}{2^p} = \frac{1}{2^p}\Big[t\Big]_{2^{p-1}}^{2^p} = \frac{2^p - 2^{p-1}}{2^p} = \frac{1}{2}.$$

Applying this inequality to the successive integrals in the formula for $\log N$, and noting that $\int_{2^m}^{N} \frac{dt}{t}$ is positive, we obtain the relation

$$\log N > \tfrac{1}{2} + \tfrac{1}{2} + \tfrac{1}{2} + \ldots + \tfrac{1}{2} \quad (m \text{ terms}),$$

so that $\log N$, which exceeds $\frac{1}{2}m$, increases without bound.

COROLLARY. *The value of $\log x$ tends to* MINUS *infinity as x tends to zero.*

For
$$\log\left\{\left(\frac{1}{N}\right)N\right\} = \log 1 = 0,$$

so that
$$\log\left(\frac{1}{N}\right) + \log N = 0,$$

or
$$\log\left(\frac{1}{N}\right) = -\log N.$$

As N tends to infinity, $1/N$ tends to zero, and the result follows.

3. The graph of $\log x$.

If
$$y = \log x,$$

then
$$\frac{dy}{dx} = \frac{1}{x}.$$

Hence $\frac{dy}{dx}$ is positive for all positive values of x, so that $\log x$ *is a steadily increasing function of x for positive x* (Fig. 62). Also the gradient $\frac{1}{x}$ is large and positive when x is small and positive, decreases as x increases, taking the value 1 when $x = 1$, and tends to the value zero as x increases indefinitely. Moreover, as above, y itself tends to 'minus infinity' when x tends to zero, increases with x, taking the value 0 when $x = 1$, and 'tends to infinity' as x increases indefinitely.

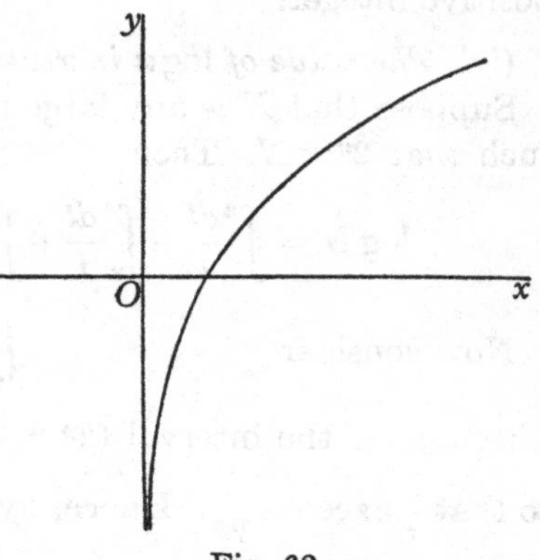

Fig. 62.

The general shape is therefore that shown in the diagram.

The value $x = 0$ imposes a downward barrier on the logarithm, and *the function* $\log x$ *is undefined for negative values of* x.

Note. If $x_1 = x_2$, being positive, then $\log x_1 = \log x_2$; and, what is more important, the converse property holds, that, *if* $\log x_1 = \log x_2$, *then* $x_1 = x_2$. In fact, if $x_1 > x_2$, then $\log x_1 > \log x_2$, since the logarithm is an increasing function; and if $x_1 < x_2$, then $\log x_1 < \log x_2$.

It is also clear from the graph that, *if* c *is a given number, then the relation*

$$\log x = c$$

defines x *uniquely.*

WARNING. The value of the integral

$$\int_{-2}^{-3} \frac{dt}{t}$$

is fully determinate, but we cannot use the argument:

$$\int_{-2}^{-3} \frac{dt}{t} = \Big[\log t\Big]_{-2}^{-3} = \log(-3) - \log(-2)$$

$$= \log\left(\frac{-3}{-2}\right) = \log(3/2),$$

since $\log(-3)$ and $\log(-2)$ are non-existent. We must proceed as follows:

Substitute $$t = -u,$$

so that $$dt = -du.$$

Then $$\int_{-2}^{-3} \frac{dt}{t} = \int_{2}^{3} \frac{-du}{-u} = \int_{2}^{3} \frac{du}{u}$$

$$= \Big[\log u\Big]_{2}^{3} = \log(3/2).$$

UNDER NO CIRCUMSTANCES may we evaluate an integral such as

$$\int_{-2}^{+3} \frac{dt}{t}$$

where the variable of integration t runs through the value zero at which $(1/t)$ has no meaning.

We give two typical examples to show how logarithms arise in physical applications.

ILLUSTRATION 1. *An electric circuit contains a resistance R, a coil of self-inductance L, and a battery of electromotive force E, supposed constant. To find an equation to determine the current t seconds after a switch in the circuit has been closed.*

The equation for the current x is known to be

$$L\frac{dx}{dt}+Rx=E,$$

so that $$L\frac{dx}{dt}=E-Rx.$$

Hence the differentials dt, dx are connected by the relation

$$dt=\frac{Ldx}{E-Rx},$$

and so $$t=L\int\frac{dx}{E-Rx}.$$

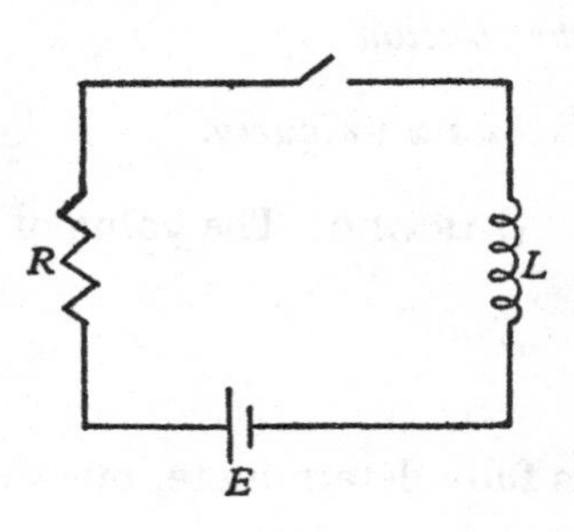

Fig. 63.

[*Note.* We are changing from the conception of x as a function of t to that of t as a function of x.]

The value of the integral may be written down at once, but the beginner may prefer to use the substitution

$$E-Rx=u,$$

so that $$-Rdx=du,$$

giving $$t=-\frac{L}{R}\int\frac{du}{u}=-\frac{L}{R}\log u+C,$$

assuming that we are dealing with a case in which u is positive.

Hence $$t=-\frac{L}{R}\log(E-Rx)+C,$$

where C is an arbitrary constant. Now $x=0$ when $t=0$, and so

$$0=-\frac{L}{R}\log E+C,$$

or $$C=\frac{L}{R}\log E.$$

Hence
$$t = \frac{L}{R}\{\log E - \log(E - Rx)\}$$
$$= \frac{L}{R}\log\frac{E}{E-Rx}.$$

An alternative method for dealing with the relation $L\frac{dx}{dt} + Rx = E$ will be given later (p. 22).

ILLUSTRATION 2. *To find the work done when a given quantity of a perfect gas expands from volume v_1 to volume v_2 at constant absolute temperature T.*

It is known that the volume v and pressure p are connected by the relation
$$pv = RT,$$
where R is constant. Also the work done is known to be
$$W \equiv \int_{v_1}^{v_2} p\,dv.$$

Hence
$$W = \int_{v_1}^{v_2} \frac{RT\,dv}{v}$$
$$= RT\Big[\log v\Big]_{v_1}^{v_2}$$
$$= RT(\log v_2 - \log v_1)$$
$$= RT\log(v_2/v_1).$$

The following illustrations are typical of integrals involving logarithms.

ILLUSTRATION 3. *To find*
$$I \equiv \int \frac{4x\,dx}{x^4-1}.$$
We have
$$I = \int\left(\frac{2x}{x^2-1} - \frac{2x}{x^2+1}\right)dx = \log(x^2-1) - \log(x^2+1)$$
$$= \log\left(\frac{x^2-1}{x^2+1}\right).$$

ILLUSTRATION 4. *To find*

$$I \equiv \int \tan x \, dx.$$

Write $$u = \cos x,$$

so that $$du = -\sin x \, dx.$$

Hence $$I = -\int \frac{du}{u} = -\log u = \log(1/u)$$

$$= \log \sec x.$$

ILLUSTRATION 5. *To find*

$$I \equiv \int x^n \log x \, dx \quad (n \neq -1).$$

On integration by parts (Vol. I, p. 103), we have

$$= \frac{x^{n+1}}{n+1} \log x - \int \frac{x^{n+1}}{n+1} \cdot \frac{1}{x} dx$$

$$= \frac{x^{n+1}}{n+1} \log x - \int \frac{x^n}{n+1} dx$$

$$= \frac{x^{n+1}}{n+1} \log x - \frac{x^{n+1}}{(n+1)^2}.$$

ILLUSTRATION 6. *To find*

$$I \equiv \int \frac{dx}{\cos x}.$$

We have $$I = \int \frac{\cos x \, dx}{\cos^2 x} = \int \frac{\cos x \, dx}{1 - \sin^2 x}.$$

Let $$u = \sin x,$$

so that $$du = \cos x \, dx.$$

Then $$I = \int \frac{du}{1-u^2} = \frac{1}{2} \int \left\{ \frac{du}{1-u} + \frac{du}{1+u} \right\}$$

$$= \tfrac{1}{2}\{-\log(1-u) + \log(1+u)\}$$

$$= \tfrac{1}{2} \log \frac{1+u}{1-u}$$

$$= \tfrac{1}{2} \log \frac{1+\sin x}{1-\sin x}.$$

This may also be expressed in the form

$$I = \tfrac{1}{2}\log\frac{(1+\sin x)^2}{1-\sin^2 x}$$

$$= \tfrac{1}{2}\log\left(\frac{1+\sin x}{\cos x}\right)^2 = \log\left(\frac{1+\sin x}{\cos x}\right)$$

$$= \log(\sec x + \tan x).$$

Another form for the answer is

$$I = \log\tan\left(\frac{\pi}{4}+\frac{x}{2}\right).$$

ILLUSTRATION 7.* *To find*

$$I \equiv \int\frac{(4x+7)\,dx}{x^2+4x+13}.$$

Notice that the differential coefficient of the denominator is

$$2x+4,$$

and express the numerator in the form

$$2(2x+4)-1.$$

Then $$I = 2\int\frac{(2x+4)\,dx}{x^2+4x+13}-\int\frac{dx}{x^2+4x+13}$$

$$= 2\int\frac{(2x+4)\,dx}{x^2+4x+13}-\int\frac{dx}{(x+2)^2+9}$$

$$= 2\log(x^2+4x+13)-\tfrac{1}{3}\tan^{-1}\left(\frac{x+2}{3}\right).$$

ILLUSTRATION 8.* *To find*

$$I \equiv \int\frac{(5x+8)\,dx}{x^2-6x+25}.$$

Notice that the differential coefficient of the denominator is

$$2x-6,$$

and express the numerator in the form

$$\tfrac{5}{2}(2x-6)+23.$$

Then $$I = \frac{5}{2}\int\frac{(2x-6)\,dx}{x^2-6x+25}+23\int\frac{dx}{(x-3)^2+16}$$

$$= \frac{5}{2}\log(x^2-6x+25)+\frac{23}{4}\tan^{-1}\left(\frac{x-3}{4}\right).$$

* An important type.

EXAMPLES I

Find the following integrals:

1. $\int \frac{dx}{x+1}$. 2. $\int \frac{dx}{2x+1}$. 3. $\int \frac{dx}{2-3x}$.

4. $\int \frac{x^2+1}{x} dx$. 5. $\int \frac{dx}{x^2-1}$. 6. $\int \left(x^2+\frac{1}{x^3}\right)^2 dx$.

Evaluate the following integrals:

7. $\int_2^4 \frac{dx}{x}$. 8. $\int_{-3}^{-5} \frac{dx}{x+1}$. 9. $\int_0^1 \frac{dx}{3x+2}$.

10. $\int_{-2}^{-3} \frac{dx}{4x+7}$. 11. $\int_3^4 \frac{dx}{x^2-4}$. 12. $\int_1^2 \frac{dx}{9x^2-1}$.

Differentiate the following functions with respect to x:

13. $\log(3x+2)$. 14. $\log \tan x$. 15. $\log \operatorname{cosec} x$.

16. $x^2 \log x$. 17. $x^n \log x$. 18. $\log(1+x^2)$.

Find the following integrals:

19. $\int \log x \, dx$. 20. $\int \frac{\log x}{x} dx$. 21. $\int \cot x \, dx$.

22. $\int x \log x \, dx$. 23. $\int \frac{dx}{\sin x}$. 24. $\int \frac{\cos^2 x}{\sin x} dx$.

25. $\int \frac{(2x+5)\,dx}{x^2+5x+12}$. 26. $\int \frac{(2x-3)\,dx}{x^2-3x+7}$. 27. $\int \frac{(2x+5)\,dx}{x^2-2x+17}$.

28. $\int \frac{(2x-6)dx}{x^2+6x+10}$. 29. $\int \frac{(5x+7)\,dx}{x^2-8x+25}$. 30. $\int \frac{(7x-2)\,dx}{x^2+10x+34}$.

4. The use of logarithms in differentiation. The differentiation of a fraction (in which the numerator and the denominator may themselves be products of factors) is often made easier by the method known as *logarithmic differentiation*, illustrated in the following examples.

ILLUSTRATION 9. *To differentiate the function*

$$y = \frac{x^3(1+x^2)}{(1-x)^4(1+2x)^2}.$$

Take logarithms. Then

$$\log y = 3\log x + \log(1+x^2) - 4\log(1-x) - 2\log(1+2x).$$

Differentiate. Then

$$\frac{1}{y}\frac{dy}{dx} = \frac{3}{x} + \frac{2x}{1+x^2} + \frac{4}{1-x} - \frac{4}{1+2x},$$

and the value of $\frac{dy}{dx}$ follows at once.

With a little practice, the two steps may be taken together:

ILLUSTRATION 10. *To differentiate the function*

$$y = \frac{(1-2x)^2 \sin^3 x}{(1+4x^2)^2}.$$

Take logarithms and differentiate. Then

$$\frac{1}{y}\frac{dy}{dx} = \frac{-4}{1-2x} + \frac{3\cos x}{\sin x} - \frac{16x}{1+4x^2}.$$

EXAMPLES II

Use the method of logarithmic differentiation to differentiate the following functions:

1. $\frac{(1+x)^2}{(1-x)^3}$. 2. $\frac{\cos^2 x}{1+x^2}$. 3. $\frac{x \sin^2 x}{1-2x^3}$.

4. $\frac{x^2(1+x)^2}{(1+x^4)^2}$. 5. $\frac{x \sin x}{(1+x)^3(1-x)}$. 6. $\frac{(1+x^2)^2}{2x \cos^2 x}$.

7. $\frac{5x^4(1-x)^3}{\tan^2 2x}$. 8. $\frac{(1+\cos x)^2}{(1+x+x^2)}$. 9. $\frac{(1-x)(1+2x)^2}{(1-3x)^3(1+4x)^4}$.

5. The use of logarithms in integrating simple rational functions. A *rational function of x* is an expression of the form

$$\frac{u(x)}{v(x)},$$

where $u(x), v(x)$ are *polynomials* in x. We shall later (p. 200) give a detailed treatment of the integration of such functions; here we give a preliminary account of the simpler cases.

If the degree of $u(x)$ is higher than that of $v(x)$, we can divide $u(x)$ by $v(x)$, and obtain an expression of the form

$$p(x)+\frac{w(x)}{v(x)},$$

where $p(x)$ is a polynomial, and $w(x)$ *is a polynomial whose degree is less than that of* $v(x)$.

The integration of the polynomial $p(x)$ is immediate. We may therefore confine our attention to the form

$$\frac{w(x)}{v(x)},$$

where $w(x), v(x)$ are polynomials in x, the degree of $w(x)$ being less than that of $v(x)$. The method is to express this quotient in *partial fractions*; details may be found in a text-book on algebra, but, for convenience, a brief account of the calculations involved is inserted for reference.

In order to explain what is required, we consider some typical examples. (Different mathematicians use varying methods. Those which follow have the advantage of giving independent checks of accuracy in some of the more complicated cases.)

(i) $$f(x)\equiv\frac{2x}{(x-2)(x+3)}.$$

The denominator consists of the two linear factors $(x-2), (x+3)$, each occurring to the first degree only. We seek to express $f(x)$ in the form

$$\frac{A}{x-2}+\frac{B}{x+3}.$$

We have $$\frac{A}{x-2}+\frac{B}{x+3}\equiv\frac{2x}{(x-2)(x+3)}.$$

Multiply throughout by $x-2$. Then

$$A+\frac{B(x-2)}{x+3}\equiv\frac{2x}{x+3}.$$

This holds for all values of x; in particular, for $x = 2$. Then

$$A+0=\frac{2.2}{2+3}=\frac{4}{5}.$$

Hence $$A=\frac{4}{5}.$$

In practice, these steps are usually telescoped, as we now illustrate in finding B. Multiply throughout by $x+3$ and *then* put $x = -3$. Thus

$$B = \frac{2(-3)}{(-3-2)} = \frac{6}{5}$$

Hence
$$f(x) \equiv \frac{4}{5(x-2)} + \frac{6}{5(x+3)}.$$

(ii)
$$f(x) \equiv \frac{x+1}{(x+2)(x-2)^3}.$$

The denominator consists of two linear factors $(x+2)$, $(x-2)$, of which $(x-2)$ occurs to degree 3. We seek to express $f(x)$ in the form

$$\frac{A}{x+2} + \frac{B}{(x-2)^3} + \frac{C}{(x-2)^2} + \frac{D}{x-2},$$

so that
$$\frac{A}{x+2} + \frac{B}{(x-2)^3} + \frac{C}{(x-2)^2} + \frac{D}{x-2} \equiv \frac{x+1}{(x+2)(x-2)^3}.$$

Multiply throughout by $x+2$ and *then* put $x = -2$. Thus

$$A = \frac{-2+1}{(-2-2)^3} = \frac{-1}{-64} = \frac{1}{64}.$$

Multiply throughout by $(x-2)^3$ and *then* put $x = 2$. Thus

$$B = \frac{2+1}{2+2} = \frac{3}{4}.$$

In order to find C, D, we use these values of A, B:

$$\frac{C}{(x-2)^2} + \frac{D}{(x-2)} \equiv \frac{x+1}{(x+2)(x-2)^3} - \frac{1}{64(x+2)} - \frac{3}{4(x-2)^3}$$

$$\equiv \frac{64(x+1) - (x-2)^3 - 48(x+2)}{64(x+2)(x-2)^3}$$

$$\equiv \frac{64x + 64 - x^3 + 6x^2 - 12x + 8 - 48x - 96}{64(x+2)\ (x-2)^3}$$

$$\equiv -\frac{x^3 - 6x^2 - 4x + 24}{64(x+2)(x-2)^3}.$$

At this point, we are able to check accuracy for the highest common factor of the denominators on the left-hand side is $(x-2)^2$.

Hence $x+2$ and $x-2$ MUST be factors of the numerator on the right. By division, we find that

$$x^3-6x^2-4x+24\equiv(x+2)(x-2)(x-6).$$

Hence
$$\frac{C}{(x-2)^2}+\frac{D}{x-2}\equiv-\frac{x-6}{64(x-2)^2}.$$

Multiply by $(x-2)^2$ and *then* put $x=2$. Thus

$$C=-\frac{(-4)}{64}=\frac{1}{16}.$$

Hence
$$\frac{D}{x-2}\equiv-\frac{(x-6)}{64(x-2)^2}-\frac{1}{16(x-2)^2}$$

$$\equiv\frac{-x+2}{64(x-2)^2}\equiv\frac{-1}{64(x-2)},$$

again checking accuracy by the cancelling of $x-2$.

Finally,
$$D=-\frac{1}{64}.$$

Hence

$$f(x)\equiv\frac{1}{64(x+2)}+\frac{3}{4(x-2)^3}+\frac{1}{16(x-2)^2}-\frac{1}{64(x-2)}.$$

(iii)
$$f(x)\equiv\frac{4x-1}{(x-1)^2(x^2+x+1)}.$$

The denominator consists of the linear factor $(x-1)$, repeated, and the quadratic factor (x^2+x+1). We require to express $f(x)$ in the form

$$\frac{A}{(x-1)^2}+\frac{B}{x-1}+\frac{Cx+D}{x^2+x+1},$$

the numerator above the quadratic factor being of the form $Cx+D$. We thus have

$$\frac{A}{(x-1)^2}+\frac{B}{x-1}+\frac{Cx+D}{x^2+x+1}\equiv\frac{4x-1}{(x-1)^2(x^2+x+1)}.$$

Multiply throughout by $(x-1)^2$, and then put $x=1$. Thus

$$A=\frac{4-1}{1+1+1}=1.$$

Hence $$\frac{B}{x-1}+\frac{Cx+D}{x^2+x+1}\equiv\frac{4x-1}{(x-1)^2(x^2+x+1)}-\frac{1}{(x-1)^2}$$

$$\equiv\frac{4x-1-(x^2+x+1)}{(x-1)^2(x^2+x+1)}$$

$$\equiv-\frac{x^2-3x+2}{(x-1)^2(x^2+x+1)}.$$

Hence $$\frac{B}{x-1}+\frac{Cx+D}{x^2+x+1}\equiv-\frac{x-2}{(x-1)(x^2+x+1)},$$

the cancelling of the factor $x-1$ providing a check of accuracy.

Multiply throughout by $x-1$, and then put $x=1$. Thus

$$B=-\frac{1-2}{1+1+1}=\frac{1}{3}.$$

Hence $$\frac{Cx+D}{x^2+x+1}\equiv-\frac{x-2}{(x-1)(x^2+x+1)}-\frac{1}{3(x-1)}$$

$$\equiv\frac{-3(x-2)-(x^2+x+1)}{3(x-1)(x^2+x+1)}$$

$$\equiv-\frac{x^2+4x-5}{3(x-1)(x^2+x+1)}$$

$$\equiv-\frac{x+5}{3(x^2+x+1)}$$

on cancelling the factor $x-1$. Hence

$$C\equiv-\tfrac{1}{3},\quad D\equiv-\tfrac{5}{3},$$

and so $$f(x)\equiv\frac{1}{(x-1)^2}+\frac{1}{3(x-1)}-\frac{x+5}{3(x^2+x+1)}.$$

The following illustrations exhibit some further points about the calculation of partial fractions, and also show how the integration of rational functions is carried out.

ILLUSTRATION 11. *To find*

$$I\equiv\int\frac{x^4\,dx}{(x^2-1)^2}.$$

The numerator is not of less degree than the denominator, so we begin by dividing out:

$$\frac{x^4}{(x^2-1)^2}=1+\frac{2x^2-1}{(x^2-1)^2}.$$

Consider, then, the function

$$g(x) \equiv \frac{2x^2-1}{(x^2-1)^2}.$$

Factorize the denominator, so that

$$g(x) \equiv \frac{2x^2-1}{(x-1)^2\,(x+1)^2}.$$

We therefore have to find constants A, B, C, D such that

$$\frac{A}{(x-1)^2}+\frac{B}{x-1}+\frac{C}{(x+1)^2}+\frac{D}{x+1} \equiv \frac{2x^2-1}{(x-1)^2\,(x+1)^2}.$$

Multiply throughout by $\begin{cases}(x-1)^2\\(x+1)^2\end{cases}$ and then put $\begin{cases}x=1\\x=-1\end{cases}$. Thus

$$A = \frac{2(1)^2-1}{(1+1)^2} = \frac{1}{4},$$

$$C = \frac{2(-1)^2-1}{(-1-1)^2} = \frac{1}{4}.$$

Hence

$$\frac{B}{x-1}+\frac{D}{x+1} \equiv \frac{2x^2-1}{(x-1)^2\,(x+1)^2}-\frac{1}{4(x-1)^2}-\frac{1}{4(x+1)^2}$$

$$\equiv \frac{8x^2-4-(x^2+2x+1)-(x^2-2x+1)}{4(x-1)^2\,(x+1)^2}$$

$$\equiv \frac{6x^2-6}{4(x-1)^2\,(x+1)^2}$$

$$\equiv \frac{3}{2(x-1)\,(x+1)}.$$

Hence, by the usual process,

$$B = \tfrac{3}{4}, \quad D = -\tfrac{3}{4}.$$

It follows that

$$I = \int\left\{1+\frac{1}{4(x-1)^2}+\frac{3}{4(x-1)}+\frac{1}{4(x+1)^2}-\frac{3}{4(x+1)}\right\}dx$$

$$= x-\frac{1}{4(x-1)}+\tfrac{3}{4}\log(x-1)-\frac{1}{4(x+1)}-\tfrac{3}{4}\log(x+1).$$

ILLUSTRATION 2. *To find*

$$I \equiv \int \frac{13\,dx}{x^3+x-10}.$$

We must first factorize the denominator. It vanishes when $x = 2$, so that $x-2$ is a factor, and, after division, we find that it is $(x-2)(x^2+2x+5)$. We therefore seek to express

$$f(x) \equiv \frac{13}{x^3+x-10} \equiv \frac{13}{(x-2)(x^2+2x+5)}$$

in the form $$\frac{A}{x-2}+\frac{Bx+C}{x^2+2x+5},$$

so that $$\frac{A}{x-2}+\frac{Bx+C}{x^2+2x+5} \equiv \frac{13}{(x-2)(x^2+2x+5)}.$$

Following a routine which should now be familiar, we have

$$A = \frac{13}{4+4+5} = 1;$$

$$\frac{Bx+C}{x^2+2x+5} \equiv \frac{13}{(x-2)(x^2+2x+5)} - \frac{1}{x-2}$$

$$\equiv \frac{13-x^2-2x-5}{(x-2)(x^2+2x+5)}$$

$$\equiv \frac{8-2x-x^2}{(x-2)(x^2+2x+5)}$$

$$\equiv \frac{-x-4}{x^2+2x+5}.$$

Hence

$$I = \int \left\{\frac{1}{x-2} - \frac{x+4}{x^2+2x+5}\right\} dx$$

$$= \int \left\{\frac{1}{x-2} - \frac{x+1}{x^2+2x+5} - \frac{3}{x^2+2x+5}\right\} dx$$

$$= \log(x-2) - \tfrac{1}{2}\log(x^2+2x+5) - \tfrac{3}{2}\tan^{-1}\left(\frac{x+1}{2}\right).$$

EXAMPLES III

Integrate the following rational functions:

1. $\dfrac{1}{x^2-9}$.
2. $\dfrac{x^2}{(x-1)^2}$.
3. $\dfrac{x}{x^2-3x+2}$.
4. $\dfrac{x^3}{x-1}$.
5. $\dfrac{1}{x^3+x^2+x+1}$.
6. $\dfrac{x^2}{x^2+5x+4}$.
7. $\dfrac{1}{(x-1)^2(x+1)}$.
8. $\dfrac{x}{(x-1)^2(x^2+1)}$.
9. $\dfrac{1}{(x-2)(x^2+4x+9)}$.
10. $\dfrac{x^2}{(x-2)^3}$.
11. $\dfrac{x^3}{(x^2-4)^2}$.
12. $\dfrac{1}{(x-1)^3(x^2+4)}$.
13. $\dfrac{x}{(x-1)(x-2)(x-3)}$.
14. $\dfrac{x^2}{(x-1)(x-2)^2(x-3)}$.
15. $\dfrac{1}{x^2(x^2-6x+13)}$.
16. $\dfrac{1}{x^2(x+4)}$.
17. $\dfrac{x-2}{x^3(x+3)}$.
18. $\dfrac{2x+5}{4x+7}$.
19. $\dfrac{x^2}{x^2+6x+25}$.
20. $\dfrac{(x+1)^2}{x^2+1}$.
21. $\dfrac{(x-1)^2}{x(x^2+4)}$.
22. $\dfrac{2x}{x^3-x^2-8x+12}$.
23. $\dfrac{1}{x^4-2x^3+2x^2-2x+1}$.
24. $\dfrac{x^4}{(x+2)^2(x^2+2x+17)}$.

6. The exponential function. Imagine the graph $y = \log x$ (Fig. 62) to be turned, as it were, through a right angle and viewed through a mirror, and the axes then renamed to give the curve shown in the diagram (Fig. 64). Then y is a certain function of x with the property that

$$x = \log y.$$

Thus y is an 'inverse' function of $\log x$ in a sense similar to that in which $\sin^{-1} x$ is (Vol. I, p. 38) an inverse function of $\sin x$.

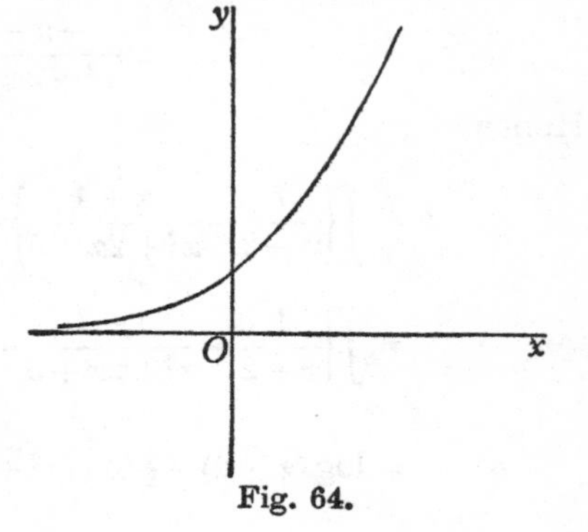

Fig. 64.

We write the relation $x = \log y$

to give y in terms of x, in the form

$$y = \exp x,$$

where $\exp x$, whose properties we now study, is called the *exponential function*. It is defined, as the graph implies, for all values of x, increasing steadily from zero to 'infinity' as x increases from 'minus infinity' to 'infinity'.

The exponential function (of a real variable x) is necessarily POSITIVE.

From the relation $\log y = x$,

we have, by differentiation with respect to x,

$$\frac{1}{y}\frac{dy}{dx} = 1,$$

so that

$$\frac{dy}{dx} = y.$$

Hence *the differential coefficient of* $\exp x$ is $\exp x$ itself.

It is convenient to have a name for the value of the function when $x = 1$, and for this we use the letter e.

Thus

$$\exp 1 = e,$$

or, in equivalent form, $\log e = 1$.

From the graph, we have the relation

$$e > 1.$$

[The value of e, to four significant figures, is 2·718.]

We now seek to identify the function $\exp x$ in terms of the constant e and the variable x. If

$$y = \exp x,$$

then

$$\log\{\exp x\} = \log y$$
$$= x.$$

Also the relation $\log(x^n) = n \log x$ leads, on replacing x, n by the letters e, x respectively, to the relation

$$\log(e^x) = x \log e$$
$$= x,$$

since $\log e = 1$. Hence

$$\log\{\exp x\} = \log(e^x).$$

But we have proved (p. 5) that, if the logarithms of two numbers are equal, then the numbers themselves are equal, and so

$$\exp x = e^x.$$

The exponential function $\exp x$ is therefore identified as the number e raised to the power x.

7. The relations $\dfrac{dy}{dx} = \dfrac{1}{x}, \dfrac{dy}{dx} = y.$

(i) The Logarithm.

It follows from the definition of a logarithm that the relation

$$\frac{dy}{dx} = \frac{1}{x}$$

yields for positive x the result

$$y = \log x + C$$

where C is an arbitrary constant.

If x is negative, say

$$x = -u,$$

where u is positive, then

$$\frac{dy}{du} = \frac{dy}{dx}\frac{dx}{du} = -\frac{dy}{dx};$$

hence

$$\frac{dy}{du} = -\frac{1}{x}$$

$$= \frac{1}{u},$$

so that

$$y = \log u + C$$

$$= \log(-x) + C.$$

We may therefore conclude that, *whether x is positive or negative, the equation*

$$\frac{dy}{dx} = \frac{1}{x}$$

leads to the relation $\quad y = \log|x| + C,$

where $|x|$ is the numerical value of x.

In practice, it is customary to use the form

$$y = \log x + C$$

with the tacit assumption that x is positive; but this needs care.

The relation $y = \log x + C$ may be put into an alternative form by writing $C = \log a$, where a is also an arbitrary constant. Then

$$\begin{aligned} y &= \log x + \log a \\ &= \log ax \quad (ax \text{ assumed positive}). \end{aligned}$$

By the definition of the exponential function, we then have

$$ax = e^y,$$

or

$$x = b\,e^y,$$

where b is likewise an arbitrary constant, assumed to have the same sign as x.

(ii) THE EXPONENTIAL FUNCTION.

We turn now to the equation

$$\frac{dy}{dx} = y.$$

Writing this in the form $\quad \dfrac{dx}{dy} = \dfrac{1}{y},$

we see that interchange of x, y in the above relation $x = b\,e^y$ leads to the result

$$y = b\,e^x.$$

Hence *the equation* $\quad \dfrac{dy}{dx} = y$

leads to the relation $\quad y = b\,e^x,$

where b is an arbitrary constant, assumed to have the same sign as y.

More generally, *the relation*

$$\frac{dy}{dx} = ky$$

leads to the relation $\quad y = be^{kx}:$

For the substitution $x = u/k$

gives
$$\frac{dy}{du} = \frac{dy}{dx}\frac{dx}{du} = \frac{1}{k}\frac{dy}{dx},$$

so that
$$\frac{dy}{du} = \frac{1}{k}(ky)$$
$$= y.$$

Hence
$$y = be^u$$
$$= be^{kx}.$$

We give two typical examples to show how exponential functions arise in physical applications.

ILLUSTRATION 13. We return to Illustration 1 (p. 6) of a circuit with resistance R, self-inductance L and electromotive force E. The equation for the current x at time t is

$$L\frac{dx}{dt} + Rx = E.$$

Hence
$$L\frac{dx}{dt} = E - Rx.$$

Write
$$E - Rx = u;$$

then
$$\frac{du}{dt} = \frac{du}{dx}\frac{dx}{dt}$$
$$= (-R)(u/L)$$
$$= (-R/L)u.$$

Hence
$$u = Ae^{-(R/L)t},$$

where A is an arbitrary constant, so that

$$E - Rx = Ae^{-(R/L)t}.$$

If $x = 0$ when $t = 0$, then

$$E = Ae^0 = A,$$

and so
$$E - Rx = Ee^{-(R/L)t},$$

or
$$x = \frac{E}{R}\{1 - e^{-(R/L)t}\}.$$

ILLUSTRATION 14. *To find the variation of pressure with height in an atmosphere obeying the law*

$$pv = \text{constant},$$

where p, v denote pressure and volume respectively.

Consider a vertical filament of air whose cross-sections have area δA (Fig. 65). Let the pressures at heights $x, x+\delta x$ be $p, p+\delta p$. Then the element of volume (shaded in the diagram) of height δx and base δA is in equilibrium under pressure round its sides, which does not concern us, and also under the following vertical forces:

(i) $p\delta A$ upwards;

(ii) $(p+\delta p)\delta A$ downwards;

(iii) $\rho\delta A\delta x$ downwards,

where ρ is the weight per unit volume at height x. Hence

$$p\delta A-(p+\delta p)\delta A-\rho\delta x\delta A = 0,$$

or

$$\delta p+\rho\delta x = 0.$$

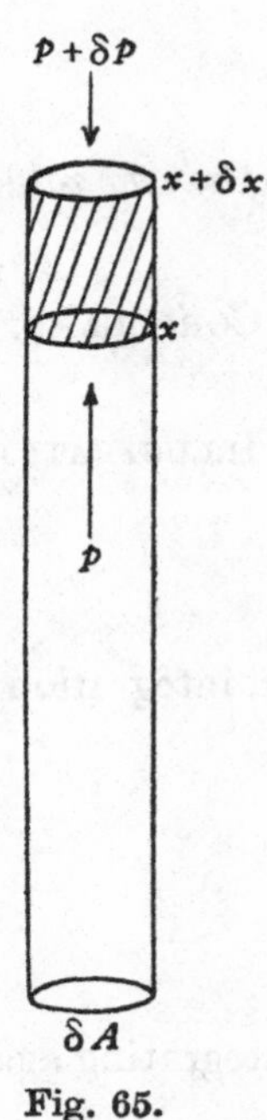

Fig. 65.

Now let p_0, ρ_0, v_0 be the values of p, ρ, v at ground level. Since ρ is the weight per unit volume, the relation

$$pv = p_0v_0$$

is equivalent to

$$\frac{p}{\rho} = \frac{p_0}{\rho_0},$$

and so

$$\delta p+\frac{\rho_0}{p_0}p\,\delta x = 0.$$

In the limit, this is

$$\frac{dp}{dx} = -\frac{\rho_0}{p_0}p,$$

and so

$$p = Ae^{-(\rho_0/p_0)x},$$

where A is an arbitrary constant. But $p = p_0$ when $x = 0$, so that

$$p_0 = Ae^0 = A.$$

Hence the pressure at height x is given by the relation

$$p = p_0e^{-(\rho_0/p_0)x}.$$

8. The integration of e^x.

To find
$$\int e^x dx,$$
we have merely to note that the relation (p. 19)
$$\frac{d}{dx}(e^x) = e^x$$
leads at once to the result
$$e^x = \int e^x dx,$$
so that *the value of* $\int e^x dx$ *is* e^x *itself*.

COROLLARY.
$$\int e^{ax} dx = \frac{1}{a} e^{ax}.$$

ILLUSTRATION 15. *To find*
$$I \equiv \int e^{ax} \sin bx\, dx.$$

On integration by parts, we have
$$I = \frac{1}{a} e^{ax} \sin bx - \int \frac{1}{a} e^{ax} . b \cos bx\, dx$$
$$= \frac{1}{a} e^{ax} \sin bx - \frac{b}{a} \int e^{ax} \cos bx dx.$$

Integrating again by parts, we have
$$I = \frac{1}{a} e^{ax} \sin bx - \frac{b}{a^2} e^{ax} \cos bx + \frac{b}{a^2} \int e^{ax}(-b \sin bx) dx$$
$$= \frac{1}{a} e^{ax} \sin bx - \frac{b}{a^2} e^{ax} \cos bx - \frac{b^2}{a^2} I.$$

Hence
$$\left(1 + \frac{b^2}{a^2}\right) I = \frac{1}{a} e^{ax} \sin bx - \frac{b}{a^2} e^{ax} \cos bx,$$
so that
$$I = \frac{e^{ax}}{a^2 + b^2}(a \sin bx - b \cos bx).$$

ILLUSTRATION 16. *To prove that, if*

$$y = e^{ax} \sin bx,$$

then
$$\frac{d^2y}{dx^2} - 2a\frac{dy}{dx} + (a^2+b^2)y = 0.$$

We have
$$\frac{dy}{dx} = a e^{ax} \sin bx + b e^{ax} \cos bx$$
$$= ay + b e^{ax} \cos bx.$$

Hence
$$\frac{d^2y}{dx^2} = a\frac{dy}{dx} + ab\, e^{ax} \cos bx - b^2 e^{ax} \sin bx$$
$$= a\frac{dy}{dx} + a\left\{\frac{dy}{dx} - ay\right\} - b^2 y,$$

so that
$$\frac{d^2y}{dx^2} - 2a\frac{dy}{dx} + (a^2+b^2)y = 0.$$

EXAMPLES IV

Find the differential coefficients of the following functions:

1. e^{2x}.
2. e^{x^2}.
3. $x e^{5x}$.
4. $e^x \cos x$.
5. $e^{\sin x}$.
6. $(1+x^2)e^{-x}$.
7. $\frac{1}{2}(e^x + e^{-x})^2$.
8. $x e^x \sin x$.
9. $\dfrac{e^x}{1-x^2}$.
10. $(1+e^x)\sin x$.
11. $e^{3x} \cos 4x$.
12. $e^x \tan x$.

Find the following integrals:

13. $\int e^{2x}\,dx$.
14. $\int e^{-5x}\,dx$.
15. $\int 2x e^{x^2}\,dx$.
16. $\int x^2 e^{-x^3}\,dx$.
17. $\int x e^x\,dx$.
18. $\int x^2 e^x\,dx$.
19. $\int e^{\sin x} \cos x\,dx$.
20. $\int \sin x \cos x\, e^{\sin^2 x}\,dx$.
21. $\int \sec^2 x\, e^{\tan x}\,dx$.
22. $\int e^x \cos x\,dx$.
23. $\int e^{3x} \cos 4x\,dx$.
24. $\int (1+x) e^{2x}\,dx$.

ILLUSTRATION 17. *To find a formula of reduction for*

$$I_n \equiv \int x^n e^x dx.$$

On integration by parts, we have the relation

$$I_n = e^x . x^n - \int e^x . nx^{n-1} dx$$
$$= x^n e^x - n I_{n-1}.$$

This is the required formula.

ILLUSTRATION 18. *To find a formula of reduction for*

$$I_n = \int e^x \sin^n x \, dx.$$

On integration by parts, we have the relation

$$I_n = e^x . \sin^n x - \int e^x . n \sin^{n-1} x \cos x \, dx$$
$$= e^x \sin^n x - n . e^x \sin^{n-1} x \cos x$$
$$+ n \int e^x \{(n-1) \sin^{n-2} x \cos^2 x - \sin^n x\} dx$$
$$= e^x \sin^n x - n e^x \sin^{n-1} x \cos x$$
$$+ n \int e^x \{(n-1) \sin^{n-2} x (1 - \sin^2 x) - \sin^n x] dx$$
$$= e^x \sin^n x - n e^x \sin^{n-1} x \cos x$$
$$+ n \{(n-1) I_{n-2} - n I_n\}.$$

Hence

$$(n^2+1) I_n = e^x \sin^n x - n e^x \sin^{n-1} x \cos x + n(n-1) I_{n-2}.$$

This is the required formula.

EXAMPLES V

Obtain formulæ of reduction for the following integrals:

1. $\int x^n e^{ax} dx.$ 2. $\int e^{ax} \sin^n x \, dx.$ 3. $\int e^{ax} \cos^n bx \, dx.$

Evaluate the following integrals:

4. $\int_0^1 x^5 e^x dx.$ 5. $\int_0^{\frac{1}{2}\pi} e^x \sin^4 x \, dx.$ 6. $\int_0^{\pi} e^x \cos^3 x \, dx.$

9. The reconciliation of $\log_e x$ and $\log_{10} x$. The reader will recall the elementary definition:

The logarithm of a number N to the base a is the index of the power to which a must be raised to give N.

If $$N = a^k,$$

then $$\log_a N = k.$$

In particular, if $$y = e^x,$$

then $$x = \log_e y.$$

By this relationship the work which we have just done is reconciled to the more elementary approach, and our use of the word 'logarithm' is justified.

Note. The relation $$\frac{d}{dx}(\log x) = \frac{1}{x}$$

is true only for the base e.

For other bases we must proceed as follows:

Let $$y = \log_a x.$$

Then $$x = a^y.$$

Take logarithms of each side to the base e. Then

$$\log_e x = y \log_e a.$$

Differentiate. Then $$\frac{1}{x} = \log_e a \frac{dy}{dx},$$

so that $$\frac{dy}{dx} = \frac{1}{x \log_e a}.$$

Thus, if $$y = \log_{10} x,$$

then $$\frac{dy}{dx} = \frac{1}{x \log_e 10}.$$

REVISION EXAMPLES III

'Advanced' Level

1. Differentiate $(1+x)^n \log(1+x)$ (i) with respect to x, (ii) with respect to $(1+x)^n$.

If
$$y = A\sqrt{(1+x)} + B\sqrt{x},$$
prove that
$$4x(1+x)\frac{d^2y}{dx^2} + 2(1+2x)\frac{dy}{dx} - y = 0.$$

2. Prove that, if $y = A\sin^2 x + B\cos^2 x$, then
$$\tan 2x \frac{d^2y}{dx^2} = 2\frac{dy}{dx}.$$

Given that $z = Ae^{2x} + Be^{-2x} + C$, find a differential equation satisfied by z and not containing the constants A, B, C.

3. Find dy/dx in terms of t when
$$x = \frac{1+t}{1-2t}, \quad y = \frac{1+2t}{1-t}.$$

Prove that $\frac{dy}{dx} = 1$ when $t = 0$, and find a second value of t for which $\frac{dy}{dx} = 1$.

Prove that
$$\frac{d^2y}{dx^2} = -\frac{2}{3}\left(\frac{1-2t}{1-t}\right)^3.$$

4. Differentiate with respect to x:
$$\log\sin^2 x, \quad \sqrt{\left(\frac{1-x^2}{1+x^2}\right)}.$$

5. Differentiate with respect to x:
$$\sin^2 x\cos x, \quad e^{ax}(\cos ax + \sin ax),$$
$$\log\{x + \sqrt{(x^2+1)}\}, \quad \tan^{-1}\left(\frac{2}{x}\right).$$

6. Find the differential coefficient of $\sqrt{x}$ from first principles. Differentiate with respect to x:
$$x, \quad \frac{1}{\sqrt{(x^2+1)}}, \quad \cos^{-1}(x^2), \quad \log_{10} x.$$

7. Differentiate the following with respect to x:

$$\frac{1}{x},\quad \sin 2x,\quad \tan^{-1}\left(\frac{2x}{1-x^2}\right),\quad e^{x^2}.$$

8. Differentiate the following with respect to x:

$$\frac{1}{1-x},\quad \sin^2 3x,\quad x\log_e x - x,\quad \sin^{-1}\left(\frac{x}{\sqrt{(1+x^2)}}\right).$$

9. Differentiate the following with respect to x, expressing your results as simply as possible:

$$\text{(i)}\ \frac{x}{\sqrt{(1-x^2)}},\quad \text{(ii)}\ \log_e \tan(\tfrac{1}{2}x+\tfrac{1}{4}\pi),\quad \text{(iii)}\ \sin^{-1}\left(\frac{3+5\cos x}{5+3\cos x}\right).$$

10. Find, from first principles, the differential coefficient of $1/x^2$ with respect to x.

Differentiate the following with respect to x, expressing the results as simply as possible:

$$\frac{x}{(x+1)^2},\quad \tan^{-1}(x^2),\quad \log_e \tan 2x.$$

11. Differentiate the following expressions with respect to x, giving your results as simply as possible:

$$\frac{x^2+1}{x},\quad (a^2-x^2)^{\frac{3}{2}},\quad \sin^{-1}\left(\frac{x}{1+x}\right).$$

12. Differentiate the following functions of x with respect to x:

$$\left(1-\frac{3}{x^2}\right)^2,\quad \sqrt{\left(\frac{1+x}{1-x}\right)},\quad \frac{\sin x-\cos x}{\sin x+\cos x}.$$

13. Prove from first principles that

$$\frac{d}{dx}\tan x = \sec^2 x$$

and deduce the values of $\dfrac{d}{dx}\tan^{-1} x$ and $\dfrac{d}{dx}\cot^{-1} x$.

Differentiate with respect to x:

$$\frac{1+x^2}{1-x^2},\quad e^{-\frac{1}{2}x}\sin 2x,\quad \log_e(\tan x+\cot x).$$

14. Differentiate the following expressions with respect to x, simplifying your results as much as you can:

$$\frac{x^3}{1+x}, \quad \sin^2 x \cos^3 x, \quad \sin^{-1}(\sqrt{x}).$$

Prove that, if $y = e^{-2x} \cos 4x$, then

$$\frac{d^2y}{dx^2}+4\frac{dy}{dx}+20y = 0.$$

15. Find from first principles the differential coefficient of $1/x^4$ with respect to x.

Differentiate the following expressions with respect to x, simplifying your results as much as you can:

$$\frac{(x^2-1)^{\frac{1}{2}}}{x}, \quad \frac{3+4\tan x}{4+3\tan x}, \quad x \log_e(x^2+1).$$

16. A particle moves along the x-axis so that its displacement x from O at time t is $e^t \cos^2 t$. Find its velocity and acceleration at time $t = \pi$.

Prove that the values of t for which the particle is at rest form two arithmetic progressions, each with common difference π, and that the successive maximum displacements from O form a geometric progression

$$\tfrac{4}{5}e^{\alpha}, \quad \tfrac{4}{5}e^{\alpha+\pi}, \quad \tfrac{4}{5}e^{\alpha+2\pi}, \quad \ldots,$$

where α is an acute angle such that $\tan\alpha = \frac{1}{2}$.

17. Two circles, with centres O and P, radii a ft. and b ft. respectively, intersect at A and B; the chord AB subtends angles 2θ and 2ϕ at O and P respectively; the area common to the two circles is denoted by Δ and you may assume that

$$2\Delta = a^2(2\theta - \sin 2\theta) + b^2(2\phi - \sin 2\phi).$$

Prove that, if P moves towards O with a speed of u ft. per sec., then

$$\frac{d\Delta}{dt} = 2au \sin\theta.$$

18. Find the equations of the tangent and normal at any point on the curve $x = a\cos^3 t, y = a \sin^3 t$, when t is a variable parameter.

Show that the axes intercept a length a on the tangent and a length $2a \cot 2t$ on the normal.

19. A rod AB of length a is hinged at A to a horizontal table and turns about A in a vertical plane with angular velocity ω. A luminous point is situated vertically above A at a height $h(>a)$. Find the length of the shadow when the rod makes an angle θ with the vertical, and prove that the length of the shadow is altering at the rate $ha\omega(h\cos\theta - a)/(h - a\cos\theta)^2$.

20. Prove that, if a, b are positive and $9b > a$, then

$$a\sin x + b\sin 3x$$

will have a maximum value for some value of x between 0 and $\frac{1}{2}\pi$.

Find this maximum value when $a = 3, b = 0{\cdot}5$, proving that it is a maximum and not a minimum.

21. Given that $\quad y = x^5 - 5x^3 + 5x^2 + 1,$

find the stationary values of y. Determine whether these values are maximum or minimum values or neither.

22. The perpendicular from the vertex A to the base BC of a triangular lamina cuts BC at D; $CD = q, DB = p$ (where $q < p$) and $AD = h$. The lamina lies in the quadrant XOY with B on OX, C on OY, and A on the side of BC remote from O. It moves so that B, C slide on OX, OY. Prove that, if $\angle OBC = \theta$, then OA is maximum (not minimum) when

$$\tan 2\theta = \frac{2h}{q-p}.$$

Prove also that $OA = CA$ when

$$\tan\theta = \frac{2h}{q-p}.$$

23. On a fixed diameter of a circle of radius 6 in., and on opposite sides of the centre O, points A, B are taken such that $AO = 3$ in., $OB = 2$ in. The points A, B are joined to any point P on the circle. Prove that, as P moves round the circle, $AP + BP$ takes minimum values 13 in. and 11 in., and takes a maximum value $5\sqrt{7}$ in. twice.

24. Prove that, for real values of x, the function

$$\frac{3\sin x}{2+\cos x}$$

cannot have a value greater than $\sqrt{3}$ or a value less than $-\sqrt{3}$.

Sketch the graph of this function for values of x from $-\pi$ to π.

25. A particle P falls vertically from a point A, its depth below A after t seconds being at^2, where a is constant. B is a fixed point at the same level as A, and at a distance b from A. Prove that the rate of increase of $\angle ABP$ at time t is

$$\frac{2abt}{a^2t^4+b^2},$$

and show that this rate of increase is greatest when

$$\angle ABP = 30^\circ.$$

26. The angle between the bounding radii of a sector of a circle of radius r is θ. Both r and θ vary, but the area of the sector remains constant and equal to c^2. Prove that the perimeter of the sector is a minimum when $r = c$ and $\theta = 2$ radians.

27. If a variable rectangle has a diagonal of constant length 10 inches, prove that its maximum area is 50 square inches.

28. A straight line with variable slope passes through the fixed point (a, b), where a, b are positive, so as to meet the positive part of the x-axis at A and the positive part of the y-axis at B. If O is the origin, prove that the minimum area of the triangle OAB is $2ab$.

Find also the minimum value of the sum of the lengths of OA and OB.

29. Prove that of all isosceles triangles with a given constant perimeter the triangle whose area is greatest is equilateral.

30. A variable line passes through the point (2, 1) and meets the positive axes OX, OY at A, B respectively. If θ denotes the angle OAB $(0 < \theta < \frac{1}{2}\pi)$, express the area of the triangle OAB in terms of θ, and prove that the area is a minimum when $\tan\theta = \frac{1}{2}$.

Find also the value of θ if the hypotenuse of the triangle is a minimum.

31. Find the turning points on the graph of the function

$$\frac{2x}{x^2+1},$$

stating (with proof) which is a maximum and which is a minimum.

Sketch the graph and find the equation of the tangent at the point on the curve where $x = \frac{1}{2}$.

32. The slant height of a right circular cone is constant and equal to l. Prove that the volume of the cone is a maximum when the radius of the base is $l\sqrt{(\frac{2}{3})}$.

33. A lighthouse AB of height c ft. stands on the edge of a vertical cliff OA of height b ft. above sea level. From a small boat at a variable distance x from O the angle subtended by AB is θ. Prove that

$$\tan\theta = \frac{cx}{x^2+b(b+c)}.$$

Prove also that, if α is the maximum value of θ, then

$$\tan\alpha = \frac{c}{2(b^2+bc)^{\frac{1}{2}}}.$$

Integrate with respect to x:

34. (i) $\dfrac{2x-7}{2x^2+x-3}$, (ii) $\sin^3 x$, (iii) $x^2\cos x$.

35. (i) $\cos^2 2x$, (ii) $\dfrac{4x}{3x^2-2x-1}$, (iii) $x^3\log_e x$.

36. (i) $\dfrac{\sin^3 x}{\cos^4 x}$, (ii) $x\tan^{-1}x$, (iii) $\dfrac{5}{4x^2+3x-1}$.

37. (i) $\dfrac{10}{2x^2+3x-2}$, (ii) $\sin^2 x\cos^3 x$, (iii) $\dfrac{1}{x^n}\log_e x$ $(n \neq 1)$.

38. (i) $\dfrac{7x}{3x^2-11x+6}$, (ii) $x\sin x$, (iii) $(a^2-x^2)^{\frac{3}{2}}$.

39. Integrate with respect to x:

$$\frac{\cos^3 x}{\sin^2 x}, \quad e^x\sin x.$$

Evaluate
$$\int_2^{12}\frac{2x+5}{(2x-3)(2x+1)}\,dx$$

to three significant figures, given that $\log_{10}e = 0{\cdot}4343$.

40. Interpret by means of a sketch the definite integral

$$\int_0^1 \sqrt{(4-x^2)}\,dx,$$

and evaluate this integral.

41. Integrate the following functions with respect to x:

$$\frac{x}{(1-x^2)^{\frac{3}{2}}}, \quad \frac{(x-1)(x-2)}{x+1}.$$

Evaluate $$\int_1^2 x \log_e x\,dx.$$

42. By a suitable change of variable, prove that

$$\int_0^{\frac{1}{2}\pi} \frac{dx}{1+\sin x} = \int_0^{\frac{1}{2}\pi} \frac{dx}{1+\cos x},$$

and by means of the substitution $t = \tan \frac{1}{2}x$, or otherwise, evaluate one of these integrals.

43. Integrate with respect to x:

$$x^2 \cos x, \quad \frac{5+3x}{1-9x^2}.$$

44. Integrate with respect to x:

$$x\sqrt{(1+x^2)}, \quad \sqrt{(1+x^2)}, \quad x^2/\sqrt{(1+x^2)}.$$

Evaluate $$\int_0^{\frac{1}{2}\pi} \sin 3x \cos 2x\,dx.$$

45. Explain the method of integration by parts, and employ it to integrate $\sqrt{(1-x^2)}$ with respect to x.

Integrate with respect to x

$$\frac{\sin^{-1} x}{\sqrt{(1-x^2)}}, \quad \frac{1}{(1-x^2)^{\frac{3}{2}}}, \quad e^x(\cos x - \sin x).$$

46. Integrate with respect to x:

$$\frac{1}{x+\sqrt{x}}, \quad \sin^5 x, \quad \frac{x^2+1}{x^2-11x+30}.$$

47. Integrate with respect to x:

$$\frac{1}{\sin^2 x + 2\cos^2 x}, \quad \frac{1}{x^3+1}.$$

By integration by parts, or otherwise, integrate $\sin^{-1} x$.

48. Integrate with respect to x:

$$\frac{2-x}{1-x}, \quad (1-\cos^2 x)^2 \sin x, \quad x e^x.$$

Prove that
$$\int_{\frac{3}{5}}^{\frac{4}{5}} \frac{1-x}{\sqrt{(1-x^2)}}\,dx = \sin^{-1}\tfrac{7}{25} - \tfrac{1}{5}.$$

49. By integration by parts, show that

$$\int_0^\alpha x \sin(\alpha - x)\,dx$$

is equal to
$$\alpha - \sin\alpha,$$

and also to
$$\frac{\alpha^3}{3!} - \frac{1}{3!}\int_0^\alpha x^3 \sin(\alpha - x)\,dx.$$

From graphical or other considerations prove that, if $0 < \alpha < \pi$, then

$$\int_0^\alpha x^3 \sin(\alpha - x)\,dx < \int_0^\alpha x^3\,dx,$$

and deduce that
$$\int_0^\alpha x^3 \sin(\alpha - x)\,dx < \tfrac{1}{4}\alpha^4.$$

50. Integrate with respect to x:

$$\frac{1}{x^2(1+x^2)}, \quad \tan^4 x.$$

Evaluate
$$\int_0^{\frac{1}{2}\pi} x \sin x\,dx.$$

51. Integrate with respect to x:

$$\frac{x+1}{x+2}, \quad x\sec^2 x.$$

Evaluate
$$\int_0^{\frac{1}{2}\pi} \sin 5x \cos 3x\,dx.$$

52. Integrate with respect to x:

$$\frac{4x-1}{2x^2-x-3}, \quad \cos^3 x \sin^2 x.$$

Evaluate
$$\int_0^{\frac{1}{2}\pi} x^2 \sin x\,dx.$$

53. Integrate $\left(1+\frac{3}{x^2}\right)^2$ with respect to x, and evaluate

$$\int_0^{\frac{1}{2}\pi} \sin^3 x \cos^2 x\,dx, \quad \int_0^1 \frac{(1-x)\,dx}{(x+1)(x^2+1)}.$$

54. Evaluate the definite integrals:

$$\int_0^{\pi} \cos^2 2x\,dx, \quad \int_1^4 x \log x\,dx.$$

If $\quad u_n = \int_0^{\frac{1}{2}\pi} x^n \sin x\,dx, \quad$ and $n > 1$, show that

$$u_n = n(\tfrac{1}{2}\pi)^{n-1} - n(n-1)\,u_{n-2},$$

and evaluate u_4.

55. (i) Prove that, if

$$I_n = \int \sec^n x\,dx,$$

then $\quad (n-1)\,I_n = \tan x \sec^{n-2} x + (n-2)\,I_{n-2}.$

Use the formula to evaluate

$$\int_0^{\frac{1}{4}\pi} \sec^5 x\,dx.$$

(ii) Find the positive value of x for which the definite integral

$$\int_0^x \frac{1-t}{\sqrt{(1+t)}}\,dt$$

is greatest, and evaluate the integral for this value of x.

56. Prove that

$$\frac{d}{dx}(\sin^{m+1} x \cos^{n-1} x)$$
$$= (m+n)\sin^m x \cos^n x - (n-1)\sin^m x \cos^{n-2} x\,dx,$$

and deduce a formula connecting

$$\int_0^{\frac{1}{2}\pi} \sin^m x \cos^n x\,dx, \quad \int_0^{\frac{1}{2}\pi} \sin^m x \cos^{n-2} x\,dx.$$

Evaluate

$$\int_0^{\frac{1}{2}\pi} \sin^3 x \cos^5 x\,dx, \quad \int_{-\frac{1}{2}\pi}^{\frac{1}{2}\pi} \sin^3 x \cos^5 x\,dx.$$

57. Prove the rule for integration by parts and use it in finding

$$\int e^{ax}\cos cx\,dx, \quad \int e^{ax}\sin^2 bx\,dx.$$

If I_n denotes $\int_0^a (a^2-x^2)^n\,dx$, prove that, if $n>0$,

$$I_n = \frac{2na^2}{2n+1} I_{n-1}.$$

58. (i) Find a reduction formula for

$$\int_0^1 (1+x^2)^{n+\frac{1}{2}}\,dx,$$

and evaluate the integral when $n = 2$.

(ii) By integration by parts, show that, if $0 < m < n$, and

$$I = \int_0^1 x^m \frac{d^n}{dx^n}\{x^n(1-x)^n\}\,dx,$$

then $$I = -m\int_0^1 x^{m-1}\frac{d^{n-1}}{dx^{n-1}}\{x^n(1-x)^n\}\,dx.$$

Deduce that $I = 0$.

CHAPTER VIII

TAYLOR'S SERIES AND ALLIED RESULTS

1. A series giving $\sin x$. By repeated application of the method given in Volume I (p. 53), we may establish that, when x is positive, $\sin x$ lies between the following pairs of functions:

(i) $$x \quad \text{and} \quad x-\frac{x^3}{3!},$$

(ii) $$x-\frac{x^3}{3!} \quad \text{and} \quad x-\frac{x^3}{3!}+\frac{x^5}{5!},$$

(iii) $$x-\frac{x^3}{3!}+\frac{x^5}{5!} \quad \text{and} \quad x-\frac{x^3}{3!}+\frac{x^5}{5!}-\frac{x^7}{7!},$$

(iv) $$x-\frac{x^3}{3!}+\frac{x^5}{5!}-\frac{x^7}{7!} \quad \text{and} \quad x-\frac{x^3}{3!}+\frac{x^5}{5!}-\frac{x^7}{7!}+\frac{x^9}{9!},$$

... ...

A simple inductive step completes the argument. Moreover, the results as stated are equally true when x is negative, though the directions of the inequalities must then be reversed; for example, if x is positive, then

$$x > \sin x > x-\frac{x^3}{3!},$$

whereas, if x is negative,

$$x < \sin x < x-\frac{x^3}{3!}.$$

In either case $\sin x$ is between x and $x-\dfrac{x^3}{3!}$.

There is, however, a more significant form in which the statements may be cast:

(i) $\sin x$ differs from x by not more than $\dfrac{|x|^3}{3!}$, where $|x|$ stands for the numerical value of x;

(ii) $\sin x$ differs from $$x-\frac{x^3}{3!}$$

by not more than $\dfrac{|x|^5}{5!}$;

(iii) $\sin x$ differs from $x-\frac{x^3}{3!}+\frac{x^5}{5!}$

by not more than $\frac{|x|^7}{7!}$;

(iv) $\sin x$ differs from $x-\frac{x^3}{3!}+\frac{x^5}{5!}-\frac{x^7}{7!}$

by not more than $\frac{|x|^9}{9!}$;

and so on.

We are therefore led to a series

$$x-\frac{x^3}{3!}+\frac{x^5}{5!}-\frac{x^7}{7!}+\frac{x^9}{9!}-\dots,$$

whose n^{th} term is $(-1)^{n-1}\frac{x^{2n-1}}{(2n-1)!}$, with the property that $\sin x$ *differs from the sum of the first n terms by less than* $\frac{|x|^{2n+1}}{(2n+1)!}$.

Let us examine this 'difference' term $\frac{|x|^{2n+1}}{(2n+1)!}$, writing it in the form

$$\frac{|x|}{1}\cdot\frac{|x|}{2}\cdot\frac{|x|}{3}\cdot\frac{|x|}{4}\cdots\frac{|x|}{2n}\cdot\frac{|x|}{(2n+1)}.$$

Suppose that x has some definite value, positive or negative. If we 'watch' n increase a step at a time, there will come a point when $2n+1$ exceeds $|x|$. Thereafter, the later factors in the product are less than 1; moreover the factor

$$\frac{|x|}{(2n+1)}$$

tends to zero as n continues to increase. By taking n sufficiently large, we may thus ensure that the value of $\sin x$ differs from that of the sum of the first n terms of the series

$$x-\frac{x^3}{3!}+\frac{x^5}{5!}-\frac{x^7}{7!}+\frac{x^9}{9!}-\dots+(-1)^{n-1}\frac{x^{2n-1}}{(2n-1)!}+\dots$$

by as little as we please. In that case, $\sin x$ is called the 'sum to infinity' of the series, and we describe the series as an *expansion* of $\sin x$ in ascending powers of x.

EXAMPLES I

1. Complete the inductive step in the argument to prove that $\sin x$ lies between

$$x-\frac{x^3}{3!}+\frac{x^5}{5!}-\ldots+(-1)^{n-1}\frac{x^{2n-1}}{(2n-1)!}$$

and
$$x-\frac{x^3}{3!}+\frac{x^5}{5!}-\ldots+(-1)^{n}\frac{x^{2n+1}}{(2n+1)!}.$$

2. Obtain the expansion for $\cos x$ as a series of ascending powers of x in the form

$$1-\frac{x^2}{2!}+\frac{x^4}{4!}-\frac{x^6}{6!}+\ldots+(-)^{n-1}\frac{x^{2n-2}}{(2n-2)!}+\ldots.$$

2. A series giving $1/(1+x)$. Consider the sum

$$S_n \equiv 1-x+x^2-x^3+\ldots+(-1)^{n-1}x^{n-1},$$

consisting of n terms. By direct multiplication, we have

$$xS_n = \quad x-x^2+x^3-\ldots+(-1)^{n-2}x^{n-1}+(-1)^{n-1}x^n,$$

so that
$$(1+x)S_n = \quad 1+(-1)^{n-1}x^n,$$

or
$$S_n = \frac{1}{1+x}+\frac{(-1)^{n-1}x^n}{1+x}.$$

Hence

$$\frac{1}{1+x} = 1-x+x^2-x^3+\ldots+(-1)^{n-1}x^{n-1}+\frac{(-1)^n x^n}{1+x},$$

as is probably familiar.

We have therefore obtained a series

$$1-x+x^2-x^3+\ldots,$$

whose n^{th} term is $(-1)^{n-1}x^{n-1}$, with the property that $1/(1+x)$ *differs from the sum of the first n terms in this series by precisely the amount* $|x^n/(1+x)|$.

Consider, then, the 'difference' term $|x^n/(1+x)|$. When x lies between $-1, 1$, so that $-1<x<1$, this difference can be made as small as we please by taking n sufficiently large, and so we may ensure that, when $|x|<1$, the value of $1/(1+x)$ differs from that of the sum of the n terms

$$1-x+x^2-x^3+\ldots+(-1)^{n-1}x^{n-1}$$

by as little as we please.

On the other hand, when $|x|>1$, the difference $|x^n/(1+x)|$ becomes larger and larger with increasing n, and the value of

$$1-x+x^2-x^3+\ldots+(-1)^{n-1}x^{n-1},$$

so far from approximating to $1/(1+x)$, oscillates wildly as the number of terms increases when x is positive, and increases beyond all bounds when x is negative.

We have therefore obtained a series

$$1-x+x^2-x^3+\ldots,$$

which represents the function $1/(1+x)$ for a certain range of values of x (namely $-1<x<1$), but whose sum has no value when $|x|>1$; in contrast to the series

$$x-\frac{x^3}{3!}+\frac{x^5}{5!}-\frac{x^7}{7!}+\ldots$$

which was shown (p. 39) to be an expansion for $\sin x$ for all values of x.

The intermediate values $x=+1$, $x=-1$ require separate consideration:

When $x=+1$, the series is

$$1-1+1-1+\ldots,$$

whose sum to n terms is oscillating, being 1 when n is odd and 0 when n is even.

When $x=-1$, the series is

$$1+1+1+1+\ldots,$$

and the sum of the first n terms increases indefinitely as n increases.

In neither of these cases can we assign a meaning to the sum 'to infinity'.

3. Expansion in series. The two examples given in §§ 1, 2 illustrate the way in which a function $f(x)$ can be expanded as a series of ascending powers of x in the form

$$a_0+a_1x+a_2x^2+a_3x^3+\ldots+a_nx^n+\ldots,$$

possibly for a restricted range of values of x. They suffer, however, by referring to very particular functions, and the treatment given for $\sin x$ and $1/(1+x)$ leaves us with no idea of how to proceed in more general cases.

In the next paragraph we shall give a formula for the coefficients $a_0, a_1, a_2, \ldots$ in terms of $f(x)$ and its differential coefficients, and later we proceed to a more detailed discussion. First, however, we must say a few words about the meaning of the 'sum to infinity' in general; for a fuller treatment, a text-book on analysis should be consulted.

Suppose that we have a series whose successive terms are, say, $u_1, u_2, u_3, \ldots, u_n, \ldots$. It may happen that the sum of the first n terms

$$S_n \equiv u_1 + u_2 + \ldots + u_n$$

tends to a limit S as n tends to infinity.

(For example, if $x \neq 1$,

$$1 + x + x^2 + \ldots + x^{n-1} = \frac{1}{1-x} - \frac{x^n}{1-x},$$

so that, for this series, $S_n = \dfrac{1}{1-x} - \dfrac{x^n}{1-x}$,

and, if $|x| < 1$, $\quad S = \lim_{n \to \infty} S_n = \dfrac{1}{1-x}$.)

In this case we call S the *sum to infinity* of the series. We write

$$S = u_1 + u_2 + \ldots + u_n + \ldots,$$

and say that the series *converges* to S.

On the other hand, if S_n does not tend to a limit as n tends to infinity, the series has no 'sum to infinity', and the expression

$$u_1 + u_2 + \ldots + u_n + \ldots$$

has no arithmetical meaning.

If the terms $u_1, u_2, \ldots$ of the series depend on x (as in the particular example just quoted) so also does the sum S_n of the first n terms, and the sum to infinity when there is one. We shall be concerned exclusively with series of the form

$$a_0 + a_1 x + a_2 x^2 + \ldots + a_n x^n + \ldots,$$

where the coefficients $a_0, a_1, \ldots$ are constants, and we have seen (§§ 1, 2) that such a series may converge for all values of x or for some values only.

Note. It is important to realize that the word 'converge' implies definite tending to a limit. A series such as

$$1-1+1-1+1-1+\ldots,$$

for which $S_n = 1$ when n is odd and 0 when n is even, has a finite sum for all values of n, but does not converge. The series is said to *oscillate boundedly*.

4. The coefficients in an infinite series. We first assume that expansion in an infinite series is possible, and seek a formula to determine the coefficients:

To prove that, *if $f(x)$ is a given function which* CAN *be expanded in the form*

$$f(x) \equiv a_0 + a_1 x + a_2 x^2 + \ldots + a_n x^n + \ldots,$$

then

$$a_n = \frac{f^{(n)}(0)}{n!},$$

where $f^{(n)}(0)$ is the value of $f^{(n)}(x)$ when $x = 0$.

We assume without proof that (in normal cases) we can differentiate the sum of an infinite series by differentiating the terms separately and adding the results, as we should for a finite number of terms. Then

$$f'(x) = a_1 + 2a_2 x + 3a_3 x^2 + \ldots + na_n x^{n-1} + \ldots,$$

$$f''(x) = 2a_2 + 3.2a_3 x + 4.3a_4 x^2 + \ldots + n(n-1)a_n x^{n-2} + \ldots,$$

$$f'''(x) = 3.2a_3 + 4.3.2a_4 x + \ldots + n(n-1)(n-2)a_n x^{n-3} + \ldots,$$

and so on. Putting $x = 0$, we have successively

$$f(0) = a_0,$$

$$f'(0) = a_1,$$

$$f''(0) = 2a_2,$$

$$f'''(0) = 3.2a_3,$$

$$f^{(iv)}(0) = 4.3.2.a_4,$$

and, generally,

$$f^{(n)}(0) = n(n-1)\ldots 3.2a_n = n!\,a_n.$$

Hence

$$f(x) = f(0) + xf'(0) + \frac{x^2}{2!}f''(0) + \frac{x^3}{3!}f'''(0) + \ldots + \frac{x^n}{n!}f^{(n)}(0) + \ldots.$$

Note. This work gives no help about whether the function CAN be expanded; for that we must go to the next paragraph.

ILLUSTRATION 1. *A particle falls from rest under gravity in a medium whose resistance is proportional to the speed. To find an expression for the distance fallen in time t.*

If x is the distance fallen, then the acceleration is $\ddot{x}$ downwards; the forces acting per unit mass are (i) gravity, of magnitude g downwards, (ii) resistance of magnitude $k\dot{x}$ upwards. Hence

$$\ddot{x} = g - k\dot{x}.$$

The formula just given, when adapted to this notation, is

$$x = x_0 + t\dot{x}_0 + \frac{t^2}{2!}\ddot{x}_0 + \frac{t^3}{3!}\dddot{x}_0 + \ldots,$$

where $x_0, \dot{x}_0, \ddot{x}_0, \ldots$ are the values of $x, \dot{x}, \ddot{x}, \ldots$ when $t = 0$.

From the initial conditions,

$$x_0 = 0, \quad \dot{x}_0 = 0,$$

so that

$$\ddot{x}_0 = g - k\dot{x}_0 = g.$$

By successive differentiation of the equation of motion, we have

$$\dddot{x}_0 = -k\ddot{x}_0 = -kg,$$

$$\ddddot{x}_0 = -k\dddot{x}_0 = k^2 g,$$

and so on. Hence

$$x = \frac{t^2}{2!}g + \frac{t^3}{3!}(-kg) + \frac{t^4}{4!}(k^2 g) + \frac{t^5}{5!}(-k^3 g) + \ldots$$

$$= gt^2\left\{\frac{1}{2!} - \frac{(kt)}{3!} + \frac{(kt)^2}{4!} - \frac{(kt)^3}{5!} + \ldots\right\}.$$

The series converges for all values of t.

EXAMPLES II

Use the formula of § 4 to find expansions for the following functions:

1. $\sin x$. 2. $\cos x$. 3. $1/(1+x)$.

5. Taylor's theorem. We come to a somewhat difficult theorem on which the validity of expansion in series can be based.

Suppose that $f(x)$ is a given function of x, possessing as many differential coefficients as are required in the subsequent work.

To prove that, if a, b are two given values of x, then there exists a number ξ between a, b such that, for given n,

$$f(b) = f(a) + (b-a)f'(a) + \frac{(b-a)^2}{2!}f''(a) + \ldots$$
$$+ \frac{(b-a)^{n-1}}{(n-1)!}f^{(n-1)}(a) + \frac{(b-a)^n}{n!}f^{(n)}(\xi).$$

Write

$$F(x) \equiv f(b) - f(x) - (b-x)f'(x) - \frac{(b-x)^2}{2!}f''(x) - \ldots$$
$$- \frac{(b-x)^{n-1}}{(n-1)!}f^{(n-1)}(x) - \frac{(b-x)^k}{(b-a)^k}R_n,$$

where R_n is a number whose properties will be described as required, and k is a positive integer to be specified later. We propose to use Rolle's theorem (Vol. I, p. 60) for $F(x)$ exactly as we did (Vol. I, p. 61) for the mean value theorem, of which this is, indeed, a generalization; we therefore want the relations $F(a) = F(b) = 0$.

By direct substitution of b for x in the expression for $F(x)$, we have

$$F(b) = 0.$$

In order to obtain the relation $F(a) = 0$, we substitute a for x on the right-hand side and equate the result to zero; thus

$$0 = f(b) - f(a) - (b-a)f'(a) - \frac{(b-a)^2}{2!}f''(a) - \ldots$$
$$- \frac{(b-a)^{n-1}}{(n-1)!}f^{(n-1)}(a) - R_n.$$

We must therefore give to R_n the value

$$R_n \equiv f(b) - f(a) - (b-a)f'(a) - \frac{(b-a)^2}{2!}f''(a) - \ldots$$
$$- \frac{(b-a)^{n-1}}{(n-1)!}f^{(n-1)}(a).$$

We have now ensured the relations

$$F(b) = F(a) = 0,$$

and so, by Rolle's theorem, there exists a value ξ of x between a, b at which $F'(x) = 0$.

The next step is to evaluate $F'(x)$. This will involve a number of terms of which

$$\frac{(b-x)^p}{p!}f^{(p)}(x)$$

is typical, and the differential coefficient of this term is

$$\frac{(b-x)^p}{p!}f^{(p+1)}(x)+\frac{p(b-x)^{p-1}(-1)}{p!}f^{(p)}(x)$$

$$=\frac{(b-x)^p}{p!}f^{(p+1)}(x)-\frac{(b-x)^{p-1}}{(p-1)!}f^{(p)}(x).$$

Hence, remembering that R_n is a constant, we have

$$F'(x) = -f'(x)-\{(b-x)f''(x)-f'(x)\}$$

$$-\left\{\frac{(b-x)^2}{2!}f'''(x)-(b-x)f''(x)\right\}-\dots$$

$$-\left\{\frac{(b-x)^{n-1}}{(n-1)!}f^{(n)}(x)-\frac{(b-x)^{n-2}}{(n-2)!}f^{(n-1)}(x)\right\}+\frac{k(b-x)^{k-1}}{(b-a)^k}R_n$$

$$=-\frac{(b-x)^{n-1}}{(n-1)!}f^{(n)}(x)+\frac{k(b-x)^{k-1}}{(b-a)^k}R_n,$$

after cancelling like terms of opposite signs. But there is a number ξ between a, b for which $F'(\xi) = 0$, and so

$$R_n = \frac{(b-a)^k(b-\xi)^{n-k}}{k.(n-1)!}f^{(n)}(\xi).$$

Equating this to the value of R_n obtained above, we have

$$f(b) = f(a)+(b-a)f'(a)+\frac{(b-a)^2}{2!}f''(a)+\dots$$

$$+\frac{(b-a)^{n-1}}{(n-1)!}f^{(n-1)}(a)+\frac{(b-a)^k(b-\xi)^{n-k}}{k.(n-1)!}f^{(n)}(\xi).$$

The value of k is still at our disposal. If we put $k = n$ (as we might have done from the start, of course, had we so desired) we have

$$R_n = \frac{(b-a)^n}{n!}f^{(n)}(\xi),$$

and the result enunciated at the head of the paragraph follows at once. This gives us the most usual form for Taylor's theorem, and we have adopted it in the formal statement; but there are advantages in keeping k more general, as we see below.

The theorem is therefore established.

The expression

$$R_n \equiv \frac{(b-a)^k (b-\xi)^{n-k}}{k.(n-1)!} f^{(n)}(\xi)$$

is called *the remainder after n terms.* It is, in the first instance, what it says, namely a remainder, the difference between

$$f(b)$$

and $$f(a)+(b-a)f'(a)+\frac{(b-a)^2}{2!}f''(a)+\dots\frac{(b-a)^{n-1}}{(n-1)!}f^{(n-1)}(a).$$

The theorem just proved enables us, however, to express this remainder in the suggestive form (with $k=n$)

$$\frac{(b-a)^n}{n!}f^{(n)}(\xi)$$

by choosing ξ suitably. This expression is known as *Lagrange's form* of the remainder.

By giving k other values, we obtain various forms for the remainder. In particular, when $k=1$, we have

$$R_n \equiv \frac{(b-a)(b-\xi)^{n-1}}{(n-1)!} f^{(n)}(\xi).$$

This may be expressed alternatively by writing ξ, which lies between a, b, as $a+\theta(b-a)$, where θ lies between 0 and 1. Then

$$\begin{aligned} R_n &\equiv \frac{(b-a)\{b-a-\theta(b-a)\}^{n-1}}{(n-1)!} f^{(n)}\{a+\theta(b-a)\} \\ &= \frac{(b-a)^n(1-\theta)^{n-1}}{(n-1)!} f^{(n)}\{a+\theta(b-a)\}. \end{aligned}$$

This is called *Cauchy's form* of the remainder. Though less 'in sequence' than Lagrange's form, it enables us to deal with some series for which Lagrange's form does not work.

ALTERNATIVE TREATMENT. There is an alternative treatment of Taylor's theorem, based on integration by parts, which leads to yet another form of the remainder. The method is essentially a continued application of the formula

$$\int_0^b \frac{(b-t)^{k-1}}{(k-1)!} f^{(k)}(a+t)\,dt$$

$$= \left[-\frac{(b-t)^k}{k!} f^{(k)}(a+t)\right]_0^b + \int_0^b \frac{(b-t)^k}{k!} f^{(k+1)}(a+t)\,dt$$

$$= \frac{b^k}{k!} f^{(k)}(a) + \int_0^b \frac{(b-t)^k}{k!} f^{(k+1)}(a+t)\,dt,$$

the term $\dfrac{(b-t)^k}{k!} f^{(k)}(a+t)$ vanishing for $t = b$ when $k > 0$.

This formula may be expressed more concisely by writing

$$u_k \equiv \int_0^b \frac{(b-t)^{k-1}}{(k-1)!} f^{(k)}(a+t)\,dt \quad (k \geqslant 1),$$

so that

$$u_k = \frac{b^k}{k!} f^{(k)}(a) + u_{k+1}.$$

Hence

$$\begin{aligned} u_1 &= bf'(a) + u_2 \\ &= bf'(a) + \frac{b^2}{2!} f''(a) + u_3 \\ &= bf'(a) + \frac{b^2}{2!} f''(a) + \frac{b^3}{3!} f'''(a) + u_4 \\ &= bf'(a) + \frac{b^2}{2!} f''(a) + \ldots + \frac{b^{n-1}}{(n-1)!} f^{(n-1)}(a) + u_n. \end{aligned}$$

Moreover,

$$u_1 = \int_0^b f'(a+t)\,dt = f(a+b) - f(a).$$

Equating the two values of u_1, we obtain the 'Taylor' relation (with our original 'b' now replaced by '$a+b$')

$$f(a+b) = f(a) + bf'(a) + \frac{b^2}{2!} f''(a) + \ldots + \frac{b^{n-1}}{(n-1)!} f^{(n-1)}(a) + R_n,$$

where

$$R_n = \int_0^b \frac{(b-t)^{n-1}}{(n-1)!} f^{(n)}(a+t)\,dt.$$

EXAMPLE III

1. Prove that, if $R_n(x)$ is the function of x defined by the relation

$$R_n(x) = \left\{f(x) + (b-x)f'(x) + \ldots + \frac{(b-x)^{n-1}}{(n-1)!}f^{(n-1)}(x)\right\} - f(b),$$

then
$$R'_n(x) = \frac{(b-x)^{n-1}}{(n-1)!}f^{(n)}(x),$$

and deduce that
$$R_n(x) = \int_b^x \frac{(b-t)^{n-1}}{(n-1)!}f^{(n)}(t)\,dt.$$

6. Maclaurin's theorem. If we write $b = a+h$, Taylor's theorem (with the Lagrange remainder) becomes

$$f(a+h) = f(a) + hf'(a) + \frac{h^2}{2!}f''(a) + \ldots + \frac{h^{n-1}}{(n-1)!}f^{(n-1)}(a) + \frac{h^n}{n!}f^{(n)}(\xi),$$

where ξ is a certain number between $a, a+h$.

A convenient form is found by putting $a = 0$ and then renaming h to be the current variable x:

$$f(x) = f(0) + xf'(0) + \frac{x^2}{2!}f''(0) + \ldots + \frac{x^{n-1}}{(n-1)!}f^{(n-1)}(0) + \frac{x^n}{n!}f^{(n)}(\xi),$$

where ξ is a certain number between $0, x$. This important result is known as Maclaurin's theorem. Compare p. 43.

With the Cauchy form of remainder, the corresponding result is

$$f(x) = f(0) + xf'(0) + \frac{x^2}{2!}f''(0) + \ldots + \frac{x^{n-1}}{(n-1)!}f^{(n-1)}(0) + \frac{x^n(1-\theta)^{n-1}}{(n-1)!}f^{(n)}(\theta x),$$

where θ is a certain number between 0, 1.

7. Maclaurin's series. The remainder R_n in Maclaurin's theorem appears in the form

$$\frac{x^n}{n!}f^{(n)}(\xi) \quad \text{or} \quad \frac{x^n(1-\theta)^{n-1}}{(n-1)!}f^{(n)}(\theta x),$$

where ξ lies between $0, x$ and θ between $0, 1$. Then

$$f(x) \equiv f(0) + xf'(0) + \ldots + \frac{x^{n-1}}{(n-1)!} f^{(n-1)}(0) + R_n.$$

It may be possible to prove that, as n becomes larger and larger (x having a definite value for a particular problem) the remainder R_n tends to the limit zero. When this happens, the sum of the first n terms of the series

$$f(0) + xf'(0) + \frac{x^2}{2!} f''(0) + \ldots$$

tends to the limit $f(x)$, and so the sum to infinity of the series exists, and is $f(x)$.

The condition for the remainder to tend to zero may involve x, so that it is fulfilled for some values of x but not for others.

It is on this basis that the possibility of obtaining an expansion rests. The succeeding paragraphs give the details for a number of important functions.

8. The series for $\sin x$ and $\cos x$.

Let

$$f(x) \equiv \sin x.$$

Then

$$f'(x) = \cos x, \quad f''(x) = -\sin x, \quad f'''(x) = -\cos x, \quad \ldots,$$

so that

$$f(0) = 0, \quad f'(0) = 1, \quad f''(0) = 0, \quad f'''(0) = -1, \quad \ldots.$$

Hence the Maclaurin series is

$$0 + x.1 + \frac{x^2}{2!}.0 + \frac{x^3}{3!}(-1) + \ldots,$$

or

$$x - \frac{x^3}{3!} + \frac{x^5}{5!} - \frac{x^7}{7!} + \ldots.$$

To see whether the series converges to $\sin x$, we consider R_n, the remainder after n terms, where

$$R_n \equiv \frac{x^n}{n!} f^{(n)}(\xi).$$

The numerical value $|f^{(n)}(\xi)|$ is certainly not greater than 1, since $f^{(n)}(\xi)$ is a sine or a cosine. Hence

$$|R_n| \leqslant \frac{|x^n|}{n!} = \frac{|x|^n}{n!}.$$

But (see p. 39) $$\lim_{n\to\infty}\frac{|x|^n}{n!}=0,$$

for all values of x, and so the series converges to $\sin x$ for all values of x.

By similar argument, we obtain the expansion

$$\cos x = 1-\frac{x^2}{2!}+\frac{x^4}{4!}-\frac{x^6}{6!}+\ldots,$$

convergent for all values of x.

9. The binomial series. Let

$$f(x)\equiv(1+x)^p,$$

where p may be positive or negative, and not necessarily an integer. Then

$$f'(x) = p(1+x)^{p-1},$$
$$f''(x) = p(p-1)(1+x)^{p-2},$$
$$f'''(x) = p(p-1)(p-2)(1+x)^{p-3},$$

and, generally,

$$f^{(n)}(x) = p(p-1)\ldots(p-n+1)(1+x)^{p-n}.$$

Hence

$$f(0) = 1,$$
$$f'(0) = p,$$
$$f''(0) = p(p-1),$$
$$f'''(0) = p(p-1)(p-2),$$
$$\ldots\ldots\ldots\ldots\ldots\ldots$$
$$f^{(n)}(0) = p(p-1)\ldots(p-n+1).$$

The Maclaurin series is thus

$$1+px+\frac{p(p-1)}{2!}x^2+\frac{p(p-1)(p-2)}{3!}x^3+\ldots+\frac{p(p-1)\ldots(p-n+1)}{n!}x^n+\ldots.$$

If p is a positive integer, the series terminates, giving the expansion, familiar from any text-book on algebra,

$$(1+x)^p = 1+c_1x+c_2x^2+\ldots+c_px^p,$$

where $$c_n=\frac{p(p-1)\ldots(p-n+1)}{n!}=\frac{p!}{n!(p-n)!}.$$

If p is NOT *a positive integer*, we obtain an infinite series, and the conditions for convergence become important. We prove that *the expansion*

$$(1+x)^p = 1+px+\frac{p(p-1)}{2!}x^2+\ldots+\frac{p(p-1)\ldots(p-n+1)}{n!}x^n+\ldots,$$

where p is not a positive integer, is valid for values of x in the interval

$$-1<x<1.$$

The Cauchy form of remainder* gives

$$R_n = \frac{x^n(1-\theta)^{n-1}}{(n-1)!}p(p-1)\ldots(p-n+1)(1+\theta x)^{p-n}$$

$$= x^n\left(\frac{1-\theta}{1+\theta x}\right)^{n-1}\frac{p(p-1)\ldots(p-n+1)}{(n-1)!}(1+\theta x)^{p-1}.$$

Now $$1-\theta<1+\theta x$$

whether x is positive or negative, since $-1<x<1, 0<\theta<1$.

Hence $$\left(\frac{1-\theta}{1+\theta x}\right)^{n-1}<1.$$

Also, if $p>1$, $$(1+\theta x)^{p-1}<(1+|x|)^{p-1},$$

and, if $p<1$, $$(1+\theta x)^{p-1}=\frac{1}{(1+\theta x)^{1-p}}$$

$$<\frac{1}{(1-|x|)^{1-p}}.$$

For any given x in the interval $-1<x<1$, the product

$$\left(\frac{1-\theta}{1+\theta x}\right)^{n-1}(1+\theta x)^{p-1}$$

is therefore less than an ascertainable positive number A.

Consider next the product

$$\frac{p(p-1)\ldots(p-n+1)}{(n-1)!}x^n,$$

which we denote by the symbol u_n. Then

$$\frac{u_{n+1}}{u_n}=\frac{p(p-1)\ldots(p-n+1)(p-n)}{n!}x^{n+1}\cdot\frac{(n-1)!}{p(p-1)\ldots(p-n+1)}\cdot\frac{1}{x^n}$$

$$=\frac{p-n}{n}x.$$

* The proof which follows may be postponed, if desired.

Take n so large as to be greater than p. Then

$$\left|\frac{u_{n+1}}{u_n}\right| = \frac{n-p}{n}|x|.$$

Write $y = \frac{1}{2}(1+|x|)$, so that $0 < y < 1$. Let n_0 be the first positive integer such that $(n-p)\,|x|/n < y$, that is, such that

$$n(1-|x|) > -2p|x|.$$

Then
$$\left|\frac{u_{n+1}}{u_n}\right| . \left|\frac{u_n}{u_{n-1}}\right| . \left|\frac{u_{n-1}}{u_{n-2}}\right| \ldots \left|\frac{u_{n_0+2}}{u_{n_0+1}}\right| . \left|\frac{u_{n_0+1}}{u_{n_0}}\right|$$
$$< y . y . y \ldots . y . y,$$

so that
$$|u_{n+1}| < |u_{n_0}| y^{n-n_0+1} \quad (n > n_0).$$

Since u_{n_0} is a definite ascertainable number, and since $|y| < 1$, it follows that

$$|u_n| \to 0$$

as $n \to \infty$.

But
$$|R_n| < A|u_n|,$$

when A is the positive number already defined. Hence

$$|R_n| \to 0$$

as $n \to \infty$, and the validity of the expansion is established.

10. The logarithmic series.

Let

$$f(x) \equiv \log(1+x).$$

Then
$$f'(x) = \frac{1}{1+x},$$
$$f''(x) = -\frac{1}{(1+x)^2},$$
$$f'''(x) = \frac{2}{(1+x)^3},$$
$$f^{(\mathrm{iv})}(x) = -\frac{3.2}{(1+x)^4},$$

and generally,
$$f^{(n)}(x) = (-1)^{n-1}\frac{(n-1)!}{(1+x)^n}.$$

Hence
$$f(0) = \log 1 = 0,$$
$$f^{(n)}(0) = (-1)^{n-1}(n-1)!$$

so that the Maclaurin series is

$$0 + x.1 + \frac{x^2}{2!}(-1) + \frac{x^3}{3!}(2!) + \frac{x^4}{4!}(-3!) + \ldots,$$

or
$$x - \tfrac{1}{2}x^2 + \tfrac{1}{3}x^3 - \tfrac{1}{4}x^4 + \ldots + (-1)^{n-1}\frac{1}{n}x^n + \ldots.$$

This expansion for $\log(1+x)$ *is valid for all values of* x *in the range*
$$-1 < x \leqslant 1.$$

When $x = 1$, we have the result

$$\log 2 = 1 - \tfrac{1}{2} + \tfrac{1}{3} - \tfrac{1}{4} + \ldots.$$

The case $x = -1$ is reflected in the graph (Fig. 62, p. 4) where $y \to -\infty$ as $x \to 0$.

EXAMPLE IV

1. Use the method given for the binomial series to prove the validity of the logarithmic series when $-1 < x < 1$.

11. The exponential series.

Let
$$f(x) = e^x.$$

Then
$$f'(x) = e^x,$$
$$f''(x) = e^x,$$

and, generally,
$$f^{(n)}(x) = e^x.$$

Hence
$$f(0) = f'(0) = f''(0) = \ldots = 1.$$

We therefore have the Maclaurin series

$$1 + x + \frac{x^2}{2!} + \frac{x^3}{3!} + \ldots + \frac{x^n}{n!} + \ldots.$$

It may be proved that *this expansion for* e^x *is valid for all values of* x. [See Examples V (1) below.]

COROLLARY. The work of this paragraph enables us to fill a gap in the discussion of the exponential function (pp. 18–20) by obtaining an expression for the number e. Putting $x = 1$, we have

$$e = 1 + 1 + \frac{1}{2!} + \frac{1}{3!} + \ldots + \frac{1}{n!} + \ldots.$$

EXAMPLES V

1. Use the method given for the sine series to prove the validity of the exponential series for all values of x.

By direct calculation of the differential coefficients and substitution in Maclaurin's formula, obtain expansions for the following functions as series of ascending powers of x:

2. e^{3x}.
3. $\log(1+2x)$.
4. $\sin 2x$.
5. $1/(1+x)$.
6. $1/(1-x)^2$.
7. $\cos 4x$.
8. e^{-2x}.
9. $\sqrt{(1+2x)}$.
10. $\log(1-3x)$.

12. Approximations.

If the successive terms in the expansion

$$f(x) \equiv a_0 + a_1 x + a_2 x^2 + \ldots$$

become rapidly smaller, a good approximation to the value of the function $f(x)$ may be found by taking the first few terms.

ILLUSTRATION 2. *To estimate* $\sqrt{(3\cdot 98)}$.

Writing the expression in the form

$$(4 - \cdot 02)^{\frac{1}{2}} = 2(1 - \cdot 005)^{\frac{1}{2}},$$

we may expand by the binomial series (p. 51) to obtain

$$2\left\{1 + \tfrac{1}{2}(-\cdot 005) + \frac{\frac{1}{2}(-\frac{1}{2})}{2!}(-\cdot 005)^2 + \frac{\frac{1}{2}(-\frac{1}{2})(-\frac{3}{2})}{3!}(-\cdot 005)^3 + \ldots\right\}$$

$$= 2\{1 - \cdot 0025 - \tfrac{1}{8}(\cdot 000025) + \tfrac{1}{16}(\cdot 000000125)\ldots\},$$

and a good approximation is

$$2(1 - \cdot 0025)$$

$$= 1\cdot 9950.$$

A limit may be set to the error by means of Taylor's theorem, as follows:

Since, for a general function $f(x)$,

$$f(x) = f(0) + xf'(0) + \frac{x^2}{2!}f''(\theta x) \quad (0 < \theta < 1),$$

the error cannot exceed the greatest value of

$$\left|\frac{x^2}{2!}f''(\theta x)\right|,$$

or, here,
$$2 \times \frac{(\cdot 005)^2}{2!} \cdot \frac{1}{4(1 - \cdot 005\theta)^{\frac{3}{2}}} \equiv \frac{\cdot 00000625}{(1 - \cdot 005\theta)^{\frac{3}{2}}}.$$

In the most unfavourable case, with $\theta = 1$, this gives a value less than $\cdot 0000063$ for the error.

EXAMPLES VI

Estimate the values of the following expressions:

1. $\sqrt{(4\cdot 02)}$.
2. $\sqrt{(8\cdot 97)}$.
3. $\sqrt[3]{(8\cdot 02)}$.
4. $\sqrt[3]{(26\cdot 98)}$.
5. $\sqrt[5]{(31\cdot 97)}$.
6. $1/\sqrt{(9\cdot 03)}$.

13. Newton's approximation to a root of an equation.

Suppose that we are given an equation

$$f(x) = 0$$

and know that there is a root somewhere near the value $x = a$. Newton's method, which we now describe, shows that, *under suitable circumstances, a better approximation to the root is*

$$a - \frac{f(a)}{f'(a)}.$$

If the correct root being sought is $\xi \equiv a + h$, then

$$f(a+h) = f(\xi) = 0,$$

so that, by Taylor's theorem for $n = 2$, with Lagrange's form of the remainder,

$$f(a) + hf'(a) + \frac{h^2}{2!}f''(\eta) = 0$$

for a value of η between a and $a+h$. If h is reasonably small, then the term involving h^2 may be regarded as negligible for practical purposes. We therefore have

$$f(a)+hf'(a) = 0,$$

or

$$h = -\frac{f(a)}{f'(a)},$$

so that the root is approximately

$$a-\frac{f(a)}{f'(a)}.$$

This crude statement, however, should be supplemented by more careful analysis, and the following graphical treatment shows the precautions which ought to be taken.

We begin with an examination of the curve

$$y = f(x)$$

near the point ξ, taking first the case where the gradient is positive and the concavity 'upwards' near that point, so that $f'(x), f''(x)$ are both positive.

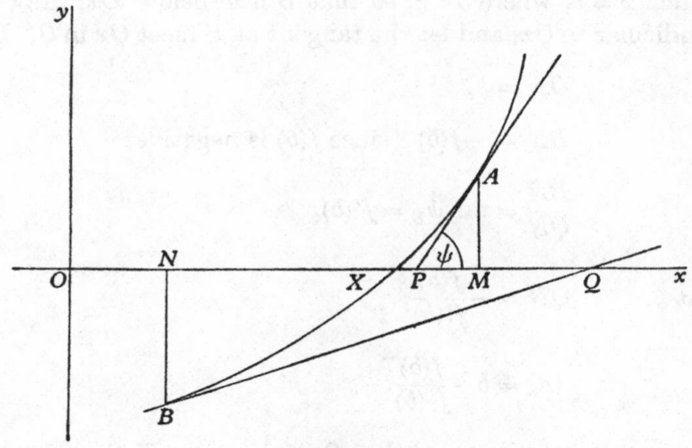

Fig. 66.

(i) Let X be the point $(\xi, 0)$ on the curve, and let A be the point for which $x = a$, where $a > \xi$; draw AM perpendicular to Ox, and let the tangent at A meet Ox in P.

Then
$$OM = a,$$
$$AM = f(a),$$
$$\frac{AM}{PM} = \tan\psi_A = f'(a),$$
so that
$$PM = \frac{f(a)}{f'(a)},$$
and
$$OP = a - \frac{f(a)}{f'(a)}.$$

But under our assumptions that $f'(x), f''(x)$ are positive near X (so that the gradient is positive and the concavity upwards) the tangent AP lies between AM and the curve, so that P lies between X and M. Hence, under these conditions, *OP is a better approximation than OM to OX.* That is, $a - \dfrac{f(a)}{f'(a)}$ is a better approximation than a to the root.

It is assumed that $f'(a)$ is not zero, and, indeed, that $f'(x)$ is not zero near the required root.

(ii) Suppose next that, with the same diagram, B is the point for which $x = b$, where $b < \xi$, so that B lies 'below' Ox; draw BN perpendicular to Ox, and let the tangent at B meet Ox in Q. Then
$$ON = b,$$
$$BN = -f(b) \quad \text{since } f(b) \text{ is negative},$$
$$\frac{BN}{QN} = \tan\psi_B = f'(b),$$
so that
$$QN = -\frac{f(b)}{f'(b)},$$
and
$$OQ = b - \frac{f(b)}{f'(b)}.$$

But now we cannot be sure that Q is nearer to X than N, for Q may be anywhere to the right of X according to the shape of the curve. Hence *we cannot be sure whether $b - \dfrac{f(b)}{f'(b)}$ is, or is not, a better approximation than b to the root.*

Similar argument applied to the accompanying diagram (Fig. 67) shows that the results (i), (ii) are also true if (the concavity still being 'upwards') the gradient is negative near X; P is closer than M to X, whereas Q may or may not be closer than N to X.

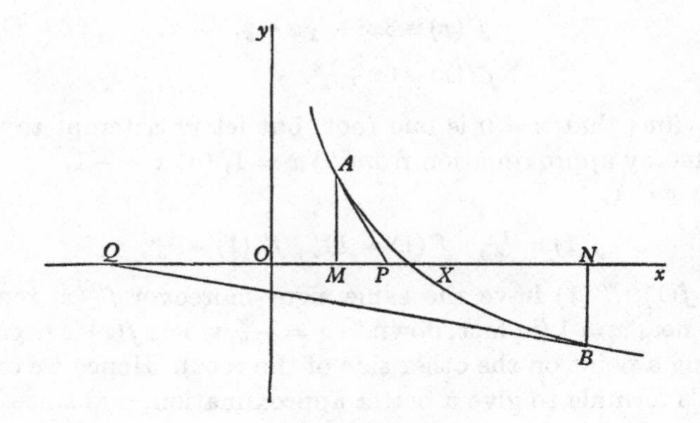

Fig. 67.

We have therefore proved, so far, that, *if* $f''(x)$ *is positive near* ξ, *the approximation* $a-\dfrac{f(a)}{f'(a)}$ *is better than* a *itself when* $f(a)$ *is positive.*

It is easy to verify in the same way that, *if* $f''(x)$ *is negative near* ξ, *the approximation* $a-\dfrac{f(a)}{f'(a)}$ *is better than* a *itself when* $f(a)$ *is negative.* (The whole diagram is merely 'turned upside down'.)

In other words, *we can be sure of a better approximation if* $f''(x)$ *retains the same sign near* $x=a$, *that sign being also the sign of* $f(a)$.

In more complicated cases, it is wise to draw sketches such as we have shown in order to determine how the tangent at A cuts the x-axis. It is, however, clear from the graphs that, for ordinary functions, the approximation $a-\dfrac{f(a)}{f'(a)}$ is ALWAYS better than a itself once we come sufficiently close to the correct answer. In other words, once it is ascertained that an approximation is reasonably good, it may confidently be expected that Newton's approximation will make it better still.

ILLUSTRATION 3. An example where the answer is apparent from the outset may help to make the principle clear. Consider the equation

$$x^3+\tfrac{8}{3}x^2+\tfrac{8}{3}x=0,$$

so that

$$f(x)\equiv x^3+\tfrac{8}{3}x^2+\tfrac{8}{3}x,$$

$$f'(x)\equiv 3x^2+\tfrac{16}{3}x+\tfrac{8}{3},$$

$$f''(x)\equiv 6x+\tfrac{16}{3}.$$

It is obvious that $x=0$ is one root, but let us attempt to reach that value by approximation from (i) $x=1$, (ii) $x=-1$.

When $x=1$,

$$f(1)=\tfrac{19}{3},\quad f'(1)=11,\quad f''(1)=\tfrac{34}{3},$$

so that $f(1)$, $f''(1)$ have the same sign; moreover $f''(x)$ remains positive near $x=1$ (in fact, down to $x=-\frac{8}{9}$, where $f(x)$ is negative, indicating a point on the other side of the root). Hence we expect Newton's formula to give a better approximation; and since

$$1-\frac{f(1)}{f'(1)}=1-\frac{19}{33}=\frac{14}{33},$$

this is actually the case.

On the other hand, when $x=-1$,

$$f(-1)=-1,\quad f'(-1)=\tfrac{1}{3},\quad f''(-1)=-\tfrac{2}{3}.$$

Here, again, $f(-1)$, $f''(-1)$ have the same sign; but $f''(x)$ changes sign from negative to positive at $x=-\frac{8}{9}$, which is near -1; we are therefore in a doubtful region; and since

$$-1-\frac{f(-1)}{f'(-1)}=-1+3=2,$$

the approximation is actually worse.

ILLUSTRATION 4. *To find an approximation to that root of the equation*

$$x^3-3x+1=0$$

which lies between 0, 1.

We have

$$f(x)\equiv x^3-3x+1,$$

$$f'(x)\equiv 3x^2-3,$$

$$f''(x)=6x.$$

Since $f''(x)$ is positive when x lies in the given interval 0, 1, it is advisable to begin with an approximation which makes $f(x)$ also positive.

Now $$f(0) = 1,\quad f(\tfrac{1}{4}) = \tfrac{17}{64},\quad f(\tfrac{1}{2}) = -\tfrac{3}{8}.$$

We should therefore like an approximation between $\frac{1}{4}$ and $\frac{1}{2}$ which keeps $f(x)$ positive, and inspection shows that $\frac{1}{3}$ appears very suitable. We then have

$$f(\tfrac{1}{3}) = \tfrac{1}{27},\quad f'(\tfrac{1}{3}) = -\tfrac{8}{3},$$

so that $$-\frac{f(\frac{1}{3})}{f'(\frac{1}{3})} = +\frac{1}{72},$$

and the corresponding approximation is

$$\tfrac{1}{3} + \tfrac{1}{72} = \tfrac{25}{72} = \cdot 3472.$$

The correct root is $\cdot 3472\ldots$, so we have already obtained four correct figures.

ILLUSTRATION 5. *If η is a small positive number, to find an approximation to that root of the equation*

$$\sin x = \eta x$$

which lies near to $x = \pi$.

(The intersection of the graphs $y = \sin x, y = \eta x$ shows that there *is* a root near to π.)

Since $\sin\pi = 0$ and η is small, the approximation $x = \pi$ is reasonably good.

Write $$f(x) \equiv \sin x - \eta x,$$

so that $$f'(x) = \cos x - \eta,$$

$$f''(x) = -\sin x.$$

Then $$f(\pi) = -\eta\pi,\quad f'(\pi) = -1-\eta,\quad f''(\pi) = 0.$$

Although $f''(\pi)$ is actually zero, so that the curve (Fig. 68) $y = \sin x - \eta x$ has an inflexion at $x = \pi$, the concavity is 'downwards' in the interval $0, \pi$ and $f(\pi)$ is negative; moreover the gradient $f'(\pi)$ is also negative. The accompanying sketch shows that the tangent lies between the ordinate $x = \pi$ and the curve, so that Newton's method will improve the approximation.

The corresponding solution is

$$\pi - \frac{f(\pi)}{f'(\pi)}$$

$$= \pi - \frac{(-\eta\pi)}{(-1-\eta)} = \pi - \frac{\eta\pi}{1+\eta} = \frac{\pi}{1+\eta}.$$

To the first order in η this is, on expansion of $(1+\eta)^{-1}$,

$$\pi(1-\eta).$$

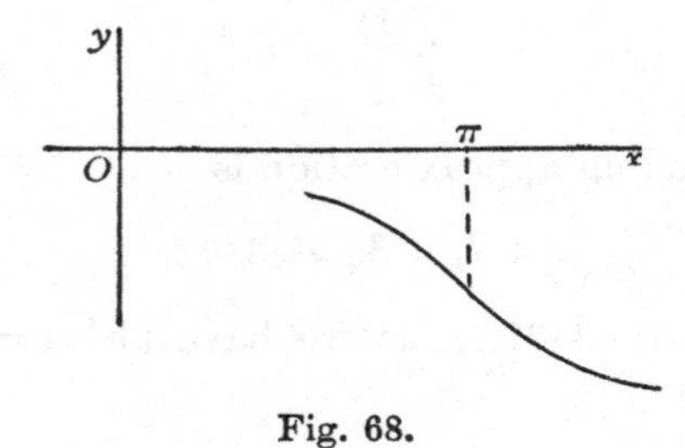

Fig. 68.

Note. This solution is less than π, as we expected from the diagram.

14. Leibniz's theorem. The theorem which follows is useful in calculating the higher differential coefficients necessary for a Maclaurin expansion.

To prove that, *if* $f(x)$ *is the product of two functions* u, v, *so that*

$$f(x) = uv,$$

then

$$f^{(n)}(x) = u^{(n)}v + {}_nc_1 u^{(n-1)}v' + {}_nc_2 u^{(n-2)}v'' + \dots$$
$$+ {}_nc_p u^{(n-p)}v^{(p)} + \dots + {}_nc_{n-1}u'v^{(n-1)} + {}_nc_n uv^{(n)},$$

where ${}_nc_p$ *is the binomial coefficient*

$${}_nc_p \equiv \frac{n!}{p!(n-p)!}.$$

We use the method of mathematical induction, assuming the result to be true for a certain integer N, so that

$$f^{(N)}(x) = u^{(N)}v + \dots + {}_Nc_p u^{(N-p)}v^{(p)} + \dots + {}_Nc_N uv^{(N)}.$$

Now differentiate this expression to obtain $f^{(N+1)}(x)$. The differential coefficient of a product such as $u^{(N-p)}v^{(p)}$ is

$$u^{(N-p+1)}v^{(p)}+u^{(N-p)}v^{(p+1)}.$$

As we write down these terms for the series on the right, we put the answer in two lines, the top line consisting of terms such as $u^{(N-p+1)}v^{(p)}$ and the lower of $u^{(N-p)}v^{(p+1)}$; also we displace the lower line one place to the right, thus:

$$u^{(N+1)}v+\dots\dots\dots+{}_N c_{p-1}u^{(N-p+2)}v^{(p-1)}+{}_N c_p\,u^{(N-p+1)}v^{(p)}+\dots\dots\dots\dots+u'v^{(N)}$$

$$+u^{(N)}v'+\dots\dots\dots\dots\dots+{}_N c_{p-1}u^{(N-p+1)}v^{(p)}+{}_N c_p\,u^{(N-p)}v^{(p+1)}+\dots\dots+uv^{(N+1)}$$

Now the coefficient of $u^{(N-p+1)}v^{(p)}$ is

$$\begin{aligned}&{}_N c_p+{}_N c_{p-1}\\ &=\frac{N!}{p!(N-p)!}+\frac{N!}{(p-1)!\,(N-p+1)!}\\ &=\frac{N!}{p!(N-p+1)!}\{(N-p+1)+p\}\\ &=\frac{(N+1)!}{p!(N+1-p)!}\\ &={}_{N+1}c_p.\end{aligned}$$

Hence

$$f^{(N+1)}(x)=u^{(N+1)}v+\dots+{}_{N+1}c_p\,u^{(N+1-p)}v^{(p)}+\dots+uv^{(N+1)}.$$

It follows that, if the theorem is true for any particular value N, then it is true for $N+1$, $N+2$, and all subsequent values. But it is easily established when $N=1$, being merely the result

$$f'(x)=u'v+uv'.$$

It is therefore true generally.

Note. The expression is symmetrical when regarded from the two ends, and will equally well be written in the form

$$f^{(n)}(x)=uv^{(n)}+\dots+{}_n c_p\,u^{(p)}v^{(n-p)}+\dots+u^{(n)}v.$$

EXAMPLES VII

Use Leibniz's theorem to find the following differential coefficients:

1. $\dfrac{d^4}{dx^4}(x^8 \sin x)$.
2. $\dfrac{d^4}{dx^4}(x^2 \sin x)$.
3. $\dfrac{d^5}{dx^5}(e^{2x} \cos 3x)$.
4. $\dfrac{d^6}{dx^6}(x^3 e^{3x})$.
5. $\dfrac{d^n}{dx^n}(x^3 \cos x) \quad (n > 3)$.
6. $\dfrac{d^n}{dx^n}(x^3 e^{2x}) \quad (n > 3)$.
7. $\dfrac{d^5}{dx^5}\{x^2(1-2x)^{10}\}$.
8. $\dfrac{d^8}{dx^8}\{x^2(3x+1)^{12}\}$.

ILLUSTRATION 6. *To apply Leibniz's theorem in finding a Maclaurin expansion for*

$$f(x) \equiv \log\{x + \sqrt{(x^2+1)}\}.$$

We have

$$f'(x) = \frac{1}{x+\sqrt{(x^2+1)}} \cdot \left\{1 + \tfrac{1}{2} \cdot \frac{2x}{\sqrt{(x^2+1)}}\right\}$$
$$= \frac{1}{\sqrt{(x^2+1)}}$$

on simplification. Hence

$$(x^2+1)\{f'(x)\}^2 = 1.$$

Differentiate. Then

$$(x^2+1).2f'(x)f''(x) + 2x\{f'(x)\}^2 = 0,$$

or
$$(x^2+1)f''(x) + xf'(x) = 0.$$

Differentiate n times, using Leibniz's theorem. Write the expansion from $(x^2+1)f''(x)$ on the first line, and from $xf'(x)$ on the second; note that the expansions terminate since $(x^2+1)''' = 0$, $x'' = 0$. We obtain

$$(x^2+1)f^{(n+2)}(x) + n.2x.f^{(n+1)}(x) + \frac{n(n-1)}{2!}.2.f^{(n)}(x)$$
$$+ \quad x.f^{(n+1)}(x) + \quad n \quad .1.f^{(n)}(x) = 0.$$

We require values when $x = 0$, so that

$$f^{(n+2)}(0) + \left\{\frac{2n(n-1)}{2!} + n\right\} f^{(n)}(0) = 0,$$

or

$$f^{(n+2)}(0) = -n^2 f^{(n)}(0).$$

But

$$f(0) = \log 1 = 0,$$

and so

$$f''(0) = f^{(\mathrm{iv})}(0) = \ldots = f^{(2n)}(0) = 0.$$

Also

$$f'(0) = 1,$$

so that

$$f'''(0) = -1^2 . 1,$$

$$f^{(\mathrm{v})}(0) = +3^2 . 1,$$

$$f^{(\mathrm{vii})}(0) = -5^2 . 3^2 . 1,$$

and so on. Hence the Maclaurin expansion is

$$x . 1 + \frac{x^3}{3!}(-1) + \frac{x^5}{5!}(3^2 . 1) + \frac{x^7}{7!}(-5^2 . 3^2 . 1) + \ldots,$$

or

$$x - \frac{1}{2}\frac{x^3}{3} + \frac{1 . 3}{2 . 4}\frac{x^5}{5} - \frac{1 . 3 . 5}{2 . 4 . 6}\frac{x^7}{7} + \ldots,$$

converging to $\log\{x + \sqrt{(x^2+1)}\}$ for such values of x (not considered here) as make the series convergent.

EXAMPLES VIII

Use the method of Illustration 6 to obtain the following expansions:

1. $f(x) \equiv \sin^{-1} x$.

2. $f(x) \equiv \tan^{-1} x$.

3. Prove that, if $f(x) \equiv \cos(m \sin^{-1} x)$, then

$$(1-x^2) f^{(n+2)}(x) - (2n+1) x f^{(n+1)}(x) + (m^2 - n^2) f^{(n)}(x) = 0.$$

4. Prove that, if $f(x) \equiv (\sin^{-1} x)^2$, then

$$(1-x^2) f^{(n+2)}(x) - (2n+1) x f^{(n+1)}(x) - n^2 f^{(n)}(x) = 0.$$

REVISION EXAMPLES IV

'Advanced' Level

1. Differentiate

$$\text{(i) } e^{ax^2 \log bx}, \quad \text{(ii) } \sin^{-1}\left\{\frac{ax}{\sqrt{(1+a^2x^2)}}\right\}.$$

If $y = e^{a \sin bx}$, and y', y'' are the first and second differential coefficients of y with respect to x show that

$$(yy''-y'^2)^2 = b^2y^2(a^2b^2y^2-y'^2).$$

Calculate the values of y', y'' when $x = 0$.

2. Differentiate

$$\tan^{-1}\frac{1-2x}{2+x}, \quad \sin^{-1}\sqrt{\left(\frac{1-x}{1+x}\right)},$$

$$\log(\sec x + \tan x), \quad \frac{2x+1}{x^2-8x-2}.$$

Find the values of x for which

$$\frac{2x+1}{x^2-8x-2}$$

is a maximum or minimum, distinguishing the maximum from the minimum.

3. If $y = e^x \tan x$, prove that

$$\frac{d^2y}{dx^2} - 2(1+\tan x)\frac{dy}{dx} + (1+2\tan x)y = 0.$$

Prove that $y = ue^x \tan x$ also satisfies this equation when u is a function of x such that

$$\tan x \frac{d^2u}{dx^2} + 2\frac{du}{dx} = 0.$$

Verify that $u = \cot x$ is such a function.

4. Differentiate $y = \sec^m x \tan^n x$.

Deduce that the kth differential coefficient of $\sec^m x$ can be expressed in the form $\sec^m x\, P_k(\tan x)$, where $P_k(\tan x)$ is a polynomial in $\tan x$ of degree k.

Evaluate $P_4(\tan x)$ when both m and k are taken equal to 4.

5. Differentiate with respect to x:

$$\frac{1}{x-2}+2\log\frac{x-3}{x-2}, \quad x-\tan x+\tfrac{1}{3}\tan^3 x,$$

simplifying your answers as much as you can.

Prove that the first function has a maximum at $x = 1$ and that the second function (whose differential coefficient vanishes at $x = 0$) has neither a maximum nor a minimum at $x = 0$.

6. Find the first four differential coefficients of $\sin^4 x$.

Show that the function

$$x^4(x-\tfrac{1}{2}\pi)^2+\sin^4 x$$

has turning values at $x = 0$ and at $x = \frac{1}{2}\pi$, and determine whether they are maxima or minima.

7. Prove that, if $y = \sin^2(x^2)$, then

$$x\frac{d^2y}{dx^2}-\frac{dy}{dx}+16x^3y = 8x^3.$$

Prove that the result of changing the independent variable in this equation from x to ξ, where $\xi = x^2$, is

$$\frac{d^2y}{d\xi^2}+ay = b,$$

where a, b are constants to be determined.

Prove also that, if A, B are any constants, the function

$$y = \tfrac{1}{2}+A\cos(2x^2)+B\sin(2x^2)$$

satisfies the first differential equation.

8. Differentiate

$$\left(x+\frac{1}{x}\right)^n, \quad (\cos x+\sec x)^n$$

with respect to x, and find the nth differential coefficient of

$$\left(x+\frac{1}{x}\right)^2$$

for all positive integral values of n, distinguishing the cases $n \leqslant 2$ and $n > 2$.

9. Show that the polynomial

$$f(x) \equiv x - \frac{4}{\pi^3}(2\pi - 5)x^3 + \frac{16}{\pi^5}(\pi - 3)x^5$$

and its differential coefficient $f'(x)$ have the same values as $\sin x$ and its differential coefficient respectively, at the values $x = 0$, $x = \pm\frac{1}{2}\pi$.

Show that the error in using $\int_0^{\frac{1}{2}\pi} f(x)dx$ as an approximation for $\int_0^{\frac{1}{2}\pi} \sin x\, dx$ is less than 0·1 per cent.

10. (i) Given that $\dfrac{du}{dx} = e^x, \quad \dfrac{dv}{dx} = \sin x,$

and that $u = 1, v = 0$ when $x = 0$, show that

$$\frac{d^2(uv)}{dx^2} = 1$$

when $x = 0$.

(ii) Given that $y = ax^n + bx^{1-n}$, prove that

$$x\frac{dy}{dx} + (n-1)y = (2n-1)ax^n,$$

and form an equation in $x, y, \dfrac{dy}{dx}, \dfrac{d^2y}{dx^2}$ which does not contain a or b.

11. Differentiate with respect to x:

$$\sqrt{\left(\frac{x-1}{x+1}\right)}, \quad \log \sec x.$$

Prove that $$\frac{d^n}{dx^n}\left(\frac{1}{1-x}\right) = \frac{n!}{(1-x)^{n+1}},$$

and find the nth differential coefficient of

$$\frac{x}{(1-x)(1-2x)}.$$

12. Find the nth derivatives with respect to x of

(i) $$\frac{1}{x} \quad \text{and} \quad \frac{1}{x^2-1};$$

(ii) $$\sin x \quad \text{and} \quad x \sin x.$$

13. (i) Find the derivative of $y = \sin^2 x$ with respect to $z = \cos x$ by evaluating the limit of $\delta y/\delta z$.

(ii) Differentiate with respect to x:

$$\frac{\sin x}{1+\tan x}, \quad (1-x)\sqrt{(1+x^2)}, \quad \log_e(e^x+e^{-x}).$$

(iii) Prove that $\dfrac{d^2x}{dy^2} = -\dfrac{d^2y}{dx^2}\bigg/\left(\dfrac{dy}{dx}\right)^3.$

14. Find from first principles the differential coefficient of $1/x^3$ with respect to x.

What is the differential coefficient of $1/x^3$ with respect to x^2?

Differentiate with respect to x:

$$\log_e \cos x, \quad x\sqrt{(1-x^2)}, \quad \frac{e^x}{(1+x^2)^2}.$$

15. Find the nth differential coefficient of y with respect to x in each of the following cases:

(i) $y = 1/x^2$, (ii) $y = \sin 2x$, (iii) $y = e^{2x}\sin 2x$.

There are three values of k for which the function $y = x^k e^{-x}$ satisfies the equation

$$\frac{d^3y}{dx^3}+3\frac{d^2y}{dx^2}+3\frac{dy}{dx}+y = 0.$$

Find these values.

16. Differentiate with respect to x:

$$\frac{1-2x}{(1+3x)^2}, \quad e^{\tan^2 x}, \quad \sin^{-1}\frac{1}{\sqrt{(1+x^2)}}.$$

Prove that, when $y = at^2+2bt+c$, $t = ax^2+2bx+c$, and a, b, c are constants,

$$\frac{d^3y}{dx^3} = 24a^2(ax+b).$$

17. Differentiate $\quad e^{\tan x}, \quad \sqrt{\left(\dfrac{x+1}{x-2}\right)}.$

Show that, if $y = \sin\theta$, $x = \cos\theta$, then

$$\frac{d}{dx}(y^3) = -\frac{3}{2}\sin 2\theta, \quad \frac{d^2}{dx^2}(y^5) = 5\sin 3\theta.$$

18. A particle moves along the axis of x so that its distance from the origin t seconds after starting is given by the formula

$$x = a\cos kt + \tfrac{1}{2}akt,$$

where a, k are positive constants. Find expressions for the velocity and acceleration in terms of t, and prove that the particle is not always moving in the same direction along the axis.

Find the positions of the particle at the two times $t = \pi/6k$ and $t = 13\pi/6k$; also find the position of the particle at the instant between these times at which it is momentarily at rest, and deduce the *total* distance travelled between the two times.

19. A particle moves in a plane so that its coordinates at time t are given by $x = e^t \cos t, y = e^t \sin t$. Find the magnitudes of the velocity and of the acceleration* at time t and prove that the acceleration is always at right angles to the radius vector.

Draw a rough sketch of the path of the particle, from time $t = 0$ to $t = \pi$, and indicate the direction of motion at times 0, $\tfrac{1}{4}\pi, \tfrac{1}{2}\pi, \tfrac{3}{4}\pi, \pi$.

20. A particle moves in a plane so that its position at time t is given by $x = a\cos pt, y = b\sin pt$. Find expressions for (i) the magnitude v of the velocity at time t; (ii) the magnitude f of the acceleration at time t; and prove that there is no value of t for which $f = dv/dt$.

Prove also that the resultant acceleration makes an angle θ with the normal to the path, where

$$2ab\tan\theta = (a^2-b^2)\sin 2pt.$$

21. A point moves in a straight line so that its distance at time t from a given point O of the line is x, where

$$x = t^2\sin t + 6t\cos t - 12\sin t.$$

Find its velocity at time t, and prove that the acceleration is then

$$-t^2\sin t - 2t\cos t + 2\sin t.$$

Determine the times ($t>0$) at which the acceleration has a turning value, distinguishing between maxima and minima.

* If v, f are the velocity and acceleration respectively, then $v^2 = \dot{x}^2+\dot{y}^2$, $f^2 = \ddot{x}^2+\ddot{y}^2$.

22. A particle moves along the axis of x so that its distance from the origin t seconds after starting is given by the formula $x = a\cos pt$. Prove that the velocity of the particle changes direction once, and only once, between the times $t = 0$, $t = 2\pi/p$ and that the change of direction occurs at the point $x = -a$.

The distance from the origin of a second particle is given by the formula

$$x = a\cos pt + \tfrac{1}{2}a\cos 2pt.$$

Write down expressions for its velocity and acceleration at time t. Show that between $t = 0$ and $t = 2\pi/p$ the velocity of the particle changes direction three times, and find the values of x at which these changes occur.

23. Prove that the equation of the tangent to the curve given by

$$x = 3t^2 + 1, \quad y = 2t^3 - 1$$

at the point where $t = \tan\alpha$ is

$$y - x\tan\alpha + \tan^3\alpha + \tan\alpha + 1 = 0.$$

Show that the curve lies on the positive side of the line $x = 1$ and is symmetrical about the line $y = -1$, and prove that the area bounded by the line $x = 4$ and the curve is $\frac{24}{5}$.

24. Prove that there are two distinct tangents to the curve

$$y = x^4 - x + 3$$

which pass through the origin. Find their equations, and their points of contact with the curve.

Give a rough sketch of the curve.

25. A curve is defined by the parametric equations

$$x = a(1-t^2), \quad y = a(2-t)(1-t),$$

where a is a positive constant. Prove that

(i) the curve passes through the points A, O, B whose coordinates are $(0, 6a), (0, 0), (-3a, 0)$.

(ii) the point t is in the first quadrant when $-1 < t < 1$ and in the third quadrant when $1 < t < 2$.

Make a rough sketch showing the part of the curve corresponding to values of t between -1 and $+2$.

Find the equation of the tangent to the curve at O, and prove that the area bounded by the arc AB and the chord AB is $27a^2/2$.

26. Prove that the slope of the curve whose equation is

$$y = 1 + x + \tfrac{1}{2}x^2 + \tfrac{1}{6}x^3$$

is always positive.

Show that the curve has a point of inflexion where $x = -1$, the slope there being $\frac{1}{2}$.

Prove also that the tangent at the point $(0, 1)$ meets the curve again at the point $(-3, -2)$.

Sketch the curve, indicating clearly the point of inflexion and the tangents at the points $(-1, \frac{1}{3})$ and $(0, 1)$.

27. A, O, B are three fixed points in order on a straight line, and $AO = p$, $OB = q$. A fixed circle has centre O and radius a greater than p or q, and P is a point on this circle. Show that the perimeter of the triangle APB is a maximum when OP bisects the angle APB, and find the corresponding magnitude of the angle POA.

28. Prove that the maximum and minimum values of the function $y = x\cos 3x$ occur when $3\tan 3x = 1/x$, and discuss the behaviour of the function when $x = 0$.

By considering the curves $y = \tan 3x$, $y = 1/x$, show that maximum values occur near the values $x = \frac{2}{3}k\pi$, and minimum values near $x = \frac{1}{3}(2k+1)\pi$, the approximation becoming more exact as x becomes larger.

29. Given that $a^2x^4 + b^2y^4 = c^6$, where a, b, c are constants, show that xy has a stationary value $c^3/\sqrt{(2ab)}$. Is this value a maximum or a minimum?

30. (i) A right circular cylinder is inscribed in a given right circular cone. Prove that its volume is a maximum when its altitude is one-third that of the cone.

(ii) A right circular cone of height h stands on a base of radius $h\tan\alpha$. A cylinder of height $h - x$ is inscribed in the cone. Prove that S, the total surface of the cylinder, is equal to

$$2\pi\{x^2(\tan^2\alpha - \tan\alpha) + xh\tan\alpha\},$$

and prove that, when $\tan\alpha > \frac{1}{2}$, S increases steadily as x increases.

31. P is a variable point in the circumference of a fixed circle of which AB is a fixed diameter and O is the centre. Prove that,

when OP is perpendicular to AB, then the function $AP+PB$ has a maximum value and the function AP^3+PB^3 has a minimum value.

32. If $f(x)$ is a function of x and $f'(x)$ is its differential coefficient, show that, when h is small, $f(a+h)$ is approximately equal to $f(a)+hf'(a)$.

Without using trigonometrical tables, find to three significant figures (i) the value of $\cos 31°$, and (ii) the positive acute angle whose sine is 0·503. Give your answer to (ii) in degrees and tenths of a degree.

33. Calculate

(i) $\sqrt[3]{8{\cdot}05}$ to 4 significant figures,

(ii) $\cos 59°$ to 3 significant figures.

[Take π to be 3·142.]

34. Determine to two places of decimals that root of the equation

$$\frac{x^{\frac{5}{3}}}{x+2} = 3{\cdot}2104$$

whose value is nearly equal to 8.

35. Determine to 3 places of decimals the value of that root of the equation

$$x^3-3x+1=0$$

which lies between 1·5 and 1·6.

36. Prove, graphically or otherwise, that, if n is a large positive integer, there is a root of the equation $x \sin x = 1$ nearly equal to $2n\pi$. Show that a better approximation is $2n\pi+(1/2n\pi)$.

37. Prove that, if η is small, the equation

$$\theta+\sin\theta\cos\theta = 2\eta\cos\theta$$

has a small root approximately equal to

$$\eta-\tfrac{1}{6}\eta^3.$$

[You may assume the power-series expansions for $\sin\theta$ and $\cos\theta$.]

38. Prove that the result of differentiating the equation

$$(1+x^2)\frac{dy}{dx}-2x = 0$$

$n+1$ times $(n \geqslant 1)$ with respect to x is

$$(1+x^2)\frac{d^{n+2}y}{dx^{n+2}}+2(n+1)x\frac{d^{n+1}y}{dx^{n+1}}+n(n+1)\frac{d^n y}{dx^n} = 0.$$

Hence verify that, if k is a positive integer and

$$z = \frac{d^k}{dx^k}\log(1+x^2),$$

then z is a solution of the equation

$$(1+x^2)\frac{d^2z}{dx^2}+2(k+1)x\frac{dz}{dx}+k(k+1)z = 0.$$

39. By using Maclaurin's theorem, or otherwise, obtain the expansion of $\log(1+\sin x)$ in ascending powers of x as far as the term in x^4.

40. Prove that, if $y = \log_e \cos x$, then

$$\frac{d^3y}{dx^3}+2\frac{d^2y}{dx^2}\frac{dy}{dx} = 0.$$

Hence, or otherwise, obtain the Maclaurin expansion of $\log_e \cos x$ as far as the term in x^4.

Deduce the approximate relation

$$\log_e 2 = \frac{\pi^2}{16}\left(1+\frac{\pi^2}{96}\right).$$

41. Prove that, if $y = e^{\tan x}$, then

$$\frac{dy}{dx} = y(1+t^2), \quad \frac{d^2y}{dx^2} = \frac{dy}{dx}(1+t)^2,$$

where $t \equiv \tan x$.

Prove that the expansion of y as far as the term in x^3 is

$$y = 1+x+\tfrac{1}{2}x^2+\tfrac{1}{2}x^3.$$

42. Prove that, if $y = \sin(\log x)$, where $x > 0$, then

$$x^2\frac{d^2y}{dx^2}+x\frac{dy}{dx}+y = 0.$$

By induction, or otherwise, prove that

$$x^2 \frac{d^{n+2}y}{dx^{n+2}} + (2n+1)x \frac{d^{n+1}y}{dx^{n+1}} + (n^2+1)\frac{d^n y}{dx^n} = 0,$$

where n is a positive integer.

43. If $y = e^x \log x$, show that

$$x\frac{d^2y}{dx^2} - (2x-1)\frac{dy}{dx} + (x-1)y = 0.$$

Find the equation obtained by differentiating this equation n times.

44. Prove that, when $y = e^{ax} \sin bx$,

$$\frac{d^2y}{dx^2} - 2a\frac{dy}{dx} + (a^2+b^2)y = 0,$$

and that
$$\frac{dy}{dx} = y(a + b\cot bx).$$

Prove that, if
$$e^{ax}\sin bx = \sum_{n=1}^{\infty} \frac{c_n}{n!} x^n,$$

then
$$c_{n+2} - 2ac_{n+1} + (a^2+b^2)c_n = 0,$$

and find the values of c_1, c_2, c_3.

45. If $\cos y = \cos\alpha\cos x$, where x, y, α lie between 0 and $\frac{1}{2}\pi$ radians and α is constant, find the values of $y, \frac{dy}{dx}, \frac{d^2y}{dx^2}$ when $x = 0$.

Taking x to be so small that x^3 and higher powers of x are negligible, use Maclaurin's theorem to show that

$$y = \alpha + \tfrac{1}{2}x^2 \cot\alpha.$$

Hence calculate y in degrees, correct to 0·001°, if $\alpha = 45°$ and $x = 1° 48'$. [Take $\pi = 3{\cdot}142$.]

46. By using Taylor's theorem obtain the expansion of

$$\tan(x + \tfrac{1}{4}\pi)$$

in powers of x up to the term in x^3.

Hence calculate the value of tan 44° 48′ correct to four places of decimals. [Take $\pi = 3{\cdot}142$.]

47. Prove that, if $y = \log_e\left(\frac{1+\sin x}{1-\sin x}\right)$, then

$$\text{(i)}\quad \frac{dy}{dx} = 2\sec x,$$

and (ii) the expansion of y in a series of powers of x as far as the term in x^4 is $2x+\frac{1}{3}x^3$.

Find, correct to four significant figures, the value of y when $x = 1^\circ\ 48'$, taking $\pi = 3{\cdot}142$.

48. Prove that, if $y = (\sin^{-1}x)^2$, then

$$(1-x^2)\left(\frac{dy}{dx}\right)^2 = 4y,$$

and
$$(1-x^2)\frac{d^2y}{dx^2} - x\frac{dy}{dx} = 2.$$

The Maclaurin expansion of y in powers of x is taken to be

$$y = a_0 + a_1 x + \frac{a_2}{2!}x^2 + \ldots + \frac{a_n}{n!}x^n + \ldots.$$

Given that $y = 0$ when $x = 0$, prove that $a_0 = 0$ and $a_1 = 0$. Prove also that $a_{n+2} = n^2 a_n$ when $n > 0$, and hence show that

$$y = x^2 + \sum_{n=2}^{\infty} \frac{2^{2n-1}(n-1)!}{n(n+1)\ldots(2n)}x^{2n}.$$

49. If $y = \int_1^x u^u\,du$, prove that

$$\frac{d^2y}{dx^2} = (1+\log x)\frac{dy}{dx}.$$

Find the values of the first four differential coefficients of y when $x = 1$, and, by using Taylor's expansion in the form

$$f(1+h) = f(1) + hf'(1) + \tfrac{1}{2}h^2 f''(1) + \ldots,$$

deduce that the value of $\int_1^{1{\cdot}1} u^u\,du$ is approximately $0{\cdot}1053$.

50. If $y = \tan x$, prove that

$$\frac{dy}{dx} = 1 + y^2, \quad \frac{d^2y}{dx^2} = 2y + 2y^3,$$

and find the third, fourth, and fifth derivatives of y.

Hence find the expansion of $\tan x$ in a series of powers of x up to x^5.

51. Prove that, if $y = \sin(m \sin^{-1} x)$, then

$$(1-x^2)\left(\frac{dy}{dx}\right)^2 + m^2y^2 = m^2,$$

$$(1-x^2)\frac{d^2y}{dx^2} - x\frac{dy}{dx} + m^2y = 0.$$

Show that the first two terms in the expansion of the principal value of y in ascending powers of x are

$$mx + \tfrac{1}{6}m(1-m^2)x^3.$$

52. Find the indefinite integrals:

$$\int xe^{-x^2}\,dx, \quad \int \frac{\tan^{-1}x}{1+x^2}\,dx, \quad \int x^2e^x\,dx.$$

If u, v are functions of x, and dashes denote differentiation with respect to x, show that

$$\int (uv''' + u'''v)\,dx = uv'' - u'v' + u''v + \text{constant}.$$

53. Show that $$\int_0^a f(x)dx = \int_0^a f(a-x)dx.$$

Deduce that $$\int_0^{\frac{1}{4}\pi}\left(\frac{1-\sin 2x}{1+\sin 2x}\right)dx = \int_0^{\frac{1}{4}\pi}\tan^2 x\,dx,$$

and evaluate the integral.

54. Find the indefinite integrals:

$$\int x\log x\,dx, \quad \int \frac{dx}{\sin x}, \quad \int \frac{x^2}{1-x^4}\,dx.$$

Prove that, when the expression $e^x \sin x$ is integrated n times, the result is

$$2^{-\frac{1}{2}n}e^x\sin(x - \tfrac{1}{4}n\pi) + P_{n-1},$$

where P_{n-1} is a polynomial of degree $n-1$ in x.

55. Find the indefinite integrals of

$$\cos^3 x, \quad \frac{1}{x^3(x+1)}, \quad (\log x)^3, \quad \frac{1}{3-2\cos x}.$$

56. Find the indefinite integrals of

$$32\cos^4 x, \quad 27x^2(\log x)^2, \quad \frac{2(1+x^4)}{1-x^4}, \quad a^x.$$

57. Integrate with respect to x:

$$\frac{4}{x(1+x)(1+x^2)}.$$

Evaluate the definite integrals:

$$\int_1^2 \log x\,dx, \quad \int_0^{\frac{1}{2}\pi} \frac{dx}{1+3\cos^2 x}.$$

58. Find the indefinite integrals:

$$\int \frac{x^2\,dx}{1-x^6}, \quad \int \frac{dx}{\sqrt{\{x(2-x)\}}}, \quad \int x\sec^2 x\,dx, \quad \int \tan^4 x\,dx.$$

59. Find the indefinite integrals:

$$\int \frac{dx}{(1-3x)^3}, \quad \int \tan^2 2x\,dx, \quad \int 2x\tan^{-1} x\,dx, \quad \int \frac{\cos x\,dx}{2-\cos^2 x}.$$

60. Find the values of

$$\int_0^{\frac{1}{2}\pi} \sin^5 x\,dx, \quad \int_0^{\pi} \cos^4 x\,dx, \quad \int_0^{2\pi} e^x \sin^2 x\,dx.$$

61. (i) Find the indefinite integrals of

$$\frac{1}{x(x+1)^2}, \quad \frac{2x+3}{x^2+2x+2}.$$

(ii) By substitution, or otherwise, prove that

$$\int_0^{1/\sqrt{2}} x\sin^{-1} x\,dx = \tfrac{1}{8}, \quad \int_0^1 x^3\sqrt{(x^2+1)}\,dx = \frac{2(1+\sqrt{2})}{15}.$$

62. Find the indefinite integrals of

$$\cos^3 3x, \quad \frac{1}{3-2\cos x}, \quad x^3 \sin x.$$

Use the method of integration by parts to integrate

$$(1-x^2)\frac{d^2y}{dx^2}-4x\frac{dy}{dx}-2y-8$$

twice with respect to x.

63. By means of the substitution $x = \alpha\cos^2\theta+\beta\sin^2\theta$, or otherwise, prove that

$$\int_\alpha^\beta \frac{dx}{\sqrt{\{(x-\alpha)(\beta-x)\}}} = \pi.$$

Prove also that

$$\int_\alpha^\beta \{(x-\alpha)(\beta-x)\}^{n-\frac{1}{2}}dx = \left(\frac{\beta-\alpha}{2}\right)^{2n}\int_0^\pi \sin^{2n}\psi\, d\psi.$$

64. Prove that

$$\int_0^\pi \cos^2\theta\, d\theta = \tfrac{1}{2}\pi, \quad \int_0^\pi \cos^4\theta\, d\theta = \tfrac{3}{8}\pi.$$

Find the area bounded by the curve $r = a(1+\cos\theta)$ and determine the position of the centre of gravity of the area.

65. Find the equation of the normal at the point $(\xi, \sin\xi)$ to the curve whose equation is $y = \sin x$.

Prove that, if ξ lies between 0, π, the normal at P divides the area bounded by the x-axis and that arc of the curve for which $0 \leqslant x \leqslant \pi$ in the ratio

$$(2-\cos\xi-\cos^3\xi) : (2+\cos\xi+\cos^3\xi).$$

66. The ellipse

$$\frac{x^2}{a^2}+\frac{y^2}{b^2} = 1$$

is rotated through two right angles about the x-axis. Prove that the volume generated is $\frac{4}{3}\pi ab^2$.

(i) Prove that, if a and b are varied subject to the condition $a+b = \frac{1}{2}$, then the greatest volume generated is $2\pi/81$.

(ii) The volume is cut in two by the plane generated by the rotation of the y-axis. Prove that the centre of gravity of either part of the volume is at a distance $\frac{3}{8}a$ from the plane of separation.

67. The complete curve $x^2/a^2+y^2/b^2 = 1$ is rotated round the y-axis through two right angles. Find the volume generated by the area enclosed by the curve.

The semi-axes a and b of the curve are each increased by ϵ. Prove that, if ϵ is small, the increase in volume is approximately $\frac{4}{3}\pi a(a+2b)\epsilon$.

68. OA is a straight rod of length a in which the density at a point distant x from O is $b+cx$, where b and c are constants. Find the distance of the centre of gravity of the rod from O.

69. The gradient at any point (x, y) of a curve is given by

$$\frac{dy}{dx} = -3x^2+3,$$

and the curve passes through the point $(2, 0)$. Find its equation and sketch the graph, indicating the turning points.

Find the distance from the y-axis of the centre of gravity of a uniform lamina bounded by the curve and the positive halves of the x and y axes.

70. A lamina in the shape of the parabola $y^2 = 4ax$ bounded by the chord $x = a$ is rotated (i) about the axis of y, (ii) about the line $x = a$. Prove that the volumes generated in the two cases are $\frac{16}{5}\pi a^3, \frac{32}{15}\pi a^3$.

71. Integrate with respect to x:

$$\sin^2 x, \quad \sin^3 x, \quad x^2 \cos x.$$

The portion of the curve $y = \sin x$ from $x = 0$ to $x = \frac{1}{2}\pi$ revolves round the axis of y. Prove that the volume contained between the surface so formed and the plane $y = 1$ is $\frac{1}{4}\pi(\pi^2-8)$.

72. The coordinates of a point on a curve referred to rectangular axes are $(at^2, 2at)$, where t is a variable parameter which lies between 0 and 1. Make a rough graph of the curve.

Calculate (i) the area enclosed by the curve and the lines $x = a, y = 0$; and (ii) the area of the surface obtained by revolving this part of the curve about the x-axis.

73. Find the area contained between the x-axis and that part of the curve $x = 2t^2+1, y = t^2-2t$ which corresponds to values of t lying between 0 and 2.

Find also the coordinates of the centroid of this area.

74. Prove that the tangent to the curve

$$\sqrt{x}+\sqrt{y} = \sqrt{a}$$

at the point (x, y) makes intercepts $\sqrt{(ax)}, \sqrt{(ay)}$ on the axes.

A solid is generated by rotating about the x-axis the area whose complete boundary is formed by (i) the arc of this curve joining the points $(a, 0), (0, a)$, and (ii) the straight lines joining the origin to these two points. Prove that, when this solid is of uniform density ρ, its mass M is $\frac{1}{15}\pi\rho a^3$.

Prove also that the moment of inertia of the solid about Ox is $\frac{1}{6}Ma^2$.

75. Find the area bounded by the curve $x = 2a(t^3-1), y = 3at^2$ and the straight lines $x = 0, y = 0$.

Prove that the tangent to the curve at the point $t = 1$ meets the curve again at the point $t = -\frac{1}{2}$, and find the area bounded by the parts of the tangent and of the curve that lie between these points.

76. Prove that the parabola $y^2 = 2ax$ divides the area of the ellipse $4x^2+3y^2 = 4a^2$ into two parts whose areas are in the ratio $4\pi+\sqrt{3} : 8\pi-\sqrt{3}$.

77. The portion of the curve $y^2 = 4ax$ from $(a, 2a)$ to $(4a, 4a)$ revolves round the tangent at the origin. Prove that the volume bounded by the curved surface so formed and plane ends perpendicular to the axis of revolution is $\frac{62}{5}\pi a^3$, and find the square of the radius of gyration of this volume about the axis of revolution.

78. Find the coordinates of the centre of gravity of the area enclosed by the loop of the curve whose equation is $r = a\cos 2\theta$, which lies in the sector bounded by the lines $\theta = \pm\frac{1}{4}\pi$.

Find also the volume obtained by rotating this loop about the line $\theta = \frac{1}{2}\pi$.

79. Find the coordinates of the centre of gravity of the loop of the curve traced out by the point $x = 1-t^2, y = t-t^3$.

Find also the volume obtained by rotating this loop about the line $x = y$.

80. A plane uniform lamina is bounded by the curve $y^2 = 4ax$ and the straight lines $y = 0, x = a$. Find the area and the centre of gravity of the lamina.

The lamina is rotated about the axis Ox to form a (uniform) solid of revolution. Find the centre of gravity of the solid and, assuming the density of the solid to be ρ, find its moment of inertia about the axis Ox.

81. A lamina in the shape of the parabola $y^2 = 4ax$, bounded by the chord $x = a$, is rotated (i) about the axis of y, (ii) about the line $x = a$. Prove that the two volumes thus generated are in the ratio 3 : 2.

82. Prove by integration that the moment of inertia of a uniform circular disc, of mass m and radius a, about a line through its centre perpendicular to its plane is $\frac{1}{2}ma^2$.

The mass of a uniform solid right circular cone is M, and the radius of its base is a. Prove that its moment of inertia about its axis is $\frac{3}{10}Ma^2$.

83. Evaluate $\int_0^{\frac{1}{2}\pi} \cos^n\theta\, d\theta$ when $n = 1, 2, 3, 4$.

The area bounded by the axis of x, the line $x = a$, and the curve

$$x = a\sin\theta, \quad y = a(1-\cos\theta),$$

from $\theta = 0$ to $\theta = \frac{1}{2}\pi$, revolves round the axis of x. Prove that the volume generated is $\frac{1}{6}\pi a^3(10-3\pi)$.

84. The portion of the curve $x^2 = 4a(a-y)$ from $x = -2a$ to $x = 2a$ revolves round the axis of x. Prove that the volume contained by the surface so formed is $\frac{32}{15}\pi a^3$, and find its radius of gyration about the axis of revolution.

85. Sketch the curve $r = a(1+\cos\theta)$, and find the area it encloses and the volume of the surface formed by revolving it about the line $\theta = 0$.

86. Find the three pairs of consecutive integers (positive, negative, or zero) between which the roots of the equation

$$x^3 - 3x^2 + 1 = 0$$

lie, and evaluate the largest root correct to two places of decimals.

87. Show that the equation

$$x^3 - 3x - 7 = 0$$

has one real root, and find it correct to three places of decimals.

88. Show that the equation

$$x^3+2x^2+3x+5=0$$

has one real root, and find it correct to three places of decimals.

89. For a function f having an Nth differential coefficient, Taylor's theorem expresses $f(a+x)$ as a polynomial of degree $N-1$ in x, together with a remainder. State the form taken by this polynomial, and one form of the remainder.

Prove by induction, or otherwise, that

$$\frac{d^n}{dx^n}(e^x \sin x\sqrt{3}) = 2^n e^x \sin(x\sqrt{3}+\tfrac{1}{3}n\pi).$$

Hence find the coefficients a_n in the Maclaurin series $\Sigma a_n x^n$ of the function $e^x \sin x\sqrt{3}$.

By means of Taylor's theorem, show that when $x>0$ the difference between $e^x \sin x\sqrt{3}$ and $\sum_{n=0}^{N-1} a_n x^n$ is not greater than

$$\frac{(2x)^N e^x}{N!}.$$

[A proof showing that the difference is not greater than

$$\frac{(2x)^N e^x}{(N-1)!}$$

is acceptable if the form of remainder which you have quoted leads to the result.]

90. If $$y=(1+x)^{-1}\log(1+x),$$

show that $$(1+x)^2\frac{dy}{dx}+(1+x)y=1.$$

Deduce the first four terms in the Maclaurin series for y in powers of x.

91. Show that $$y=\{x+\sqrt{(1+x^2)}\}^k$$

satisfies the relations $$y'\sqrt{(1+x^2)}=ky,$$

$$y''(1+x^2)+xy'=k^2y.$$

Deduce the expansion

$$y=1+kx+\frac{k^2}{2!}x^2+\frac{k(k^2-1)}{3!}x^3+\frac{k^2(k^2-2^2)}{4!}x^4+\dots.$$

Verify that this agrees with the series derived from the binomial series when $k=1$.

CHAPTER IX

THE HYPERBOLIC FUNCTIONS

1. The hyperbolic cosine and sine. There are two functions with properties closely analogous to those of $\cos x$ and $\sin x$. They are called the *hyperbolic cosine of* x, and the *hyperbolic sine of* x, and are written as $\cosh x$ and $\sinh x$. We define them in terms of the exponential function as follows:

$$\cosh x = \tfrac{1}{2}(e^x + e^{-x}),$$

$$\sinh x = \tfrac{1}{2}(e^x - e^{-x}).$$

We establish a succession of properties similar to those of the cosine and sine:

(i) *To prove that* $\quad \cosh^2 x - \sinh^2 x = 1.$

The left-hand side is

$$\begin{aligned} &\tfrac{1}{4}\{(e^x + e^{-x})^2 - (e^x - e^{-x})^2\} \\ &\qquad = \tfrac{1}{4}\{(e^{2x} + 2 + e^{-2x}) - (e^{2x} - 2 + e^{-2x})\} \\ &\qquad = \tfrac{1}{4}(4) \\ &\qquad = 1. \end{aligned}$$

(ii) *To prove that*

$$\cosh(x+y) = \cosh x \cosh y + \sinh x \sinh y.$$

It is easier to start with the right-hand side:

$$\begin{aligned} &\tfrac{1}{4}\{(e^x + e^{-x})(e^y + e^{-y}) + (e^x - e^{-x})(e^y - e^{-y})\} \\ &\quad = \tfrac{1}{4}\{(e^{x+y} + e^{x-y} + e^{-x+y} + e^{-x-y}) + (e^{x+y} - e^{x-y} - e^{-x+y} + e^{-x-y})\} \\ &\quad = \tfrac{1}{2}(e^{x+y} + e^{-x-y}) \\ &\quad = \cosh(x+y). \end{aligned}$$

(iii) *Similarly*

$$\sinh(x+y) = \sinh x \cosh y + \cosh x \sinh y.$$

(iv) *As particular cases of* (ii), (iii),

$$\cosh 2x = \cosh^2 x + \sinh^2 x,$$

$$\sinh 2x = 2 \sinh x \cosh x.$$

COROLLARY (i). $\cosh^2 x = \frac{1}{2}(\cosh 2x + 1)$,

$$\sinh^2 x = \tfrac{1}{2}(\cosh 2x - 1).$$

COROLLARY (ii). Since

$$\cosh x - 1 = 2 \sinh^2 \frac{x}{2}$$

and $\sinh^2 \frac{x}{2}$ is positive, it follows that $\cosh x$ *is greater than unity for all* (*real*) *values of* x, i.e. $\cosh x \geqslant 1$.

Note.

$$\cosh 0 = 1,$$

$$\sinh 0 = 0.$$

(v) *To prove that* $\cosh x = \cosh(-x)$.

The right-hand side is

$$\tfrac{1}{2}(e^{-x} + e^{-(-x)}) = \tfrac{1}{2}(e^{-x} + e^{x})$$

$$= \cosh x.$$

(vi) *To prove that* $\sinh x = -\sinh(-x)$.

For

$$\sinh(-x) = \tfrac{1}{2}(e^{-x} - e^{-(-x)}) = \tfrac{1}{2}(e^{-x} - e^{x})$$

$$= -\sinh x.$$

Note. Sinh x is positive when x is positive, and negative when x is negative. For example, if x is positive, then e^x is greater than e^{-x}, since e is greater than 1. Hence $\frac{1}{2}(e^x - e^{-x})$ is positive.

(vii) *To prove that*

$$\frac{d}{dx}(\cosh x) = \sinh x,$$

$$\frac{d}{dx}(\sinh x) = \cosh x.$$

We have
$$\frac{d}{dx}(\cosh x) = \frac{1}{2}\frac{d}{dx}(e^x + e^{-x})$$
$$= \tfrac{1}{2}(e^x - e^{-x})$$
$$= \sinh x;$$

and
$$\frac{d}{dx}(\sinh x) = \frac{1}{2}\frac{d}{dx}(e^x - e^{-x})$$
$$= \tfrac{1}{2}(e^x + e^{-x})$$
$$= \cosh x.$$

(viii) *To prove that*

$$\int \cosh dx = \sinh x,$$

$$\int \sinh x dx = \cosh x.$$

These results follow at once from (vii).

(ix) *By the formulæ of* (vii), we have the relations

$$\frac{d^2}{dx^2}(\cosh x) = \cosh x,$$

$$\frac{d^2}{dx^2}(\sinh x) = \sinh x.$$

Thus $\cosh x$, $\sinh x$ both satisfy the relation

$$\frac{d^2y}{dx^2} = y.$$

(x) The following *expansions in power series* are immediate consequences of Maclaurin's theorem:

$$\cosh x = 1 + \frac{x^2}{2!} + \frac{x^4}{4!} + \frac{x^6}{6!} + \ldots,$$

$$\sinh x = x + \frac{x^3}{3!} + \frac{x^5}{5!} + \frac{x^7}{7!} + \ldots.$$

The series converge to the functions for all values of x.

2. Other hyperbolic functions. The following functions are defined by analogy with the corresponding functions of elementary trigonometry:

$$\tanh x \; = \frac{\sinh x}{\cosh x},$$

$$\coth x \; = \frac{\cosh x}{\sinh x},$$

$$\operatorname{cosech} x = \frac{1}{\sinh x},$$

$$\operatorname{sech} x \; = \frac{1}{\cosh x}.$$

The relations $\operatorname{sech}^2 x + \tanh^2 x = 1,$

$$\coth^2 x - \operatorname{cosech}^2 x = 1.$$

are found by dividing the equation

$$\cosh^2 x - \sinh^2 x = 1$$

by $\cosh^2 x$, $\sinh^2 x$ respectively.

Note the implications

$$\operatorname{sech}^2 x < 1, \quad \tanh^2 x < 1.$$

The differential coefficients are easily obtained from the definitions:

(i) If

$$y = \tanh x = (\sinh x)(\cosh x)^{-1},$$

then $$\frac{dy}{dx} = (\cosh x)(\cosh x)^{-1} - (\sinh x)(\cosh x)^{-2} \sinh x$$

$$= 1 - \tanh^2 x$$

$$= \operatorname{sech}^2 x.$$

(ii) If

$$y = \coth x = (\cosh x)(\sinh x)^{-1},$$

then $$\frac{dy}{dx} = (\sinh x)(\sinh x)^{-1} - (\cosh x)(\sinh x)^{-2} \cosh x$$

$$= 1 - \coth^2 x$$

$$= -\operatorname{cosech}^2 x.$$

(iii) If $$y = \operatorname{cosech} x = (\sinh x)^{-1},$$

then $$\frac{dy}{dx} = -(\sinh x)^{-2} \cosh x$$
$$= -\operatorname{cosech} x \coth x.$$

(iv) If $$y = \operatorname{sech} x = (\cosh x)^{-1},$$

then $$\frac{dy}{dx} = -(\cosh x)^{-2} \sinh x$$
$$= -\operatorname{sech} x \tanh x.$$

ILLUSTRATION 1. *A body of mass m falls from rest under gravity in a medium whose resistance to motion is gv^2/k^2 per unit mass when the speed is v. To prove that the speed after t seconds is* $k \tanh(gt/k)$.

Let x be the distance dropped in time t. Then the acceleration downward is $\ddot{x}$ and the forces are

(i) mg downwards due to gravity,

(ii) $mg\dot{x}^2/k^2$ upwards due to the resistance.

Hence $$m\ddot{x} = mg - mg\dot{x}^2/k^2.$$

Write $\dot{x} = v$. Then $$\frac{dv}{dt} = g - gv^2/k^2,$$

or $$k^2 \frac{dv}{dt} = g(k^2 - v^2).$$

Substitute $$v = k \tanh \theta.$$

(This substitution is possible so long as v is less than k, since (p. 87) $\tanh^2 \theta < 1$.)

Then $$k^2 . k \operatorname{sech}^2 \theta \frac{d\theta}{dt} = gk^2(1 - \tanh^2 \theta)$$
$$= gk^2 \operatorname{sech}^2 \theta.$$

Hence $$\frac{d\theta}{dt} = \frac{g}{k},$$

so that $$\theta = \frac{gt}{k} + C,$$

where C is an arbitrary constant. It follows that

$$v = k \tanh\left(\frac{gt}{k} + C\right).$$

Now we are given that $v = 0$ when $t = 0$, so that

$$0 = \tanh C.$$

Hence $$C = 0,$$

and so $$v = k \tanh(gt/k).$$

Note that
$$v = k \tanh(gt/k)$$
$$= k\frac{e^{gt/k} - e^{-gt/k}}{e^{gt/k} + e^{-gt/k}}$$
$$= k\frac{1 - e^{-2gt/k}}{1 + e^{-2gt/k}}.$$

Now as t increases, $e^{2gt/k}$ tends to infinity (see, for example, the diagram (Fig. 64) for e^x on p. 18), so that $e^{-2gt/k} \to 0$. Hence

$$v \to k$$

as t increases. In other words, v tends to a *terminal value* k as the time of falling increases.

The example which follows deserves close attention. It brings out very clearly the points of similarity between the trigonometric and the hyperbolic functions.*

ILLUSTRATION 2. Suppose that a particle P (Fig. 69), of mass m, is free to move on a fixed smooth circular wire, of radius a, whose plane is vertical. A light string, of natural length a and modulus of elasticity $2kmg$ joins P to the highest point B of the wire. We wish to examine what happens if P receives a slight displacement from the lowest point of the wire.

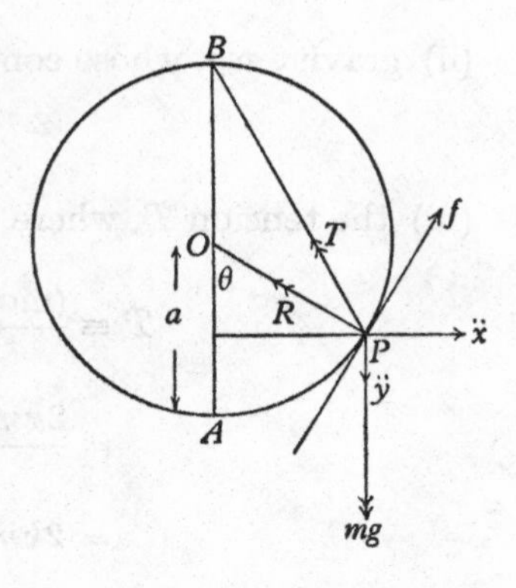

Fig. 69.

Let AB be the diameter through B, and O the centre of the circle. Denote by θ the angle $\angle AOP$.

* It may be postponed or omitted by a reader who finds the mechanics difficult.

If x denotes the horizontal distance of P to the right of the vertical diameter AB, and if y denotes the depth of P below O, then

$$x = a \sin \theta,$$

$$y = a \cos \theta.$$

Hence, differentiating with respect to time,

$$\dot{x} = a\dot{\theta} \cos \theta,$$

$$\dot{y} = -a\dot{\theta} \sin \theta,$$

and

$$\ddot{x} = a\ddot{\theta} \cos \theta - a\dot{\theta}^2 \sin \theta,$$

$$\ddot{y} = -a\ddot{\theta} \sin \theta - a\dot{\theta}^2 \cos \theta.$$

The acceleration f of P in the direction of the (upward) tangent is thus

$$\begin{aligned} f &= \ddot{x} \cos \theta - \ddot{y} \sin \theta \\ &= a\ddot{\theta}. \end{aligned}$$

[This is a standard formula of applied mathematics.]

Now the forces on the particle are

(i) the reaction R along PO, which has no component along the tangent;

(ii) gravity mg, whose component along the (upward) tangent is

$$-mg \sin \theta;$$

(iii) the tension T, where, by definition of modulus,

$$\begin{aligned} T &\equiv \frac{\text{(modulus) (extension)}}{\text{natural length}} \\ &= \frac{2kmg\{2a \cos \frac{1}{2}\theta - a\}}{a} \\ &= 2kmg(2 \cos \tfrac{1}{2}\theta - 1); \end{aligned}$$

the component of T along the (upward) tangent is thus

$$\begin{aligned} &T \sin \tfrac{1}{2}\theta \\ &\qquad = 2kmg(2 \cos \tfrac{1}{2}\theta - 1) \sin \tfrac{1}{2}\theta. \end{aligned}$$

Equating the component of the acceleration to the component of the forces, we have the equation of motion

$$a\ddot{\theta} = -g\sin\theta + 2kg(2\cos\tfrac{1}{2}\theta - 1)\sin\tfrac{1}{2}\theta.$$

Now suppose that θ is small. By the work given on p. 38, the value of $\sin\theta$ is nearly θ itself, while $\cos\theta$ is nearly equal to 1. Hence the equation is approximately

$$\begin{aligned} a\ddot{\theta} &= -g\theta + 2kg\{2(1)-1\}(\tfrac{1}{2}\theta) \\ &= -g\theta + kg\theta. \end{aligned}$$

Hence we reach the equation

$$a\ddot{\theta} = g(k-1)\theta,$$

valid during the time while θ is small.

The argument now divides, according as k is less than or greater than unity; that is, according as the string is 'fairly slack' or 'fairly tight'.

(i) *Suppose that* $k<1$.

Then
$$\begin{aligned} \ddot{\theta} &= -\frac{(1-k)g}{a}\theta \\ &= -n^2\theta, \end{aligned}$$

where
$$an^2 = (1-k)g.$$

It may be proved that, when $\ddot{\theta} = -n^2\theta$, then θ MUST be of the form

$$\theta = A\cos nt + B\sin nt,$$

where A, B are constants. In the meantime, the reader may easily verify the converse result, that this value of θ does satisfy the relation.

If we suppose that the particle is initially drawn aside so that θ has the small value α, then $\theta = \alpha$ when $t=0$, so that

$$A = \alpha.$$

If also the particle is released from rest, then $\dot{\theta} = 0$ when $t=0$, so that

$$B = 0.$$

Hence
$$\theta = \alpha\cos nt.$$

Thus if θ is initially small, it remains small, and its value oscillates between $\pm\alpha$. The equilibrium is stable at the lowest point.

(ii) *Suppose that* $k > 1$.

Then
$$\ddot{\theta} = \frac{(k-1)g}{a}\theta$$
$$= p^2\theta,$$
where
$$ap^2 = (k-1)g.$$

This equation is not satisfied by sines and cosines, but we can express the relation between θ and t in the form
$$\theta = Ae^{pt} + Be^{-pt},$$
where, again, the reader may verify the converse result that this value of θ does satisfy the equation.

With the same initial conditions as before, we have
$$\alpha = A + B,$$
$$0 = Ap - Bp,$$
so that
$$A = B = \tfrac{1}{2}\alpha.$$
Hence
$$\theta = \tfrac{1}{2}\alpha(e^{pt} + e^{-pt})$$
$$= \alpha \cosh pt.$$

As t increases, $\cosh pt$ increases steadily to 'infinity' (since e^{pt} does), so that θ ceases to be small. The equilibrium is unstable at the lowest point.

The differential equation, in fact, ceases to be accurate once θ ceases to be small.

It is instructive to consider the same problem under the alternative initial conditions that the particle is projected from the lowest point with speed v. Thus $\theta = 0, a\dot{\theta} = v$ when $t = 0$. We take the two cases in succession:

(i)
$$\theta = A\cos nt + B\sin nt,$$
where
$$0 = A,$$
$$v/a = nB.$$
Hence
$$\theta = (v/an)\sin nt.$$

(ii) $$\theta = Ae^{pt} + Be^{-pt},$$

where $$0 = A + B,$$

$$v/a = pA - pB,$$

so that $$A = \frac{v}{2ap}, \quad B = -\frac{v}{2ap}.$$

Hence $$\theta = \frac{v}{2ap}(e^{pt} - e^{-pt})$$

$$= (v/ap)\sinh pt.$$

The analogy between the pairs of solutions

$$\alpha \cos nt, \quad \alpha \cosh pt$$

and $$(v/an)\sin nt, (v/ap)\sinh pt$$

affords striking confirmation of the analogy between the two classes of functions.

EXAMPLES I

Differentiate the following functions:

1. $\sinh 3x$.
2. $\cosh^2 2x$.
3. $x \tanh x$.
4. $\sinh^2(2x+1)$.
5. $\sinh x \cos x$.
6. $\operatorname{sech} x \sin^2 x$.
7. $(1+x)^3 \cosh^3 3x$.
8. $x^2 \tanh^2 4x$.
9. $\log \sinh x$.
10. $\log(\sinh x + \cosh x)$.
11. $e^{\sinh x}$.
12. $xe^{-\tanh x}$.

Find the following integrals:

13. $\int \sinh 4x\,dx$.
14. $\int \sinh^2 x\,dx$.
15. $\int \cosh^2 x\,dx$.
16. $\int x \sinh x\,dx$.
17. $\int e^x \cosh x\,dx$.
18. $\int \sinh 3x \cosh x\,dx$.
19. $\int x \sinh^2 x\,dx$.
20. $\int \cosh^3 x\,dx$.
21. $\int \tanh^2 x\,dx$.
22. $\int x^2 \cosh x\,dx$.
23. $\int e^{2x} \sinh 5x\,dx$.
24. $\int \tanh x \operatorname{sech}^2 x\,dx$.

Establish the following formulæ:

25. $\sinh A + \sinh B = 2\sinh\dfrac{A+B}{2}\cosh\dfrac{A-B}{2}.$

26. $\sinh A - \sinh B = 2\cosh\dfrac{A+B}{2}\sinh\dfrac{A-B}{2}.$

27. $\cosh A + \cosh B = 2\cosh\dfrac{A+B}{2}\cosh\dfrac{A-B}{2}.$

28. $\cosh A - \cosh B = 2\sinh\dfrac{A+B}{2}\sinh\dfrac{A-B}{2}.$

29. Prove that, for all values of u, the point $(a\cosh u, b\sinh u)$ lies on the hyperbola whose equation is

$$\frac{x^2}{a^2}-\frac{y^2}{b^2}=1,$$

and that the tangent at that point is

$$\frac{x}{a}\cosh u-\frac{y}{b}\sinh u=1.$$

(But note that that point is restricted to the part of the hyperbola for which x is positive.)

3. The graph $y = \cosh x$. The two relations

$$y=\cosh x,$$
$$\frac{dy}{dx}=\sinh x$$

give us sufficient information to indicate the general shape of the curve:

Since

$$\cosh(-x)=\cosh(x),$$

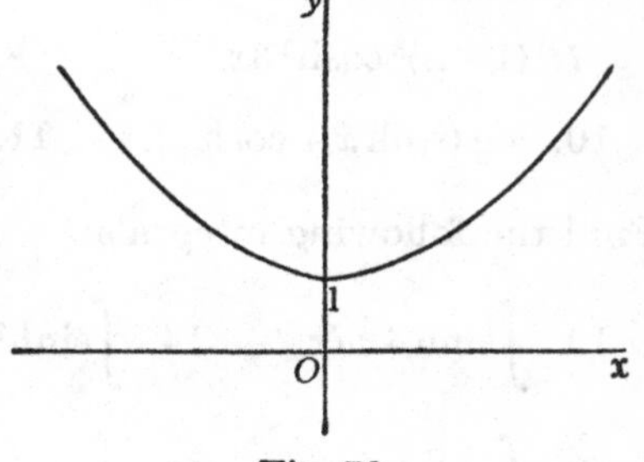

Fig. 70.

the curve is symmetrical about the y-axis; and since

$$\cosh x \geqslant 1,$$

the curve lies entirely above the line $y=1$. The value of y increases rapidly with x.

Taking x to be positive (when it is negative the corresponding part of the curve is obtained simply by reflexion in the y-axis) we have $\sinh x$ positive, so that the gradient is positive. Moreover,

$$\frac{d^2y}{dx^2} = \cosh x,$$

which is positive, and so (Vol. I, p. 54) the concavity is 'upwards'.

The general shape of the curve is therefore that indicated in the diagram (Fig. 70).

4. The graph $y = \sinh x$. We have the relations

$$y = \sinh x,$$

$$\frac{dy}{dx} = \cosh x.$$

Since $\frac{dy}{dx}$ is positive, the gradient is always positive, and y is an increasing function of x, running from $-\infty$ to $+\infty$ as x increases from $-\infty$ to $+\infty$.

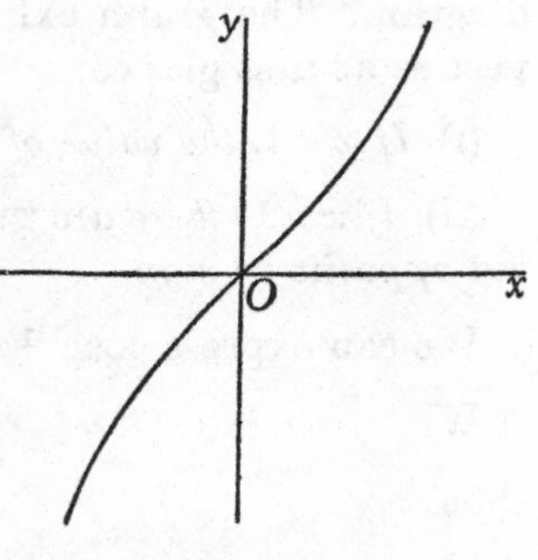

Fig. 71.

At the origin, $\frac{dy}{dx} = 1$, so that the curve crosses the x-axis there at an angle of $\frac{1}{4}\pi$. Also

$$\frac{d^2y}{dx^2} = \sinh x,$$

which is positive for positive x and negative for negative. Hence the curve lies entirely in the first and third quadrants, with (Vol. I, p. 54) concavity 'upwards' in the first and 'downwards' in the third. At the origin, $\frac{d^2y}{dx^2} = 0$, so that (Vol. I, p. 55) the curve has an inflexion there.

Since $$\sinh(-x) = -\sinh(x),$$

the curve is symmetrical about the origin.

The general shape of the curve is therefore that indicated in the diagram (Fig. 71).

5. The inverse hyperbolic cosine. The problem arises in practice to determine a function whose hyperbolic cosine has a given value x. If the function is y, then

$$x = \cosh y,$$

and we use the notation

$$y = \cosh^{-1} x$$

to denote the *inverse hyperbolic cosine.*

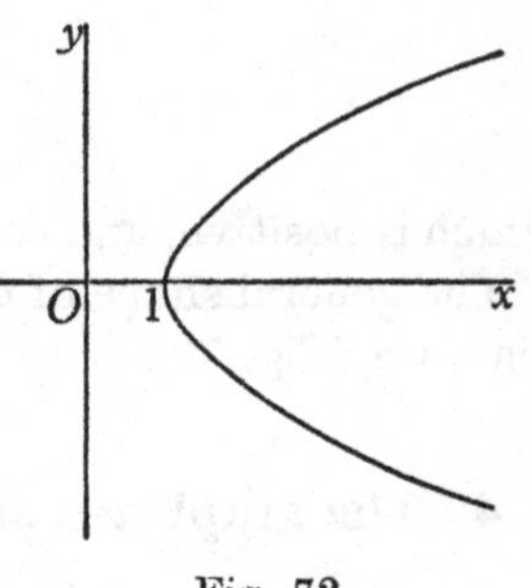

Fig. 72.

The graph (Fig. 72) $y = \cosh^{-1} x$ is found by 'turning the graph $y = \cosh x$ through a right angle' and then renaming the axes, as shown in the diagram. The graph exhibits two properties at first glance:

(i) *If* $x < 1$, *the value of* $\cosh^{-1} x$ *does not exist;*

(ii) *If* $x > 1$, *there are* TWO *values of* $\cosh^{-1} x$, *equal in magnitude but opposite in sign.*

We can express $\cosh^{-1} x$ in terms of logarithms as follows:

If $\qquad y = \cosh^{-1} x,$

then $\qquad x = \cosh y$

$$= \tfrac{1}{2}(e^y + e^{-y}),$$

so that $\qquad e^{2y} - 2xe^y + 1 = 0.$

Solving this equation as a quadratic in e^y, we have

$$e^y = x \pm \sqrt{(x^2 - 1)},$$

and so, by definition of the exponential function,

$$y = \log\{x \pm \sqrt{(x^2 - 1)}\}.$$

These are the two values of y. Moreover their sum is

$$\begin{aligned}\log\{x + \sqrt{(x^2-1)}\} &+ \log\{x - \sqrt{(x^2-1)}\} \\ &= \log[\{x + \sqrt{(x^2-1)}\}\{x - \sqrt{(x^2-1)}\}] \\ &= \log[x^2 - (x^2 - 1)] = \log 1 \\ &= 0.\end{aligned}$$

Hence the two roots are equal and opposite. The positive root is $\log\{x + \sqrt{(x^2-1)}\}$ and the negative $\log\{x - \sqrt{(x^2-1)}\}$.

To find the differential coefficient of $\cosh^{-1} x$, we differentiate the relation

$$\cosh y = x$$

with respect to x. Then

$$\sinh y \frac{dy}{dx} = 1,$$

or

$$\frac{dy}{dx} = \frac{1}{\sinh y} = \frac{1}{\pm\sqrt{(\cosh^2 y - 1)}}$$

$$= \pm\frac{1}{\sqrt{(x^2-1)}}.$$

The gradient is positive when y is positive and negative when y is negative. In particular, if we take the POSITIVE value of $\cosh^{-1} x$, then

$$\frac{dy}{dx} = \frac{1}{\sqrt{(x^2-1)}}.$$

We can also obtain the result from the formula

$$y = \log\{x + \sqrt{(x^2-1)}\},$$

taking the positive value. For if

$$u \equiv x + (x^2-1)^{\frac{1}{2}},$$

then

$$\frac{du}{dx} = 1 + x(x^2-1)^{-\frac{1}{2}}$$

$$= \frac{1}{\sqrt{(x^2-1)}}\{\sqrt{(x^2-1)} + x\} = \frac{u}{\sqrt{(x^2-1)}}.$$

Hence

$$\frac{dy}{dx} = \frac{dy}{du}\frac{du}{dx}$$

$$= \frac{1}{u}\frac{u}{\sqrt{(x^2-1)}}$$

$$= \frac{1}{\sqrt{(x^2-1)}}.$$

Note the corresponding integral

$$\int \frac{dx}{\sqrt{(x^2-1)}} = \cosh^{-1} x \quad \text{(positive value)}$$

$$= \log\{x + \sqrt{(x^2-1)}\}.$$

6. The inverse hyperbolic sine. On 'turning the graph of $\sinh x$ through a right angle and taking a mirror image', and then renaming the axes, we obtain the graph (Fig. 73)

$$y = \sinh^{-1} x$$

of the *inverse hyperbolic sine* of x, that is, of the number whose hyperbolic sine is x. The function $\sinh^{-1} x$ is a SINGLE-VALUED function, uniquely determined for all values of x.

To express $\sinh^{-1} x$ in terms of logarithms, we write

$$\sinh^{-1} x = y,$$

so that

$$x = \sinh y = \tfrac{1}{2}(e^y - e^{-y}),$$

giving $e^{2y} - 2xe^y - 1 = 0$.

Solving this equation as a quadratic in e^y, we have

$$e^y = x \pm \sqrt{(x^2+1)}.$$

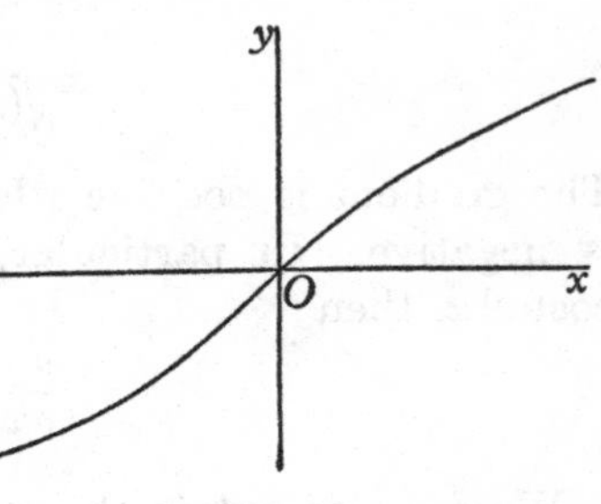

Fig. 73.

But (p. 19) the exponential function is positive, so that we must take the positive sign for the square root. Hence

$$e^y = x + \sqrt{(x^2+1)},$$

or

$$y = \log\{x + \sqrt{(x^2+1)}\}$$

without ambiguity.

To find the differential coefficient of $\sinh^{-1} x$, we differentiate the relation

$$\sinh y = x$$

with respect to x. Then

$$\cosh y \frac{dy}{dx} = 1,$$

or

$$\frac{dy}{dx} = \frac{1}{\cosh y} = \frac{1}{\pm\sqrt{(\sinh^2 y + 1)}}$$

$$= \pm \frac{1}{\sqrt{(x^2+1)}}.$$

But, from the graph (Fig. 73), the gradient is always positive, and so

$$\frac{dy}{dx} = +\frac{1}{\sqrt{(x^2+1)}}.$$

We can also obtain this result from the formula

$$y = \log\{x+\sqrt{(x^2+1)}\}.$$

For if

$$u = x+(x^2+1)^{\frac{1}{2}},$$

then

$$\frac{du}{dx} = 1+x(x^2+1)^{-\frac{1}{2}}$$

$$= \frac{1}{\sqrt{(x^2+1)}}\{\sqrt{(x^2+1)}+x\} = \frac{u}{\sqrt{(x^2+1)}}.$$

Hence

$$\frac{dy}{dx} = \frac{dy}{du}\frac{du}{dx}$$

$$= \frac{1}{u}\frac{u}{\sqrt{(x^2+1)}}$$

$$= \frac{1}{\sqrt{(x^2+1)}}.$$

Note the corresponding integral

$$\int \frac{dx}{\sqrt{(x^2+1)}} = \sinh^{-1} x$$

$$= \log\{x+\sqrt{(x^2+1)}\}.$$

EXAMPLES II

Find the differential coefficients of the following functions:

1. $x\cosh^{-1} x$.
2. $\sinh^{-1}(1+x^2)$.
3. $\tanh^{-1} x$.
4. $\operatorname{sech}^{-1} x$.
5. $\operatorname{cosech}^{-1} x$.
6. $\log(\cosh^{-1} x)$.
7. $x\cosh^{-1}(x^2+1)$.
8. $(\cosh^{-1} x)^2$.
9. $1/(\sinh^{-1} x)$.

Find the following integrals:

10. $\int \frac{dx}{\sqrt{(x^2-4)}}$.
11. $\int \frac{dx}{\sqrt{(9x^2+1)}}$.
12. $\int \frac{dx}{\sqrt{(4x^2-9)}}$.
13. $\int \frac{dx}{\sqrt{\{(x+1)^2-4\}}}$.
14. $\int \frac{dx}{\sqrt{(x^2+2x+5)}}$.
15. $\int \frac{dx}{\sqrt{(4x^2-4x-15)}}$.

CHAPTER X

CURVES

1. Parametric representation. Hitherto we have regarded a curve as defined by an equation of the form

$$y = f(x).$$

For many purposes it is more convenient to adopt a *parametric representation* whereby the coordinates x, y of a point T of the curve are expressed as functions of a *parameter* t in the form

$$x = f(t), \quad y = g(t).$$

(Of course, there is no reason why the parameter should not be x itself.) For economy of notation, however, we often write

$$x = x(t), \quad y = y(t).$$

Familiar examples from elementary coordinate geometry are the representations

$$x = at^2, \quad y = 2at$$

for the parabola $y^2 = 4ax$, and

$$x = ct, \quad y = c/t$$

for the rectangular hyperbola $xy = c^2$.

We confine ourselves to the simplest case, in which $x(t), y(t)$ are single-valued functions of t with as many continuous differential coefficients as the argument may require.

The 'dot' notation

$$\dot{x}, \dot{y}, \ddot{x}, \ddot{y}, \ldots$$

will be used to denote the differential coefficients

$$\frac{dx}{dt}, \frac{dy}{dt}, \frac{d^2x}{dt^2}, \frac{d^2y}{dt^2}, \ldots.$$

The tangent at the point 't' of the curve

$$x = f(t), \quad y = g(t)$$

may easily be found; it is the line through that point with gradient

$$\frac{dy}{dx} = \frac{\dot{y}}{\dot{x}}$$

$$= \frac{g'(t)}{f'(t)}.$$

For the reader familiar with determinants, an alternative form of equation may be given:

The equation of any straight line is

$$lx + my + n = 0.$$

If this line passes through the point $x = f(t), y = g(t)$, then

$$lf(t) + mg(t) + n = 0;$$

if it passes through the point $x = f(t+\delta t), y = g(t+\delta t)$, then

$$lf(t+\delta t) + mg(t+\delta t) + n = 0.$$

Subtracting, we have the relation

$$l\{f(t+\delta t) - f(t)\} + m\{g(t+\delta t) - g(t)\} = 0,$$

or, on division by δt,

$$l\frac{f(t+\delta t) - f(t)}{\delta t} + m\frac{g(t+\delta t) - g(t)}{\delta t} = 0.$$

For the tangent, we must take the limiting form of this relation as $\delta t \to 0$, namely

$$lf'(t) + mg'(t) = 0.$$

Hence, on eliminating the ratios $l : m : n$, we obtain the equation of the tangent in the form

$$\begin{vmatrix} x & y & 1 \\ f(t) & g(t) & 1 \\ f'(t) & g'(t) & 0 \end{vmatrix} = 0.$$

2. The sense of description of a curve. We regard that sense of description of a curve as *positive* which is followed by a variable point for *increasing* values of the defining parameter. For example, if the points A, P, Q (Fig. 74) correspond to the values a, p, q respectively, and if $a < p < q$, then the sense is $\overrightarrow{APQ}$.

It is important to realize that sense is not an inherent property; it is a man-made convention. Thus different parametric representations may give rise to different senses along the curve.

For example, the positive quadrant of the circle $x^2+y^2=1$ may be expressed with x as parameter in the form

$$x = x, \quad y = +\sqrt{(1-x^2)}.$$

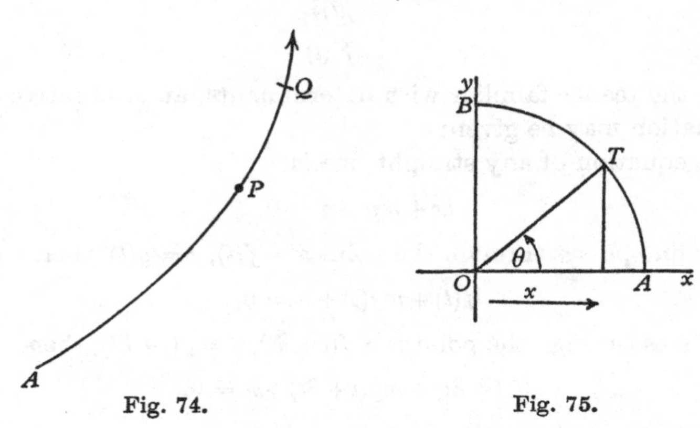

Fig. 74. Fig. 75.

As x increases from 0 to 1, the arc is described in the sense $\overrightarrow{BA}$ of the diagram (Fig. 75). On the other hand, if the polar angle θ is taken as parameter, we have

$$x = \cos\theta, \quad y = \sin\theta$$

As θ increases from 0 to $\frac{1}{2}\pi$, the arc is described in the sense $\overrightarrow{AB}$.

3. The 'length' postulate. If we confine our attention to the simplest case, where the curve has a continuously turning tangent, as in the diagram (Fig. 76), then our instinctive ideas will be satisfied if we ensure that the length of a curved arc PQ is nearly the same as that of the chord PQ when the points P, Q are very close together. For this purpose, we shall base our treatment of length on the postulate

$$\lim_{Q\to P} \frac{\text{arc } QP}{\text{chord } QP} = 1.$$

Fig. 76.

Our aim is to let the derivation of all the standard formulæ of the geometry of curves rest on this single assumption, together with the normal manipulations of algebra, trigonometry and the calculus.

The warning ought perhaps to be added that in a more advanced treatment it would be necessary to examine whether 'length' exists at all and to proceed somewhat differently. The present treatment suffices when $dx/dt, dy/dt$ are continuous; this is true for 'ordinary' cases such as we shall be considering.

4. The length of a curve. Suppose that $U\ P, P'$ are three points on the curve (Fig. 77)

$$x = x(t), \quad y = y(t)$$

given by the values $u, p, p+\delta p$ of the parameter. For convenience, we assume that

$$u < p < p + \delta p,$$

so that the curve is described in the sense $\overrightarrow{UPP'}$.

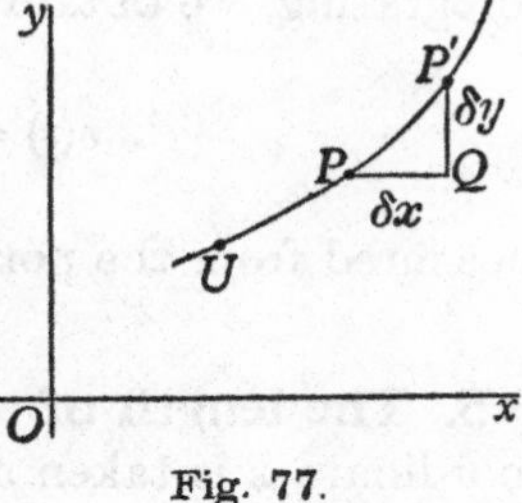

Fig. 77.

If P, P' are the points (x, y), $(x+\delta x, y+\delta y)$ respectively, then the length of the chord PP' is

$$\sqrt{\{(\delta x)^2 + (\delta y)^2\}},$$

whether δx, δy are positive or negative.

Now the length of the arc UP is a function of p, which we may call $s(p)$. Thus, since arc PP' = arc UP' − arc UP, we have

$$\text{arc } PP' = s(p+\delta p) - s(p).$$

But
$$\frac{s(p+\delta p)-s(p)}{\delta p} = \frac{s(p+\delta p)-s(p)}{\text{chord } PP'} \cdot \frac{\text{chord } PP'}{\delta p}$$

$$= \frac{\text{arc } PP'}{\text{chord } PP'} \sqrt{\left\{\left(\frac{\delta x}{\delta p}\right)^2 + \left(\frac{\delta y}{\delta p}\right)^2\right\}}.$$

If we proceed to the limit, as $\delta p \to 0$ so that $P' \to P$, then

$$\lim_{\delta p \to 0} \frac{s(p+\delta p)-s(p)}{\delta p} = s'(p),$$

$$\lim_{P' \to P} \frac{\text{arc } PP'}{\text{chord } PP'} = 1,$$

$$\lim_{\delta p \to 0} \frac{\delta x}{\delta p} = \frac{dx}{dp}, \quad \lim_{\delta p \to 0} \frac{\delta y}{\delta p} = \frac{dy}{dp}.$$

Hence
$$s'(p) = \sqrt{\left\{\left(\frac{dx}{dp}\right)^2 + \left(\frac{dy}{dp}\right)^2\right\}},$$
where the POSITIVE square root must be taken since s increases with p.

Replacing p by the current letter t, we have the relation
$$s'(t) = \sqrt{(\dot{x}^2 + \dot{y}^2)}.$$
Integrating, we obtain the formula
$$s(t) = \int_u^t \sqrt{\left\{\left(\frac{dx}{dt}\right)^2 + \left(\frac{dy}{dt}\right)^2\right\}}\, dt,$$
measured from the point U with parameter u.

5. The length of a curve in Cartesian coordinates. If the coordinate x is taken as the parameter t, then the formula of § 4 becomes
$$s'(x) = \sqrt{\left\{1 + \left(\frac{dy}{dx}\right)^2\right\}},$$
so that
$$s(x) = \int_a^x \sqrt{\left\{1 + \left(\frac{dy}{dx}\right)^2\right\}}\, dx,$$
measured from the point where $x = a$, where the positive sense of the curve is determined by x increasing.

In terms of the coordinate y, we have similarly
$$s(y) = \int_b^y \sqrt{\left\{\left(\frac{dx}{dy}\right)^2 + 1\right\}}\, dy,$$
measured from the point where $y = b$, where the positive sense of the curve is determined by y increasing.

ILLUSTRATION 1. *To find the length of the arc of the parabola $y^2 = 4ax$ from the origin to the point (x, y), where y is taken to be positive.*

The parametric representation is
$$x = at^2, \quad y = 2at,$$
so that
$$\dot{x} = 2at, \quad \dot{y} = 2a.$$
Hence
$$s = \int_0^t 2a\sqrt{(t^2 + 1)}\, dt.$$

To evaluate the integral, write

$$I = \int \sqrt{(t^2+1)}\,dt$$

$$= t\sqrt{(t^2+1)} - \int t.\frac{t}{\sqrt{(t^2+1)}}\,dt,$$

on integration by parts. Hence

$$I = t\sqrt{(t^2+1)} - \int \frac{(t^2+1)-1}{\sqrt{(t^2+1)}}\,dt$$

$$= t\sqrt{(t^2+1)} - I + \int \frac{dt}{\sqrt{(t^2+1)}}$$

$$= t\sqrt{(t^2+1)} - I + \log\{t+\sqrt{(t^2+1)}\}.$$

Hence $$I = \tfrac{1}{2}t\sqrt{(t^2+1)} + \tfrac{1}{2}\log\{t+\sqrt{(t^2+1)}\},$$

so that $$s = at\sqrt{(t^2+1)} + a\log\{t+\sqrt{(t^2+1)}\}.$$

6. The length of a curve in polar coordinates. Let the equation of the curve in polar coordinates be

$$r = f(\theta).$$

If θ is taken as the parameter, then the formula of § 4 becomes

$$s'(\theta) = \sqrt{\left\{\left(\frac{dx}{d\theta}\right)^2 + \left(\frac{dy}{d\theta}\right)^2\right\}}.$$

Now $$x = r\cos\theta, \quad y = r\sin\theta,$$

where r is a function of θ. Hence

$$\frac{dx}{d\theta} = \frac{dr}{d\theta}\cos\theta - r\sin\theta, \quad \frac{dy}{d\theta} = \frac{dr}{d\theta}\sin\theta + r\cos\theta,$$

so that $$\left(\frac{dx}{d\theta}\right)^2 + \left(\frac{dy}{d\theta}\right)^2 = \left(\frac{dr}{d\theta}\right)^2 + r^2.$$

It follows that $$s'(\theta) = \sqrt{\left\{\left(\frac{dr}{d\theta}\right)^2 + r^2\right\}},$$

so that $$s(\theta) = \int_\alpha^\theta \sqrt{\left\{\left(\frac{dr}{d\theta}\right)^2 + r^2\right\}}\,d\theta,$$

measured from the point where $\theta = \alpha$, where the positive sense of the curve is determined by θ increasing.

ILLUSTRATION 2. *To find the length of the curve* (*a* CARDIOID) *given by the equation*

$$r = a(1+\cos\theta).$$

The shape of the curve is shown in the diagram (Fig. 78).

We have $\frac{dr}{d\theta} = -a\sin\theta$, so that

$$\left(\frac{dr}{d\theta}\right)^2 + r^2 = a^2\{\sin^2\theta + (1+2\cos\theta+\cos^2\theta)\}$$
$$= 2a^2(1+\cos\theta) = 4a^2\cos^2\tfrac{1}{2}\theta.$$

O 2a

Fig. 78.

Hence the length of the curve is

$$\int_{-\pi}^{\pi} 2a\cos\tfrac{1}{2}\theta\, d\theta = 4a\left[\sin\tfrac{1}{2}\theta\right]_{-\pi}^{\pi}$$
$$= 4a[\sin(\tfrac{1}{2}\pi) - \sin(-\tfrac{1}{2}\pi)]$$
$$= 4a[1-(-1)]$$
$$= 8a.$$

Note. If we had taken the limits of integration as $0, 2\pi$, we should apparently have had the result that the length is

$$\int_0^{2\pi} 2a\cos\tfrac{1}{2}\theta\, d\theta = 4a\left[\sin\tfrac{1}{2}\theta\right]_0^{2\pi}$$
$$= 4a[\sin\pi - \sin 0]$$
$$= 0.$$

It is instructive to trace the source of error. This lies in our assumption that

$$\sqrt{(4a^2\cos^2\tfrac{1}{2}\theta)} = 2a\cos\tfrac{1}{2}\theta.$$

When θ lies between $-\pi$, π, this is true, since $\cos\frac{1}{2}\theta$ is then positive. But in the interval $\pi, 2\pi$, the value of $\cos\frac{1}{2}\theta$ is negative, so that

$$\sqrt{(4a^2\cos^2\tfrac{1}{2}\theta)} = -2a\cos\tfrac{1}{2}\theta.$$

Hence we must use the argument:

$$\int_0^{2\pi}\sqrt{(4a^2\cos^2\tfrac{1}{2}\theta)}\,d\theta$$
$$= \int_0^{\pi}\sqrt{(4a^2\cos^2\tfrac{1}{2}\theta)}\,d\theta + \int_{\pi}^{2\pi}\sqrt{(4a^2\cos^2\tfrac{1}{2}\theta)}\,d\theta$$
$$= \int_0^{\pi} 2a\cos\tfrac{1}{2}\theta\,d\theta - \int_{\pi}^{2\pi} 2a\cos\tfrac{1}{2}\theta\,d\theta$$
$$= 4a\left[\sin\tfrac{1}{2}\theta\right]_0^{\pi} - 4a\left[\sin\tfrac{1}{2}\theta\right]_{\pi}^{2\pi}$$
$$= 4a[1-0] - 4a[0-1]$$
$$= 8a.$$

7. The 'gradient angle' ψ. If the tangent at a point P of the curve

$$y = f(x)$$

makes an angle ψ with the x-axis, then

$$\frac{dy}{dx} = \tan\psi.$$

The diagrams (Fig. 79) represent the four ways (indicated by the arrows) in which a curve may 'leave' a point P on it, the parameter being such that the positive sense along the curve is that of the arrow.

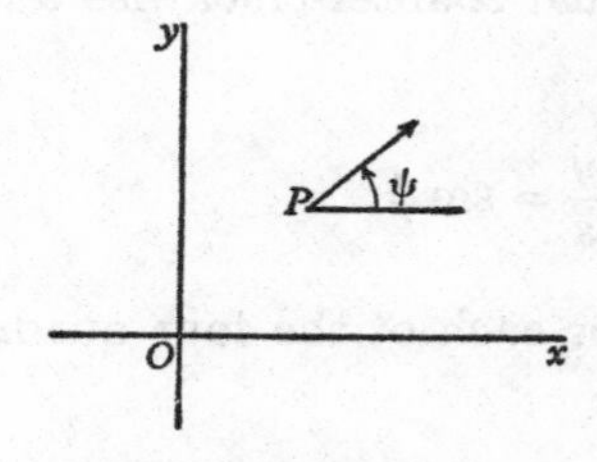

(i) FIRST QUADRANT

$dx+$; $dy+$; $\cos\psi+$; $\sin\psi+$.

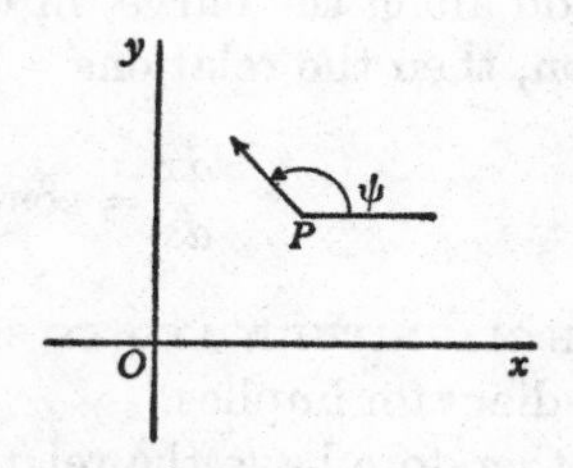

(ii) SECOND QUADRANT

$dx-$; $dy+$: $\cos\psi-$; $\sin\psi+$.

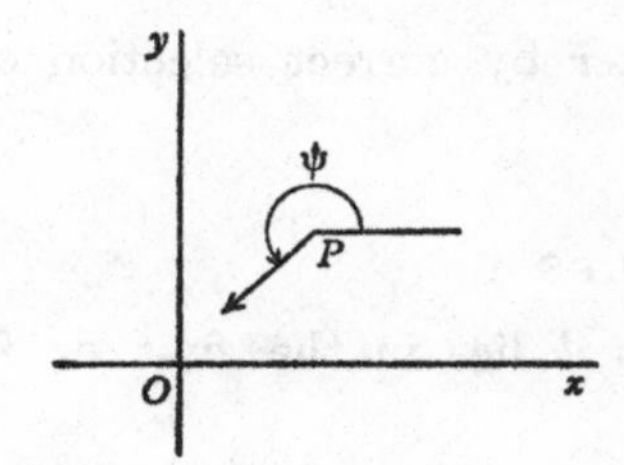

(iii) THIRD QUADRANT

$dx-$; $dy-$; $\cos\psi-$; $\sin\psi-$.

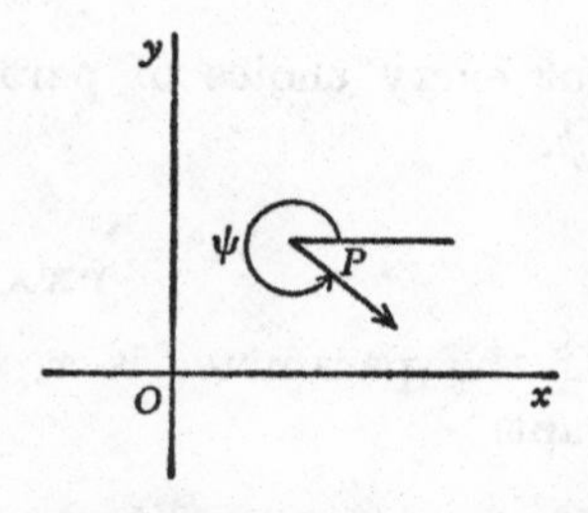

(iv) FOURTH QUADRANT

$dx+$; $dy-$; $\cos\psi+$; $\sin\psi-$.

Fig. 79.

Thus $\frac{dx}{ds}$ is positive for (i), (iv) and negative for (ii), (iii); while $\frac{dy}{ds}$ is positive for (i), (ii) and negative for (iii), (iv).

But $$\left(\frac{ds}{dx}\right)^2 = 1+\left(\frac{dy}{dx}\right)^2 = 1+\tan^2\psi = \sec^2\psi,$$

so that $$\left(\frac{dx}{ds}\right)^2 = \cos^2\psi;$$

similarly $$\left(\frac{dy}{ds}\right)^2 = \sin^2\psi.$$

Hence $\frac{dx}{ds}, \frac{dy}{ds}$ have the numerical values of $\cos\psi$, $\sin\psi$ respectively; and, if we define the angle ψ to be the angle $\left(\text{whose tangent is } \frac{dy}{dx}\right)$ from the positive direction of the x-axis round to the positive direction along the curve, in the usual counter-clockwise sense of rotation, then the relations

$$\frac{dx}{ds} = \cos\psi, \quad \frac{dy}{ds} = \sin\psi$$

hold IN MAGNITUDE AND IN SIGN for each of the four quadrants, as the diagram implies.

We therefore have the relations

$$\frac{dx}{ds} = \cos\psi, \quad \frac{dy}{ds} = \sin\psi,$$

true for every choice of parameter by correct selection of the angle ψ.

EXAMPLES I

1. If the parameter is x, then ψ lies in the first or fourth quadrant.

2. If the parameter is y, then ψ lies in the first or second quadrant.

3. If the parameter is θ, then ψ lies between θ and $\theta+\pi$ (reduced by 2π if necessary).

4. What modifications are required in the treatment given in § 7 if the curve is parallel to the y-axis? Prove that the formulæ $\frac{dx}{ds} = \cos\psi$, $\frac{dy}{ds} = \sin\psi$ are still true.

8. The angle from the radius vector to the tangent. The direction of the tangent to the curve

$$r = f(\theta)$$

may be described in terms of the angle ϕ 'behind' the radius vector. For precision, we define ϕ as follows:

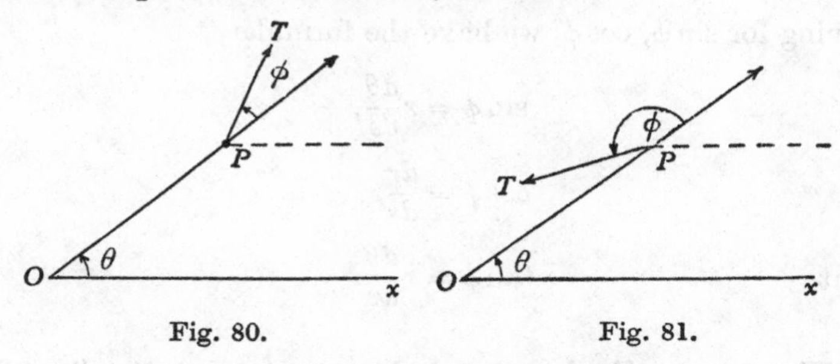

Fig. 80. Fig. 81.

A radius vector, centred on the point P (Figs. 80, 81) of the curve, starts in the direction (and sense) of the initial line Ox; after counter-clockwise rotation through an angle θ, it lies along the radius OP produced. A further counter-clockwise rotation ϕ brings it to the tangent to the curve, in the POSITIVE sense; this defines ϕ.

If we assume, as usual, that θ is the parameter, then the positive sense along the curve is that in which the length increases with θ. Hence *the angle* ϕ *lies between* 0 *and* π, as the diagrams (Figs. 80, 81) indicate.

By definition of ψ (p. 108) we have the relation

$$\psi = \theta + \phi,$$

with possible subtraction of 2π if desired.

It is important to remember when using this formula that θ is the parameter used in defining ψ.

To find expressions for $\sin\phi$, $\cos\phi$, $\tan\phi$:

From the relations

$$x = r\cos\theta, \quad y = r\sin\theta,$$

we have

$$\frac{dx}{ds} = \frac{dr}{ds}\cos\theta - r\frac{d\theta}{ds}\sin\theta,$$

$$\frac{dy}{ds} = \frac{dr}{ds}\sin\theta + r\frac{d\theta}{ds}\cos\theta,$$

so that (p. 108), when θ is the parameter,

$$\frac{dr}{ds}\cos\theta - r\frac{d\theta}{ds}\sin\theta = \cos\psi = \cos(\theta+\phi) = \cos\phi\cos\theta - \sin\phi\sin\theta,$$

$$\frac{dr}{ds}\sin\theta + r\frac{d\theta}{ds}\cos\theta = \sin\psi = \sin(\theta+\phi) = \cos\phi\sin\theta + \sin\phi\cos\theta.$$

Solving for $\sin\phi$, $\cos\phi$, we have the formulæ

$$\sin\phi = r\frac{d\theta}{ds},$$

$$\cos\phi = \frac{dr}{ds},$$

so that

$$\tan\phi = r\frac{d\theta}{dr}.$$

9. The perpendicular on the tangent. Let the line ON (Fig. 82) be drawn from the pole O on to the tangent at the point P of the curve

$$r = f(\theta).$$

Then, by elementary trigonometry,

$$p = r\sin\phi$$

$$= r^2\frac{d\theta}{ds}.$$

This formula may be cast into an alternative useful form:

$$\frac{1}{p^2} = \frac{1}{r^4}\left(\frac{ds}{d\theta}\right)^2$$

$$= \frac{1}{r^4}\left\{\left(\frac{dr}{d\theta}\right)^2 + r^2\right\}$$

$$= \frac{1}{r^4}\left(\frac{dr}{d\theta}\right)^2 + \frac{1}{r^2}.$$

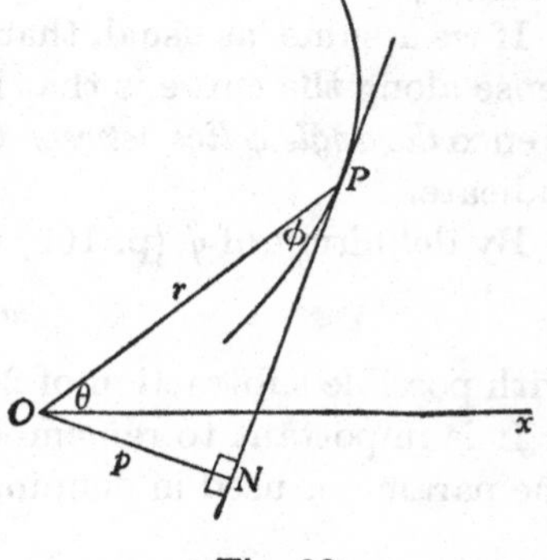

Fig. 82.

If we write

$$u = \frac{1}{r},$$

so that

$$\frac{du}{d\theta} = -\frac{1}{r^2}\frac{dr}{d\theta},$$

then

$$\frac{1}{p^2} = \left(\frac{du}{d\theta}\right)^2 + u^2.$$

Note. Since ϕ lies between $0, \pi$, the value of $\sin\phi$ is necessarily positive. We may retain the formula

$$p = r\sin\phi$$

generally if we allow p to take the sign of r. In any case, the numerical value of p is that of $r\sin\phi$.

The expression for p in terms of Cartesian coordinates follows at once; for

$$\begin{aligned} p &= r\sin\phi \\ &= r\sin(\psi-\theta) \\ &= r\sin\psi\cos\theta - r\cos\psi\sin\theta \\ &= x\sin\psi - y\cos\psi, \end{aligned}$$

where (x, y) are the coordinates of P.

It should be noticed that the step $\phi = \psi - \theta$ depends on the conventions adopted when θ is the parameter. If another parameter (for example, x) is used, the sign attached to p may require separate checking.

10. Other coordinate systems.

The Cartesian coordinates (x, y) and the polar coordinates (r, θ) are by no means the only coordinates available for defining the position of a point of a curve. Others in use are the *intrinsic* coordinates (s, ψ) and the *pedal* coordinates (p, r). The passage between Cartesian and polar coordinates is familiar. To pass from Cartesian to intrinsic coordinates, we have the relations

$$\frac{dx}{ds} = \cos\psi, \quad \frac{dy}{ds} = \sin\psi,$$

or the equivalent

$$\frac{ds}{dx} = \sqrt{\left\{1+\left(\frac{dy}{dx}\right)^2\right\}}, \quad \tan\psi = \frac{dy}{dx}.$$

To pass from polar to pedal coordinates, we have

$$\frac{1}{p^2} = \frac{1}{r^4}\left(\frac{dr}{d\theta}\right)^2 + \frac{1}{r^2}.$$

We do not propose to develop the theory of these new coordinate systems any further. The following illustration demonstrates the use of some of the formulæ.

ILLUSTRATION 3. *The* PEDAL CURVE *of a given curve with respect to a given pole.*

The locus of the foot of the perpendicular from a given point O on to the tangent at a variable point P of a given curve is called the *pedal curve* of O with respect to the given curve.

Referring to the diagram (Fig. 82) on p. 110, we see that N is the point of the pedal curve which corresponds to P. The polar coordinates of N are (r_1, θ_1), where

$$r_1 = p,$$

$$\theta_1 = \psi - \tfrac{1}{2}\pi.$$

Let us find the pedal coordinates (p_1, r_1) of N in terms of the pedal coordinates (p, r) of P. We have at once

$$r_1 = p.$$

Also, if ϕ_1 is the angle 'behind' the radius vector for the locus of N,

$$\tan\phi_1 = r_1 \frac{d\theta_1}{dr_1}$$

$$= p\frac{d\psi}{dp} = p\frac{d\theta}{dp} + p\frac{d\phi}{dp}.$$

Now $$p = r\sin\phi, \quad r\frac{d\theta}{dr} = \tan\phi,$$

so that $$\tan\phi_1 = r\sin\phi\frac{d\theta}{dp} + r\sin\phi\frac{d\phi}{dp}$$

$$= \sin\phi \,.\, r\frac{d\theta}{dr}\frac{dr}{dp} + r\sin\phi\frac{d\phi}{dp}$$

$$= \sin\phi\tan\phi\frac{dr}{dp} + r\sin\phi\frac{d\phi}{dp}$$

$$= \tan\phi\left(\frac{dr}{dp}\sin\phi + r\cos\phi\frac{d\phi}{dp}\right)$$

$$= \tan\phi\frac{d}{dp}(r\sin\phi) = \tan\phi\frac{dp}{dp}$$

$$= \tan\phi.$$

Hence
$$\phi_1 = \phi,$$
since each angle lies between $0, \pi$ and θ is being used as parameter. It follows that
$$p_1 = r_1 \sin \phi_1$$
$$= p \sin \phi$$
$$= p^2/r,$$
and so the pedal coordinates (p_1, r_1) of N are $(p^2/r, p)$.

COROLLARY. Since $\tan \phi_1 = p\dfrac{d\psi}{dp}$, and $\phi = \phi_1$, we have the relation
$$\tan \phi = p\frac{d\psi}{dp},$$
true for any curve.

11. Curvature. The instinctive idea of curvature, or bending, might be expressed in some such phrase as 'change of angle with distance', and it is just this conception which we use for our formal definition.

DEFINITION. (See Fig. 83.) *The* CURVATURE κ *at a point* P *of a curve is defined by the relation*
$$\kappa = \frac{d\psi}{ds}.$$

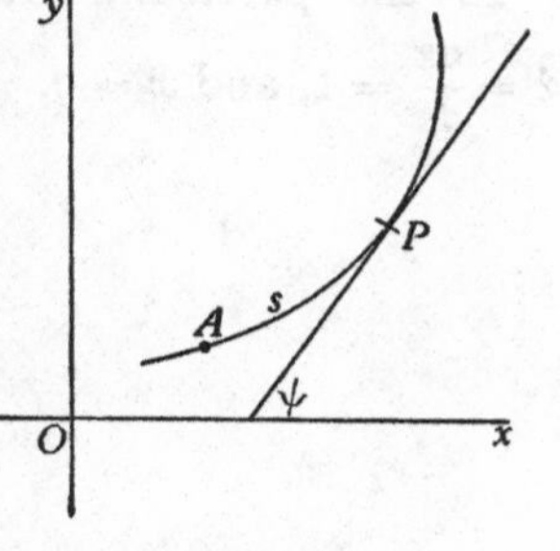

Fig. 83.

The value of κ may be positive or negative, according as ψ increases or decreases with s.

For many purposes, *the calculation of* κ *is best effected by the direct use of the definition.* It is, however, useful to be able to obtain the formulæ in the various systems of coordinates which we have described.

(i) CARTESIAN PARAMETRIC FORM.

If x, y are functions $x(t), y(t)$ of a parameter t, the relations $\dfrac{dx}{ds} = \cos\psi$, $\dfrac{dy}{ds} = \sin\psi$ assume the form
$$\dot{x} = \dot{s}\cos\psi, \quad \dot{y} = \dot{s}\sin\psi.$$

Differentiating, $\ddot{x} = \ddot{s}\cos\psi - \dot{s}\dot{\psi}\sin\psi$

$$= \ddot{s}\cos\psi - \dot{s}^2\kappa\sin\psi,$$

since $$\dot{\psi} = \frac{d\psi}{ds}\dot{s} = \kappa\dot{s};$$

similarly $$\ddot{y} = \ddot{s}\sin\psi + \dot{s}^2\kappa\cos\psi.$$

Hence $$\ddot{y}\cos\psi - \ddot{x}\sin\psi = \dot{s}^2\kappa,$$

or $$\frac{\ddot{y}\dot{x}}{\dot{s}} - \frac{\ddot{x}\dot{y}}{\dot{s}} = \dot{s}^2\kappa,$$

so that $$\kappa = \frac{\ddot{y}\dot{x} - \ddot{x}\dot{y}}{\dot{s}^3},$$

where $$\dot{s} = \sqrt{(\dot{x}^2 + \dot{y}^2)},$$

with (p. 104) POSITIVE square root.

Note. We must choose our parameter to avoid the possibility that $\dot{x} = 0, \dot{y} = 0$ simultaneously. This can be done, for example, by identifying t with x, when $\dot{x} = 1$.

(ii) CARTESIAN FORM.

In the particular case when x is the parameter, we have $\dot{x} = \dfrac{dx}{dx} = 1$, and $\ddot{x} = 0$. Hence, by (i),

$$\kappa = \frac{\dfrac{d^2y}{dx^2}}{\left(\dfrac{ds}{dx}\right)^3}$$

$$= \frac{\dfrac{d^2y}{dx^2}}{\left\{1 + \left(\dfrac{dy}{dx}\right)^2\right\}^{\frac{3}{2}}};$$

the denominator is POSITIVE since $\dfrac{ds}{dx}$ is positive when x is the parameter.

It follows that, *with our conventions, the sign of κ is the same as the sign of $\dfrac{d^2y}{dx^2}$*. Thus (Vol. I, p. 54) the sign of κ is positive when the concavity of the curve is 'upwards', and negative when the concavity is 'downwards'.

(iii) POLAR FORM.

When θ is the parameter, we have the relation (p. 109)

$$\psi = \theta + \phi,$$

so that

$$\kappa = \frac{d\psi}{ds} = \frac{d\psi}{d\theta}\frac{d\theta}{ds}$$

$$= \left(1 + \frac{d\phi}{d\theta}\right)\frac{d\theta}{ds}.$$

Now (p. 110)

$$\cot\phi = \frac{1}{r}\frac{dr}{d\theta},$$

so that

$$-\operatorname{cosec}^2\phi\,\frac{d\phi}{d\theta} = \frac{1}{r}\frac{d^2r}{d\theta^2} - \frac{1}{r^2}\left(\frac{dr}{d\theta}\right)^2,$$

or

$$-\left\{1 + \frac{1}{r^2}\left(\frac{dr}{d\theta}\right)^2\right\}\frac{d\phi}{d\theta} = \frac{1}{r}\frac{d^2r}{d\theta^2} - \frac{1}{r^2}\left(\frac{dr}{d\theta}\right)^2.$$

Hence

$$\frac{d\phi}{d\theta} = \frac{\left(\frac{dr}{d\theta}\right)^2 - r\frac{d^2r}{d\theta^2}}{r^2 + \left(\frac{dr}{d\theta}\right)^2},$$

and

$$1 + \frac{d\phi}{d\theta} = \frac{r^2 + 2\left(\frac{dr}{d\theta}\right)^2 - r\frac{d^2r}{d\theta^2}}{r^2 + \left(\frac{dr}{d\theta}\right)^2}.$$

Also (p. 105)

$$\frac{ds}{d\theta} = +\sqrt{\left\{r^2 + \left(\frac{dr}{d\theta}\right)^2\right\}},$$

the positive sign being taken since θ is the parameter.

Hence

$$\kappa = \frac{r^2 + 2\left(\frac{dr}{d\theta}\right)^2 - r\frac{d^2r}{d\theta^2}}{\left\{r^2 + \left(\frac{dr}{d\theta}\right)^2\right\}^{\frac{3}{2}}}.$$

(iv) PEDAL FORM.

We use the Corollary (p. 113)

$$\tan\phi = p\frac{d\psi}{dp},$$

giving

$$\tan\phi = r\sin\phi\,.\frac{d\psi}{ds}\frac{ds}{dr}\frac{dr}{dp}.$$

But $\frac{dr}{ds} = \cos\phi$, so that $\frac{ds}{dr} = \sec\phi$. Hence

$$\tan\phi = \kappa r \tan\phi \frac{dr}{dp},$$

so that
$$\kappa = \frac{1}{r}\frac{dp}{dr}.$$

The sign conventions used in this proof imply that θ was originally taken as parameter on the curve; otherwise, the sign of κ may require independent examination.

12. A parametric form for a curve in terms of s. Suppose that O is a given point on a curve (Fig. 84). Choose the tangent at O as x-axis and the normal at O as y-axis. We seek to express x, y in terms of s as parameter, assuming that the conditions are such that a Maclaurin expansion is possible.

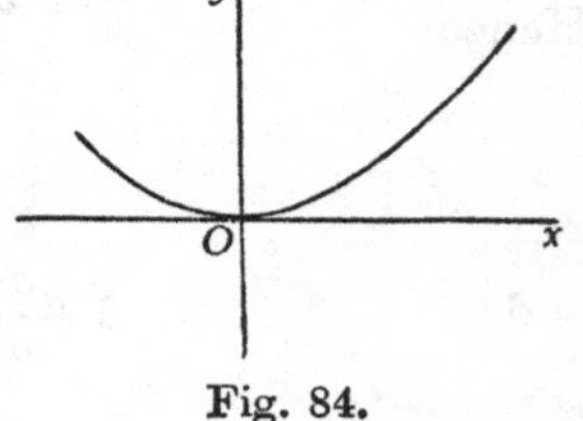

Fig. 84.

The Maclaurin formulæ (p. 49) are

$$x(s) = x(0) + sx'(0) + \frac{s^2}{2!}x''(0) + \ldots,$$

$$y(s) = y(0) + sy'(0) + \frac{s^2}{2!}y''(0) + \ldots.$$

Now
$$\frac{dx}{ds} = \cos\psi,$$

so that
$$\frac{d^2x}{ds^2} = -\sin\psi\frac{d\psi}{ds},$$

$$\frac{d^3x}{ds^3} = -\cos\psi\left(\frac{d\psi}{ds}\right)^2 - \sin\psi\frac{d^2\psi}{ds^2};$$

and
$$\frac{dy}{ds} = \sin\psi,$$

so that
$$\frac{d^2y}{ds^2} = \cos\psi\frac{d\psi}{ds},$$

$$\frac{d^3y}{ds^3} = -\sin\psi\left(\frac{d\psi}{ds}\right)^2 + \cos\psi\frac{d^2\psi}{ds^2}.$$

If we write κ_0, κ_0' to denote the values of $\kappa, \dfrac{d\kappa}{ds}$ at the origin, then, since $\psi = 0$ there,

$$x'(0) = 1; \qquad x''(0) = 0; \qquad x'''(0) = -\kappa_0^2,$$

$$y'(0) = 0; \qquad y''(0) = \kappa_0; \qquad y'''(0) = \kappa_0'.$$

We therefore have the expansions

$$x(s) = s - \tfrac{1}{6}\kappa_0^2 s^3 + \ldots,$$

$$y(s) = \tfrac{1}{2}\kappa_0 s^2 + \tfrac{1}{6}\kappa_0' s^3 + \ldots.$$

13. Newton's formula. *To prove that the curvature at the origin of a curve passing simply through it, and having $y = 0$ as tangent there, is*

$$\kappa_0 = \lim_{x \to 0} \frac{2y}{x^2}.$$

We use a method like that of the preceding paragraph to obtain the expansion, assumed possible, of y as a series of ascending powers of x. We have that, if

$$y = f(x),$$

then

$$f'(x) = \frac{dy}{dx} = \tan\psi,$$

$$f''(x) = \sec^2\psi \frac{d\psi}{dx}$$

$$= \sec^2\psi \frac{d\psi}{ds}\frac{ds}{dx}$$

$$= \sec^2\psi \,.\, \kappa \sec\psi$$

$$= \kappa \sec^3\psi.$$

Thus

$$f(0) = 0, \quad f'(0) = 0, \quad f''(0) = \kappa_0,$$

so that

$$y = f(0) + xf'(0) + \frac{x^2}{2!}f''(0) + \ldots$$

$$= \tfrac{1}{2}\kappa_0 x^2 + \ldots.$$

Hence

$$\frac{2y}{x^2} = \kappa_0 + (\text{terms involving } x),$$

and so, assuming conditions to be such that the remainder tends to zero with x,

$$\lim_{x \to 0} \frac{2y}{x^2} = \kappa_0.$$

ILLUSTRATION 4. *To find the curvature of an ellipse, of semi-axes a, b, at an end of the minor axis.*

Take the required end of the minor axis as origin (Fig. 85) and the tangent there as the x-axis. The equation of the ellipse is

$$\frac{x^2}{a^2}+\frac{(y-b)^2}{b^2}=1.$$

Hence
$$y-b=\pm b\sqrt{\left\{1-\frac{x^2}{a^2}\right\}}.$$

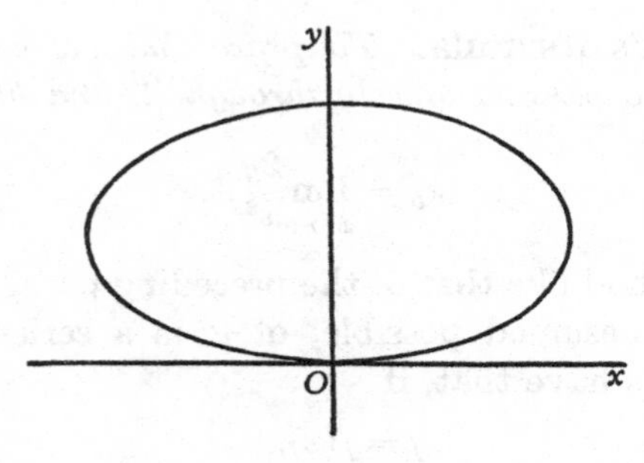

Fig. 85.

For values of y near the origin, the negative square root must be taken, giving

$$\begin{aligned}\frac{y}{b}&=1-\sqrt{\left\{1-\frac{x^2}{a^2}\right\}}\\&=1-\left[1-\frac{1}{2}\frac{x^2}{a^2}-\frac{1}{8}\frac{x^4}{a^4}+\ldots\right]\\&=\frac{x^2}{2a^2}+\frac{x^4}{8a^4}+\ldots,\end{aligned}$$

so that
$$\frac{2y}{x^2}=\frac{b}{a^2}+\frac{bx^2}{4a^4}+\ldots.$$

Hence
$$\begin{aligned}\kappa&=\lim_{x\to 0}\frac{2y}{x^2}\\&=\frac{b}{a^2}.\end{aligned}$$

14. The circle of curvature. The formulæ

$$\tan\psi = \frac{dy}{dx},$$

$$\kappa = \frac{\dfrac{d^2y}{dx^2}}{\left\{1+\left(\dfrac{dy}{dx}\right)^2\right\}^{\frac{3}{2}}}$$

show that, if the two curves

$$y = f(x),$$
$$y = g(x)$$

both pass through a point $P(\xi, \eta)$, so that

$$f(\xi) = g(\xi) = \eta,$$

and if, further, $$f'(\xi) = g'(\xi)$$

and $$f''(\xi) = g''(\xi),$$

then the two curves have the same gradient and curvature at P.

In particular, the circle which passes through P, touches the given curve $y = f(x)$ at P, and has the same curvature as the given curve at P, is called the *circle of curvature* of the given curve at P.

Denote by

$$y_P, \quad y_P', \quad y_P''$$

the values of y, y', y'' for the given curve at the point $P(x_P, y_P)$. Suppose that the equation of the circle of curvature at P is

$$(x-\alpha)^2 + (y-\beta)^2 = \rho^2,$$

where C (α, β) is the centre of the circle of curvature and ρ its radius. By differentiation of this equation with respect to x, we obtain the relations

$$(x-\alpha) + (y-\beta)y' = 0,$$
$$1 + y'^2 + (y-\beta)y'' = 0.$$

Since x, y, y', y'' are the same for the circle of curvature as for the given curve,

$$(x_P-\alpha)^2 + (y_P-\beta)^2 = \rho^2,$$
$$(x_P-\alpha) + (y_P-\beta)y_P' = 0,$$
$$(1 + y_P'^2) + (y_P-\beta)y_P'' = 0.$$

Hence
$$\beta = y_P + \frac{(1+y_P'^2)}{y_P''},$$

$$\alpha = x_P + \left\{-\frac{(1+y_P'^2)}{y_P''}\right\} y_P'$$

$$= x_P - \frac{y_P'(1+y_P'^2)}{y_P''},$$

$$\rho^2 = \frac{(1+y_P'^2)^2}{y_P''^2}(y_P'^2+1)$$

$$= \frac{(1+y_P'^2)^3}{y_P''^2}$$

$$= \frac{1}{\kappa_P^2},$$

where κ_P is the curvature of the given curve at P.

We therefore have the formulæ

$$\alpha = x_P - \frac{y_P'(1+y_P'^2)}{y_P''},$$

$$\beta = y_P + \frac{(1+y_P'^2)}{y_P''},$$

$$\rho = \pm\frac{(1+y_P'^2)^{\frac{3}{2}}}{y_P''}$$

$$= \pm\frac{1}{\kappa_P},$$

the sign being selected to make ρ positive.

If the coordinates x, y of a point on the curve are given as functions of a parameter t, then

$$\frac{dy}{dx} = \frac{dy}{dt}\bigg/\frac{dx}{dt},$$

$$\frac{d^2y}{dx^2} = \frac{\dfrac{dx}{dt}\dfrac{d^2y}{dt^2} - \dfrac{dy}{dt}\dfrac{d^2x}{dt^2}}{\left(\dfrac{dx}{dt}\right)^2}\frac{dt}{dx},$$

so that, if dots denote differentiations with respect to t,

$$\frac{dy}{dx} = \frac{\dot{y}}{\dot{x}},$$

$$\frac{d^2y}{dx^2} = \frac{\dot{x}\ddot{y} - \dot{y}\ddot{x}}{\dot{x}^3}.$$

Hence

$$\alpha = x_P - \frac{\dot{y}_P(\dot{x}_P^2 + \dot{y}_P^2)}{\dot{x}_P\ddot{y}_P - \dot{y}_P\ddot{x}_P},$$

$$\beta = y_P + \frac{\dot{x}_P(\dot{x}_P^2 + \dot{y}_P^2)}{\dot{x}_P\ddot{y}_P - \dot{y}_P\ddot{x}_P},$$

$$\rho = \pm\frac{(\dot{x}_P^2 + \dot{y}_P^2)^{\frac{3}{2}}}{\dot{x}_P\ddot{y}_P - \dot{y}_P\ddot{x}_P}$$

$$= \pm\frac{1}{\kappa_P}.$$

The point $C\ (\alpha, \beta)$ is called the *centre of curvature* of the given curve at P, and ρ its *radius of curvature*. We are adopting a convention of signs in which the radius of curvature is essentially positive.

ILLUSTRATION 5. *The curvature of a circle, and the sign of the curvature.*

Too much emphasis may easily be given to questions about the sign of curvature. Usually common sense and a diagram will settle all that is wanted. We give, however, an exposition for the case when the given curve is itself a circle, so that the reader may, if he wishes, be enabled to examine more elaborate examples.

The difficulties about sign arise, with our conventions, as a result of varying choices of the parameter used to determine the curve. We begin with the simplest case, in which the parameter is selected so that the circle is described completely in one definite sense as the parameter increases.

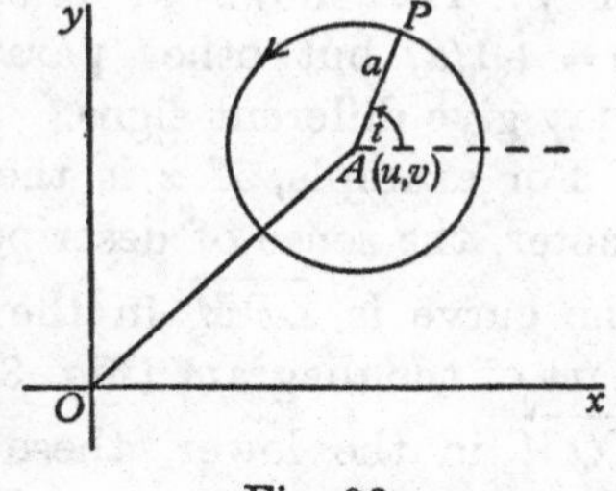

Fig. 86.

Let $A\ (u, v)$ be the centre of the given circle (Fig. 86), and a its radius. Take a variable point $P\ (x, y)$ of the circle, and denote by t

the angle which the radius vector AP makes with the positive direction of the x-axis. As t (the parameter) increases from 0 to 2π, the point P describes the circle completely in the counter-clockwise sense. Then

$$x = u + a\cos t, \quad y = v + a\sin t,$$

so that
$$\dot{x} = -a\sin t, \quad \dot{y} = a\cos t,$$
$$\ddot{x} = -a\cos t, \quad \ddot{y} = -a\sin t.$$

Hence
$$\dot{x}^2 + \dot{y}^2 = a^2,$$
$$\dot{x}\ddot{y} - \dot{y}\ddot{x} = a^2(\sin^2 t + \cos^2 t) = a^2.$$

We therefore have the relations

$$\alpha = (u + a\cos t) - \frac{(a\cos t)(a^2)}{a^2} = u,$$

$$\beta = (v + a\sin t) + \frac{(-a\sin t)(a^2)}{a^2} = v,$$

$$\rho = \pm\frac{a^3}{a^2} = a.$$

Hence for all points on a given circle, *the centre of curvature is at the centre of the circle, and the radius of curvature is equal to the radius of the circle.*

The curvature at any point is therefore $\pm 1/a$. With the present choice of parameter, the formula of p. 114 shows at once that $\kappa = +1/a$, but other parameters may give different signs.

For example, if x is the parameter, the sense of description of the curve is $\overrightarrow{LPM}$ in the upper part of the diagram (Fig. 87) and $\overrightarrow{LQM}$ in the lower, these being the directions taken by P, Q respectively as x increases. We know (p. 114) that, when x is the parameter, κ is positive when the concavity is 'upwards' and negative where it is 'downwards'. Thus $\kappa = -1/a$ in the arc

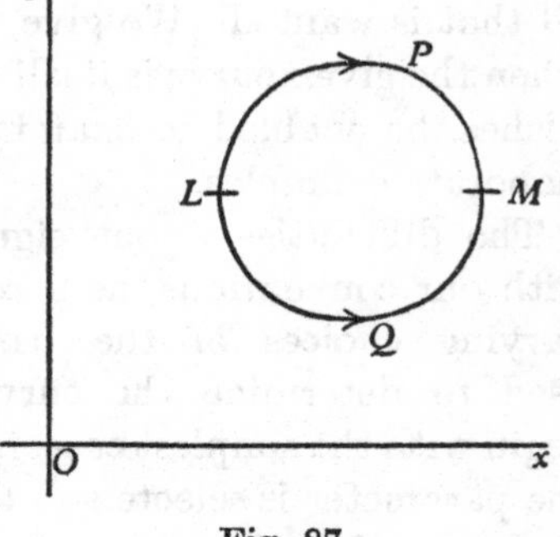

Fig. 87.

LPM, and $+1/a$ in the arc LQM. In fact, if $P(x,y)$ is in the arc LPM, then

$$y = v+\surd\{a^2-(x-u)^2\},$$

where the positive sign is attached to the square root. Hence

$$y' = \frac{-(x-u)}{\{a^2-(x-u)^2\}^{\frac{1}{2}}},$$

so that

$$1+y'^2 = \frac{a^2}{a^2-(x-u)^2},$$

and

$$y'' = \frac{-1}{\{a^2-(x-u)^2\}^{\frac{1}{2}}} - \frac{(x-u)^2}{\{a^2-(x-u)^2\}^{\frac{3}{2}}}$$

$$= -\frac{a^2}{\{a^2-(x-u^2\}^{\frac{3}{2}}}.$$

Applying the formula (p. 114) for κ, we have

$$\kappa = -\frac{a^2}{\{a^2-(x-u)^2\}^{\frac{3}{2}}} \Big/ \frac{a^3}{\{a^2-(x-u)^2\}^{\frac{3}{2}}}$$

$$= -\frac{1}{a}.$$

Similarly we may prove that $\kappa = +1/a$ in the arc LQM, where

$$y = v-\surd\{a^2-(x-u)^2\}.$$

15. Envelopes. We have been considering a curve as the path traced out by a point whose coordinates are expressed in terms of a parameter. An analogous (*dual*) problem is the study of a system of straight lines

$$lx+my+n = 0$$

when the coefficients l, m, n, instead of being constants, are given to be functions of a parameter t, say

$$l = f(t), \quad m = g(t), \quad n = h(t).$$

A familiar example is the system

$$x-yt+at^2 = 0$$

consisting of the tangents to the parabola

$$y^2 = 4ax.$$

If we take a number of values of t, we obtain correspondingly a number of lines, which lie in some such way as that indicated in the diagram (Fig. 88). They look, in fact, as if a curve could be determined to which they are all tangents. More precisely, the diagram assumes that the individual lines are numbered in an

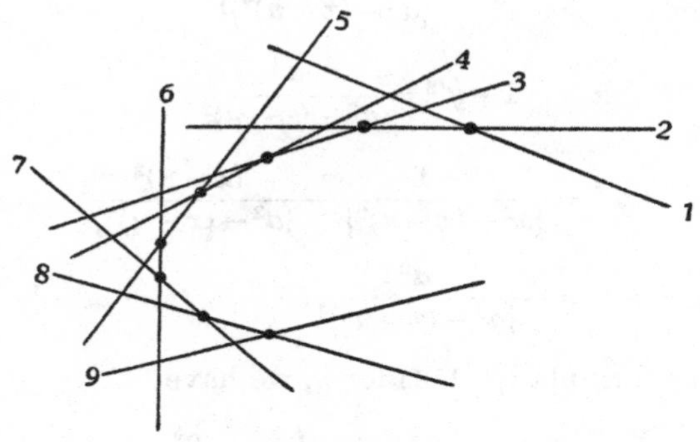

Fig. 88.

order corresponding to increasing values of the parameter, and the points of intersection of consecutive pairs (1, 2), (2, 3), (3, 4),... have been emphasized by dots. These dots appear to lie on a curve, and it is easy to conceive of the lines as becoming tangents to that curve as their number increases indefinitely. In that case, the lines are said to *envelop* the curve, and we make the following formal definition:

DEFINITION. *Given a system of lines*

$$lx+my+n=0,$$

whose coefficients l, m, n are functions of a parameter t, the locus of that point on a typical line of the system, which is the limiting position of the intersection of a neighbouring line tending to coincidence with it, is called the envelope of the system.

Consider, for example, the system of which a typical line is

$$x-yt+at^2=0.$$

Another line of the system is

$$x-yu+au^2=0.$$

They meet where $\qquad x=aut, \quad y=a(u+t).$

That point on the typical line to which the intersection tends is found by putting $u = t$ in the expressions for the coordinates, giving

$$x = at^2, \quad y = 2at.$$

This is the parametric representation of the envelope, whose equation is therefore

$$y^2 = 4ax.$$

We may now find the rule for determining parametrically the envelope of the system of which a typical line is

$$xf(t)+yg(t)+h(t) = 0.$$

Another line of the system is

$$xf(u)+yg(u)+h(u) = 0.$$

Where these lines meet, it is also true that

$$x\{f(u)-f(t)\}+y\{g(u)-g(t)\}+\{h(u)-h(t)\} = 0,$$

or, on division by $u-t$, that

$$x\frac{f(u)-f(t)}{u-t}+y\frac{g(u)-g(t)}{u-t}+\frac{h(u)-h(t)}{u-t} = 0.$$

To emphasize the limiting approach of u to t, write $u = t+\delta t$. Then the point of intersection of the two lines 'u', 't' also lies on the line

$$x\frac{f(t+\delta t)-f(t)}{\delta t}+y\frac{g(t+\delta t)-g(t)}{\delta t}+\frac{h(t+\delta t)-h(t)}{\delta t} = 0.$$

In the limit, as $\delta t \to 0$, this is the line

$$xf'(t)+yg'(t)+h'(t) = 0.$$

Hence *the envelope is the locus, as t varies, of the point of intersection of the lines*

$$xf(t)+yg(t)+h(t) = 0,$$
$$xf'(t)+yg'(t)+h'(t) = 0.$$

ILLUSTRATION 6. *To find the envelope of the system*

$$x\cos t+y\sin t+a = 0.$$

The envelope is the locus of the point of intersection of this line with the line

$$-x\sin t+y\cos t = 0.$$

That point is $x = -a\cos t, \quad y = -a\sin t,$

and so the envelope is the circle

$$x^2+y^2 = a^2.$$

Note. Envelopes exist for many families of curves as well as for families of straight lines; but we are not in a position to give a treatment of the more general case.

EXAMPLES II

Find the envelopes of the following families of straight lines:

1. $(1+t^2)x+2ty+(1-t^2)a = 0.$
2. $x\sec t-y\tan t-a = 0.$
3. $x\cosh t-y\sinh t-a = 0.$
4. $2tx+(1-t^2)y+(1+t^2)a = 0.$
5. $tx+y-a(t^3+2t) = 0.$
6. $t^3x-ty-c(t^4-1) = 0.$

16. Evolutes. Particular interest attaches to the envelope of those lines which are the normals of a given curve. Using the notation of § 14, denote by

$$y_P, \quad y_P', \quad y_P''$$

the values of y, y', y'' (where dashes denote differentiations with respect to x) for the given curve at the point $P(x_P, y_P)$. The equation of the normal at P is

$$y-y_P = (-1/y_P')(x-x_P),$$

or

$$x+y_P'y = x_P+y_P'y_P.$$

Taking x_P as the parameter, the envelope of this line is the locus of its intersection with the line

$$y_P''y = 1+y_P''y_P+y_P'^2,$$

so that

$$y = y_P+\frac{(1+y_P'^2)}{y_P''},$$

$$x = x_P-\frac{y_P'(1+y_P'^2)}{y_P''}.$$

Comparison with the results in § 14 establishes the theorem:

The centre of curvature at a point P of a given curve is that point on the normal at P which corresponds to P on the envelope of the normals.

DEFINITION. The envelope of the normals, which is also the locus of the centres of curvature, is called the *evolute* of the given curve.

ILLUSTRATION 7. *To find the evolute of the rectangular hyperbola*

$$xy = c^2.$$

Parametrically, the hyperbola is

$$x = ct, \quad y = c/t,$$

and the gradient at this point is $-1/t^2$. Hence the normal is

$$y - c/t - t^2(x - ct) = 0,$$

or

$$t^3x - ty = ct^4 - c.$$

For the envelope, we have

$$3t^2x - y = 4ct^3.$$

The centre of curvature, being the point of intersection of these two lines, is given by

$$2t^3x = 3ct^4 + c,$$

so that

$$x = \left(\frac{3t}{2} + \frac{1}{2t^3}\right)c,$$

$$y = \left(\frac{3}{2t} + \frac{t^3}{2}\right)c.$$

The evolute is the curve given parametrically by these two equations.

EXAMPLES III

Find the evolutes of the following curves:

1. The parabola $(at^2, 2at)$.
2. The ellipse $(a\cos t, b\sin t)$.
3. The rectangular hyperbola $(a\sec t, a\tan t)$.

17. The area of a closed curve. Consider the closed curve $PUQV$ shown in the diagram (Fig. 89). For simplicity, we suppose it to be oval in shape, and also, to begin with, to lie entirely in the first quadrant.

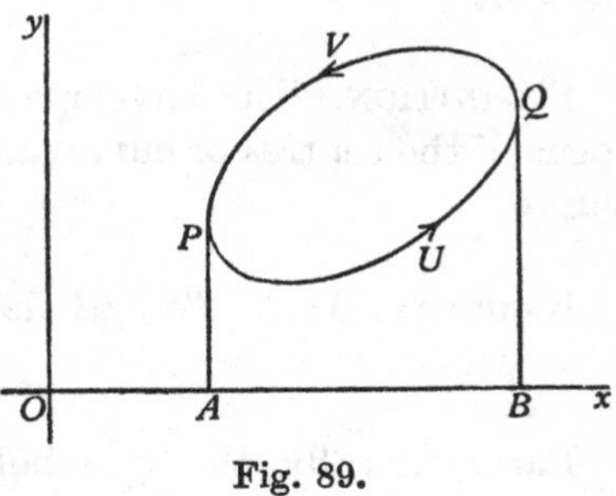

Fig. 89.

The coordinates of the points of the curve being expressed in the parametric form

$$x = x(t), \quad y = y(t),$$

we suppose that the positive sense, namely that of t increasing, is COUNTER-CLOCKWISE round the curve, as implied by the arrows in the diagram. [If this is not so, replacement of t by $-t$ will reverse the sense.] Thus the curve is described once by the point (x, y) as t increases from a value t_0 to a value t_1. Moreover, since the curve is closed, the two values t_0, t_1 give rise to the SAME point, so that

$$x(t_0) = x(t_1); \quad y(t_0) = y(t_1).$$

A simple example is the circle

$$x = 5 + 3\cos t, \quad y = 4 + 3\sin t,$$

described once in the counter-clockwise sense as t increases from 0 to 2π. The values $t = 0$, $t = 2\pi$ both give the point $(8, 4)$.

In order to calculate the area, draw the ordinates AP, BQ which just contain the curve, touching it at two points P, Q whose parameters we write as t_P, t_Q, respectively. For reference, let U be a point in the lower arc PQ and V in the upper. Then the area enclosed by the curve can be expressed in the form

$$\text{area } APVQ - \text{area } APUQ.$$

The area $APVQ$ is given by the formula

$$\text{area } APVQ = \int_{x_P}^{x_Q} y\,dx$$

integrated over points (x, y) on the arc PVQ. Let us suppose first that the 'junction' point given by t_0 or t_1, does not lie on this arc.

Then the parameter t varies steadily along the arc, and the area is

$$\int_{t_P}^{t_Q} y\frac{dx}{dt}\,dt.$$

On the other hand, the 'junction' point J must then lie on the lower arc PUQ, so we write the formula for the area $APUQ$ in the form

$$\text{area } APUQ = \int_{x_P}^{x_Q} y\,dx$$

integrated over points (x, y) on the arc PUQ, giving

$$\begin{aligned}\text{area } APUQ &= \int_{x_P}^{J} y\,dx + \int_{J}^{x_Q} y\,dx \\ &= \int_{t_P}^{t_1} y\frac{dx}{dt}\,dt + \int_{t_0}^{t_Q} y\frac{dx}{dt}\,dt,\end{aligned}$$

where the value t_1 or t_0 is taken for J according to the segment of the arc PUQ over which the respective integral is calculated, t_1 for PJ and t_0 for JQ. In all, the area enclosed by the curve is thus

$$\begin{aligned}&\int_{t_P}^{t_Q} y\frac{dx}{dt}\,dt - \left\{\int_{t_P}^{t_1} y\frac{dx}{dt}\,dt + \int_{t_0}^{t_Q} y\frac{dx}{dt}\,dt\right\} \\ &\qquad = -\left\{\int_{t_0}^{t_Q} y\frac{dx}{dt}\,dt + \int_{t_Q}^{t_P} y\frac{dx}{dt}\,dt + \int_{t_P}^{t_1} y\frac{dx}{dt}\,dt\right\} \\ &\qquad = -\int_{t_0}^{t_1} y\frac{dx}{dt}\,dt.\end{aligned}$$

Similarly, if J lies on the arc PVQ, we have the formulæ

$$\begin{aligned}\text{area } APVQ &= \int_{x_P}^{J} y\,dx + \int_{J}^{x_Q} y\,dx \\ &= \int_{t_P}^{t_0} y\frac{dx}{dt}\,dt + \int_{t_1}^{t_Q} y\frac{dx}{dt}\,dt,\end{aligned}$$

$$\begin{aligned}\text{area } APUQ &= \int_{x_P}^{x_Q} y\,dx \\ &= \int_{t_P}^{t_Q} y\frac{dx}{dt}\,dt,\end{aligned}$$

so that the area enclosed by the curve is

$$\left\{\int_{t_P}^{t_0} y\frac{dx}{dt}\,dt + \int_{t_1}^{t_Q} y\frac{dx}{dt}\,dt\right\} - \int_{t_P}^{t_Q} y\frac{dx}{dt}\,dt$$

$$= -\left\{\int_{t_0}^{t_P} y\frac{dx}{dt}\,dt + \int_{t_P}^{t_Q} y\frac{dx}{dt}\,dt + \int_{t_Q}^{t_1} y\frac{dx}{dt}\,dt\right\}$$

$$= -\int_{t_0}^{t_1} y\frac{dx}{dt}\,dt.$$

Hence, in both cases, *the area of the closed curve is*

$$-\int_{t_0}^{t_1} y\frac{dx}{dt}\,dt.$$

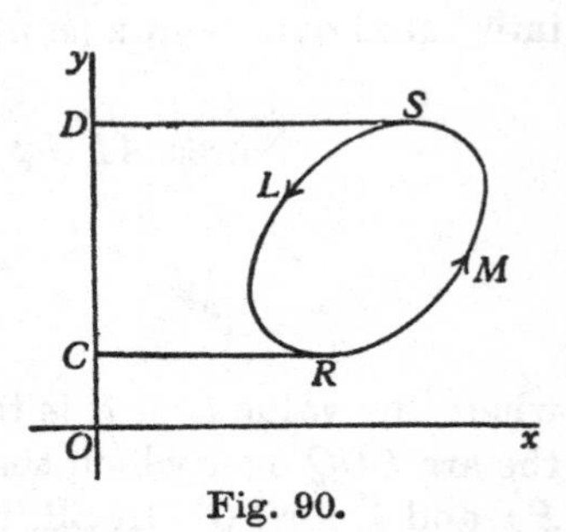

Fig. 90.

In the same way, if we draw the lines CR, DS parallel to Ox (Fig. 90), just containing the curve, to touch it at R, S, and if L, M are points on the left and right arcs RS respectively, then the area of the closed curve is

area $CRMS$ − area $CRLS$.

Now
$$\text{area } CRMS = \int_{y_R}^{y_S} x\,dy$$

integrated over points (x, y) on the arc RMS. If the 'junction' point J is not on this arc, we have

$$\int_{t_R}^{t_S} x\frac{dy}{dt}\,dt.$$

For the area $CRLS$, the 'junction' point J must then lie on the arc RLS, so we have

$$\int_{t_R}^{t_0} x\frac{dy}{dt} + \int_{t_1}^{t_S} x\frac{dy}{dt}\,dt.$$

In all, the area of the closed curve is (in brief notation)

$$\int_{t_R}^{t_S} - \left\{\int_{t_R}^{t_0} + \int_{t_1}^{t_S}\right\}$$

$$= \int_{t_0}^{t_R} + \int_{t_R}^{t_S} + \int_{t_S}^{t_1}$$

$$= \int_{t_0}^{t_1} x\frac{dy}{dt}\,dt.$$

Similarly for the 'junction' point on the arc RMS, we have

$$\left\{\int_{t_R}^{t_1}+\int_{t_0}^{t_S}\right\}-\left\{\int_{t_R}^{t_S}\right\}$$

$$=\int_{t_0}^{t_S}+\int_{t_S}^{t_R}+\int_{t_R}^{t_1}$$

$$=\int_{t_0}^{t_1} x\frac{dy}{dt}\,dt.$$

Hence *the area of the closed curve may be expressed in either of the forms*

$$-\int_{t_0}^{t_1} y\frac{dx}{dt}\,dt,$$

$$+\int_{t_0}^{t_1} x\frac{dy}{dt}\,dt,$$

where the closed curve is described completely, in the counter-clockwise sense, as t increases from t_0 to t_1.

A useful alternative form is found by taking half from each of these:

The area of the closed curve is

$$\frac{1}{2}\int_{t_0}^{t_1}\left(x\frac{dy}{dt}-y\frac{dx}{dt}\right)dt.$$

ILLUSTRATION 8. *To find the area of the ellipse*

$$\frac{x^2}{a^2}+\frac{y^2}{b^2}=1.$$

The ellipse is traced out by the point

$$x=a\cos t,\quad y=b\sin t$$

as t moves from 0 to 2π. Then

$$dx=-a\sin t\,dt,\quad dy=b\cos t\,dt,$$

and $\frac{1}{2}\int(x\,dy-y\,dx)$

$$=\frac{1}{2}\int_0^{2\pi}(a\cos t.b\cos t+b\sin t.a\sin t)\,dt$$

$$=\frac{1}{2}ab\Big[t\Big]_0^{2\pi}$$

$$=\pi ab.$$

The restriction for the curve to lie in the first quadrant is not essential. For, if it does not, a transformation of the type

$$x' = x+a, \quad y' = y+b,$$

for suitable values of a, b, can always be employed to bring it into the first quadrant of a fresh set of axes. But this shows that the area enclosed is

$$\frac{1}{2}\int_{t_0}^{t_1}\left(x'\frac{dy'}{dt}-y'\frac{dx'}{dt}\right)dt = \frac{1}{2}\int_{t_0}^{t_1}\left\{(x+a)\frac{dy}{dt}-(y+b)\frac{dx}{dt}\right\}dt$$

$$= \frac{1}{2}\int_{t_0}^{t_1}\left(x\frac{dy}{dt}-y\frac{dx}{dt}\right)dt+\tfrac{1}{2}a\int_{t_0}^{t_1}\frac{dy}{dt}\,dt-\tfrac{1}{2}b\int_{t_0}^{t_1}\frac{dx}{dt}\,dt.$$

But

$$\int_{t_0}^{t_1}\frac{dy}{dt}\,dt = \Big[y\Big]_{t_0}^{t_1} = 0,$$

$$\int_{t_0}^{t_1}\frac{dx}{dt}\,dt = \Big[x\Big]_{t_0}^{t_1} = 0,$$

since t_0, t_1 give the SAME point of the curve. Hence the area is

$$\frac{1}{2}\int_{t_0}^{t_1}\left(x\frac{dy}{dt}-y\frac{dx}{dt}\right)dt.$$

18.* Second theorem of Pappus. Suppose that a surface of revolution is obtained by rotating an arc PQ (Fig. 91) about the x-axis (assumed not to meet it). We proved (Vol. I, p. 130) that, if PQ is the curve

$$y = f(x),$$

the area S so generated is given by the formula

$$S = \int_a^b 2\pi y\,ds.$$

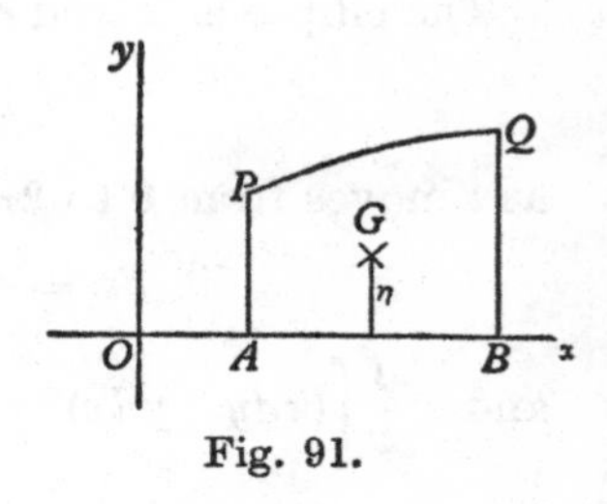

Fig. 91.

Now it is easy to prove that the y-coordinate of the centre of gravity of the arc is given by the formula

$$\eta = \frac{\int_a^b y\,ds}{l},$$

where l is the length of the arc PQ. (Compare the similar work in Chapter VI.) Hence

$$S = 2\pi\eta l.$$

* This paragraph may be postponed, if desired.

Thus *if a given curve, lying on one side of a given line, is rotated about that line as axis to form a surface of revolution, then the area of the surface so generated is equal to the product of the length of the curve and the distance rotated by its centre of gravity.*

ILLUSTRATION 9. *To find the centre of gravity of a semicircular arc.* Suppose that the circle is of radius a (Fig. 92), and that the centre of gravity, lying on the axis of symmetry, is at a distance η from the centre.

On rotating the semicircle about its bounding diameter, we obtain the surface of a sphere, whose area is known to be

$$4\pi a^2.$$

Hence $$4\pi a^2 = 2\pi\eta . \pi a,$$

so that $$\eta = \frac{2a}{\pi}.$$

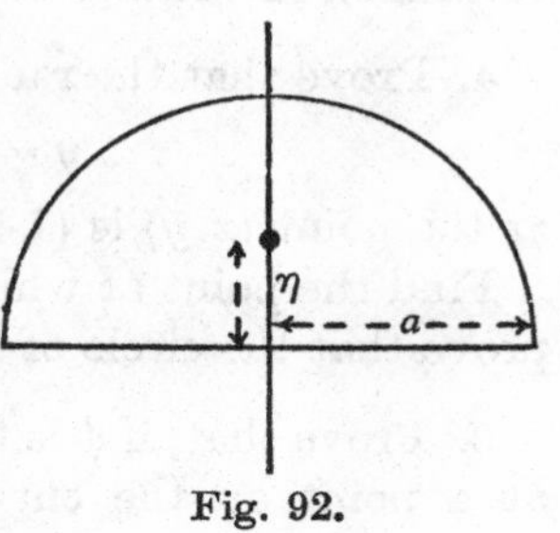

Fig. 92.

REVISION EXAMPLES V
'Advanced' Level

1. Find the equations of the tangent and normal at any point of the cycloid given by the equations

$$x = a(2\psi + \sin 2\psi), \quad y = a(1 - \cos 2\psi).$$

Prove that ψ is the angle which the tangent makes with the axis of x; and verify that, if p and q are the lengths of the perpendiculars drawn from the origin to the tangent and normal respectively, then q and $dp/d\psi$ are numerically equal.

2. Define the length of an arc of a curve and obtain an expression for it.

A curve is given in the form

$$x = \cosh t - t, \quad y = \cosh t + t.$$

Express t in terms of the length s of the arc of the curve measured from the point $(1, 1)$.

The coordinates of any point of the curve are expressed in terms of a and are then expanded in series of ascending powers of s.

Prove that the first few terms of the expansions are

$$x = 1 - \frac{s}{\sqrt{2}} + \frac{s^2}{4} + \ldots,$$

$$y = 1 + \frac{s}{\sqrt{2}} + \frac{s^2}{4} + \ldots.$$

3. Find the radius of curvature of the parabola $y^2 = 4\lambda x$ at a point for which $x = c$, and deduce that, if λ varies (c remaining constant), this radius of curvature is a minimum when $\lambda = \frac{1}{2}c$.

4. Prove that the radius of curvature of the curve

$$y = \tfrac{1}{4}x^2 - \tfrac{1}{2}\log x \quad (x > 0)$$

at the point (x, y) is $(1+x^2)^2/4x$.

Find the point at which this curve is parallel to the x-axis, and prove that the circle of curvature at this point touches the y-axis.

5. Prove that, if ψ is the inclination to the x-axis of the tangent at a point on the curve $y = a \log \sec (x/a)$, then the radius of curvature at this point is $a \sec \psi$.

6. A particle moves in a plane so that, at time t, its coordinates referred to rectangular axes are given by

$$x = a\cos 2t + 2a\cos t, \quad y = a\sin 2t + 2a\sin t.$$

Find the components of the velocity parallel to the axes and the resultant speed of the particle.

Show that ρ, the radius of curvature at a point of the path, is proportional to the speed of the particle at that point.

7. The coordinates of the points of the curve $4y^3 = 27x^2$ are expressed parametrically in the form $(2t^3, 3t^2)$. By using this parametric representation, or otherwise, prove that the length of the curve between the origin and the point P with parameter t_1 is

$$2(1+t_1^2)^{\frac{3}{2}} - 2,$$

and that the radius of curvature at P is numerically equal to

$$6t_1(1+t_1^2)^{\frac{3}{2}}.$$

8. The tangent to a curve at the point (x, y) makes an angle ψ with the x-axis. Prove that the centre of curvature at (x, y) is

$$\left(x - \frac{dy}{d\psi}, \quad y + \frac{dx}{d\psi}\right).$$

A curve is given parametrically by the equations

$$x = 2a\cos t + a\cos 2t, \quad y = 2a\sin t - a\sin 2t.$$

Prove that $\psi = -\frac{1}{2}t$.

P is a variable point on the curve, and Q is the centre of curvature at P; the point R divides PQ internally so that $PR = \frac{1}{4}PQ$. Prove that the locus of R is the circle $x^2+y^2 = 9a^2$.

Prove also that no point of the curve lies outside this circle.

Draw a rough sketch of the curve.

9. The coordinates of a point of a curve are given in terms of a parameter t by the equations $x = te^t$, $y = t^2e^t$. Find dy/dx in terms of t, and prove that

$$\frac{d^2y}{dx^2} = \frac{t^2+2t+2}{(t+1)^3}e^{-t}.$$

Prove also that the radii of the circles of curvature at the two points at which the curve is parallel to the x-axis are in the ratio $e^2 : 1$.

10. Express $\cosh^4\theta$ in the form

$$a_0 + a_1\cosh\theta + a_2\cosh 2\theta + a_3\cosh 3\theta + a_4\cosh 4\theta.$$

11. The area lying between the curve

$$y = \cosh(x/2\lambda),$$

the ordinates $x = \pm\lambda^2$, and the x-axis is rotated about this axis. Prove that the volume of the solid of revolution so formed is

$$\pi\lambda(\lambda + \sinh\lambda).$$

Show that the volumes given by $\lambda = 1, \lambda = 1+\delta$ differ by approximately $\pi(2+e)\,\delta$ when δ is small.

12. Prove that the two curves

$$y_1 = \tfrac{1}{2}\cosh 2x, \quad y_2 = \tanh 2x$$

touch at one point and have no other point in common.

Sketch the two curves in the same diagram and, with them, the curves

$$y_3 = \cosh x, \quad y_4 = \sinh x.$$

Prove that the four curves intersect in pairs for two values of x and indicate clearly in the diagram the relative positions of the four curves for all values of x.

13. If s be the length of the arc of the catenary $y = a \cosh(x/a)$ from the point $(0, a)$ to the point (x, y), show that

$$s^2 = y^2 - a^2.$$

Find the area of the surface generated when this arc is rotated about the y-axis.

14. Sketch the curve whose coordinates are given parametrically by the relations $x = a(t + \sin t), y = a(1 + \cos t)$ for values of t between $-\pi, \pi$, and find the length of the curve between these two points.

15. Find the radius of curvature of the parabola $y^2 = x$ at the point $(2, \sqrt{2})$.

16. Find the radius of curvature and the coordinates of the centre of curvature at the point $(0, 1)$ on the curve $y = \cosh x$.

17. For what value of λ does the parabola

$$y^2 = 4\lambda x$$

have the same circle of curvature as the ellipse

$$\left(\frac{x-a}{a}\right)^2 + \frac{y^2}{b^2} = 1$$

at the origin?

18. Find the radius of curvature of the curve

$$y = \sin ax^2$$

at the origin.

19. Draw a rough graph of the function $\cosh x$.

The tangent at a point $P(x, y)$ of the curve $y = c \cosh(x/c)$ makes an angle ψ with the x-axis. Show that $y = c \sec \psi$.

If the tangent at P cuts the y-axis at Q, show that

$$PQ = xy/c.$$

20. Find the radius of curvature of the curve

$$ay^2 = x^3$$

at the point (a, a).

21. Find the radius of curvature at the point $t = \frac{1}{3}\pi$ on the curve

$$x = a \sin t, \quad y = a \cos 2t.$$

22. Find the radius of curvature of the curve

$$y = x^2 + x - 1$$

at the point (x, y) and find a point on the curve for which the centre of curvature is on the y-axis.

23. Show that the radius of curvature of the epicycloid

$$x = 3 \cos t - \cos 3t, \quad y = 3 \sin t - \sin 3t$$

is given by $3 \sin t$.

24. Sketch the curve whose equation is

$$r = a \sin 3\theta \quad (a > 0).$$

Find the area of the loop in the first quadrant and show that the radius of curvature at the point (r, θ) is

$$\frac{1}{2}a \frac{(5 + 4 \cos 6\theta)^{\frac{3}{2}}}{(7 + 2 \cos 6\theta)}.$$

25. Show that the two functions

$$\sinh^{-1}(\tan x), \quad \tanh^{-1}(\sin x)$$

have equal derivatives, and hence prove that the functions themselves are equal when $-\frac{1}{2}\pi < x < \frac{1}{2}\pi$.

26. Find the point of maximum curvature on the curve

$$y = \log x,$$

and the curvature at this point.

27. If y is the function of x given by the relation

$$\sinh y = \tan x,$$

where $-\frac{1}{2}\pi < x < \frac{1}{2}\pi$, prove that

$$\cosh y = \sec x, \quad y = \log \tan(\frac{1}{2}x + \frac{1}{4}\pi).$$

Show that y has an inflexion at $x = 0$, and draw a rough sketch of the function in the given range.

REVISION EXAMPLES VI

'Scholarship' Level

1. Transform the equation

$$\frac{d^2u}{dz^2}-\frac{2z}{1-z^2}\frac{du}{dz}+\frac{u}{(1-z^2)^2}=0$$

to one in which y is the dependent and x the independent variable, where

$$u(1-z^2)^{\frac{1}{2}}=y \quad \text{and} \quad (1+z^2)/(1-z^2)=x,$$

obtaining the result in the form

$$(x^2-1)\frac{d^2y}{dx^2}+(2x-1)\frac{dy}{dx}+\tfrac{1}{2}y=0.$$

2. If $f(x)$ is a polynomial which increases as x increases, show that, when $x>0$,

$$g(x)\equiv\frac{1}{x}\int_0^x f(y)\,dy$$

is also a polynomial which increases as x increases.

Show that, when $x>0$, the expression

$$\frac{(x^2-2x+2)e^x}{x}-\frac{2}{x}$$

is an increasing function of x.

3. Prove that

$$(x+1)\left(\frac{d}{dx}\right)^{n+1}\{(x+1)^n(x-1)^{n+1}\}=(n+1)\left(\frac{d}{dx}\right)^n\{(x+1)^{n+1}(x-1)^n\}.$$

Prove also that the function

$$\left(\frac{d}{dx}\right)^n\{(x+1)^{n+1}(x-1)^n\}$$

satisfies the equation

$$(1-x^2)\frac{d^2y}{dx^2}-(1+x)\frac{dy}{dx}+(n+1)^2y=0.$$

4. If $y=\sin(m\sin^{-1}x)$, show that

$$(1-x^2)\frac{d^2y}{dx^2}-x\frac{dy}{dx}+m^2y=0.$$

Hence or otherwise show that the expansion of $\sin m\theta$ as a power series in $\sin\theta$ is

$$m\sin\theta\left\{1-\frac{(m^2-1)}{3!}\sin^2\theta+\frac{(m^2-1)(m^2-3^2)}{5!}\sin^4\theta-\ldots\right.$$
$$\left.+(-1)^n\frac{(m^2-1)\ldots[m^2-(2n-1)^2]}{(2n+1)!}\sin^{2n}\theta+\ldots\right\}.$$

5. By induction, or otherwise, prove that, if $y=\cot^{-1}x$, then

$$\frac{d^ny}{dx^n}=(-1)^n(n-1)!\sin ny\sin^n y.$$

Hence, or otherwise, prove that, if

$$z=\tan^{-1}\left(\frac{x\sin\alpha}{1+x\cos\alpha}\right),$$

then $$\frac{d^nz}{dx^n}=(-1)^{n-1}\frac{(n-1)!}{\sin^n\alpha}\sin n(\alpha-z)\sin^n(\alpha-z).$$

6. The function $f_n(x)$ is defined by the relation

$$f_n(x)=\frac{1}{y}\frac{d^ny}{dx^n},$$

where $y=e^{x^2}$. Prove that

$$\frac{dy}{dx}=2xy$$

and deduce that

$$f_{n+2}(x)-2xf_{n+1}(x)-2(n+1)f_n(x)=0,$$

and that $f_n(x)$ is a polynomial in x of degree n.

Prove that $$f_{n+1}(x)=f_n'(x)+2xf_n(x),$$

and hence express $f_{n+2}(x)$ in terms of $f_n(x)$, $f_n'(x)$ and $f_n''(x)$. Deduce that

$$f_n''(x)+2xf_n'(x)-2nf_n(x)=0.$$

7. If $y=\sin^{-1}x$, prove that

$$(1-x^2)\frac{d^2y}{dx^2}-x\frac{dy}{dx}=$$

Determine the values of y and its successive derivatives when $x=0$, and hence expand y in a series of ascending powers of x.

8. Prove that, if $y = \tan^{-1} x$, then

$$u \equiv \frac{d^n y}{dx^n} = (n-1)! \cos^n y \cos\{ny + \tfrac{1}{2}(n-1)\pi\}$$

for every positive integer n.

Deduce, or prove otherwise, that u satisfies the differential equation

$$(1+x^2)\frac{d^2u}{dx^2} + 2(n+1)x\frac{du}{dx} + n(n+1)u = 0.$$

9. If $$y_n(x) = \frac{d^n}{dx^n}\{(x^2-1)^n\},$$

prove the relations

$$\frac{dy_{n+1}}{dx} = (x^2-1)\frac{d^2y_n}{dx^2} + 2(n+2)x\frac{dy_n}{dx} + (n+1)(n+2)y_n,$$

$$\frac{dy_{n+1}}{dx} = 2(n+1)x\frac{dy_n}{dx} + 2(n+1)^2 y_n.$$

10. If $$y_n(x) = e^x \frac{d^n}{dx^n}(x^n e^{-x}),$$

prove that $$y_n = x\frac{dy_{n-1}}{dx} + (n-x)y_{n-1},$$

$$\frac{1}{n}\frac{dy_n}{dx} = \frac{dy_{n-1}}{dx} - y_{n-1}.$$

Hence show that the polynomial y_n satisfies a certain linear differential equation of the second order. $\left[\text{i.e. linear in } \frac{d^2y_n}{dx^2}, \frac{dy_n}{dx}, y_n.\right]$

11. Show that, if the substitution $u = t^{a+1}\frac{d(\log y)}{dt}$ is made in the equation $$\frac{du}{dt} = t^a - t^{-a-1}u^2,$$

then $$t\frac{d^2y}{dt^2} + (a+1)\frac{dy}{dt} - y = 0.$$

Deduce that, if a is not a negative integer, then the value of $\frac{d^n y}{dt^n}$ when $t = 0$ is $$\frac{A}{(a+n)(a+n-1)\dots(a+1)},$$

where A is the value of y when $t = 0$.

12. Prove that the polynomial $y=(x^2-2x)^n$ satisfies the equation

$$(2x-x^2)\frac{d^{n+2}y}{dx^{n+2}}-2(x-1)\frac{d^{n+1}y}{dx^{n+1}}+n(n+1)\frac{d^n y}{dx^n}=0.$$

13. A function $f(x)$ is defined for $0\leqslant x\leqslant \frac{3}{2}$ as follows:

$$0\leqslant x<1,\quad f(x)=-\tfrac{1}{6}x^3+x;$$
$$1\leqslant x\leqslant\tfrac{3}{2},\quad f(x)=-\tfrac{1}{2}x^2+\tfrac{3}{2}x-\tfrac{1}{6}.$$

Discuss the continuity of $f(x)$ and its successive derivatives. Sketch the graphs of $f(x)$ and its derivatives throughout the whole interval $(0, \frac{3}{2})$.

Prove that the function

$$f(x)-\frac{3}{\pi}\sin\frac{\pi x}{3}$$

has a stationary value at $x=1$, and determine whether it is a maximum or a minimum.

14. Prove that, if $y=e^{-x^2/2}$ and n is a positive integer, then

$$\frac{d^{n+2}y}{dx^{n+2}}+x\frac{d^{n+1}y}{dx^{n+1}}+(n+1)\frac{d^n y}{dx^n}=0.$$

The functions $f_n(x)$ are defined by the formula

$$f_n(x)=(-1)^n e^{x^2/2}\frac{d^n}{dx^n}(e^{-x^2/2}).$$

Prove that (i) $f'_{n+1}=(n+1)f_n$,

(ii) $f_{n+1}=xf_n-f'_n$,

(iii) $f_{n+2}-xf_{n+1}+(n+1)f_n=0$,

(iv) f_n is a polynomial in x of degree n.

15. Obtain the equations

$$x=s-\tfrac{1}{6}\kappa^2 s^3+\ldots,\quad y=\tfrac{1}{2}\kappa s^2+\tfrac{1}{6}\kappa' s^3+\ldots$$

for a curve C.

Prove that the equation of the general conic having 4-point contact (i.e. 4 'coincident' intersections) with C at the origin O is

$$x^2-\tfrac{2}{3}\rho' xy+\lambda y^2-2\rho y=0,$$

where $\rho=1/\kappa$ and λ is an arbitrary constant.

Deduce that the length of the latus rectum of a parabola having 4-point contact with a circle of radius a is $2a$.

16. Obtain the equations

$$x = s - \tfrac{1}{6}\kappa^2 s^3 + \ldots, \quad y = \tfrac{1}{2}\kappa s^2 + \tfrac{1}{6}\kappa' s^3 + \ldots.$$

The point Q on the curve is such that P is the middle point of the arc OQ; the tangent at the origin O cuts the chord PQ at T. Find the limiting value of the ratio $TP : TQ$ as Q tends to O.

Prove that, correct to the second order in s, the angle between the tangents to the curve at P and Q is $\kappa s\left(1 + \dfrac{3\kappa'}{2\kappa} s\right)$.

17. Show that, if $f(x)$ has a derivative throughout the interval $a \leqslant x \leqslant b$, then

$$f(b) - f(a) = (b-a) f'\{a + \theta(b-a)\},$$

where θ lies between 0 and 1.

Find the value of θ if $f(x) \equiv \tan x$, $a = \tfrac{1}{4}\pi$, $b = \tfrac{1}{3}\pi$.

Draw an accurate graph of the function $y = \tan x$ between the limits $\tfrac{1}{4}\pi, \tfrac{1}{3}\pi$, taking 1 in. to represent $\tfrac{1}{60}\pi$ as the x-axis and 1 in. to represent 0·2 on the y-axis.

Illustrate the theorem by means of your graph.

18. Given that $$e^{ay} = \cos x$$

and that y_n denotes $d^n y/dx^n$, prove that

$$a y_2 + a^2 y_1^2 + 1 = 0$$

and express y_{n+2} in terms of $y_1, y_2, \ldots, y_{n+1}$.

The expansion of y as a power series in x is

$$y = b_1 x + b_2 x^2 + \ldots + b_n x^n + \ldots.$$

Prove that b_n is zero when n is odd and that, if a is positive, b_n is negative when n is even.

19. Prove that the nth differential coefficient of $e^{x\sqrt{3}} \cos x$ is

$$2^n e^{x\sqrt{3}} \cos\left(x + \frac{n\pi}{6}\right).$$

Deduce, or obtain otherwise, the expansion of the function as a series of ascending powers of x giving the terms up to x^6 and the general term.

Find the most general function whose nth differential coefficient is $e^{x\sqrt{3}} \cos x$.

20. (i) Prove by induction that, if $x = \cot\theta$, $y = \sin^2\theta$, then

$$\frac{d^n y}{dx^n} = (-1)^n n! \sin^{n+1}\theta \sin(n+1)\theta.$$

(ii) Prove that $\dfrac{d^n}{dx^n}\left(\dfrac{1}{x^2-1}\right) = \dfrac{f_n(x)}{(x^2-1)^{n+1}}$,

where $f_n(x)$ is a polynomial of degree n in x.

Prove further that

$$f_{n+1}(x) + 2(n+1)xf_n(x) + n(n+1)(x^2-1)f_{n-1}(x) = 0.$$

21. State, without proof, Rolle's theorem, and deduce that there is a number ξ between a, b such that

$$f(b) - f(a) = (b-a)f'(\xi),$$

explaining what conditions must be satisfied by the function $f(x)$ in order that the theorem may be valid.

If $f(x) \equiv \sin x$, find all the values of ξ when $a = 0, b = \frac{3}{2}\pi$. Illustrate the result with reference to the graph of $\sin x$.

22. If α is small, the equation $\sin x = \alpha x$ has a root nearly equal to π. Show that

$$\pi\{1 - \alpha + \alpha^2 - (\tfrac{1}{6}\pi^2 + 1)\alpha^3\}$$

is a better approximation, if α is sufficiently small.

23. A plane curve is such that the tangent at any point P is inclined at an angle $(k+1)\theta$ to a fixed line Ox, where k is a positive constant and θ is the angle xOP. The greatest length of OP is a. Find a polar equation for the curve.

Sketch the curve for the cases $k = 2, k = \frac{1}{2}$.

24. Prove that

$$\frac{3\dfrac{d^2y}{dx^2}\dfrac{d^4y}{dx^4} - 5\left(\dfrac{d^3y}{dx^3}\right)^2}{\left(\dfrac{dy}{dx}\right)^4} = \frac{3\dfrac{d^2x}{dy^2}\dfrac{d^4x}{dy^4} - 5\left(\dfrac{d^3x}{dy^3}\right)^2}{\left(\dfrac{dx}{dy}\right)^4}.$$

25. Obtain the expansion of $\sin x$ in ascending powers of x. For what values of x is this series convergent.

A small arc PQ of a circle of radius 1 is of length x. The arc PQ is bisected at Q_1 and the arc PQ_1 is bisected at Q_2. The chords PQ, PQ_1, PQ_2 are of lengths $c, \frac{1}{2}c_1, \frac{1}{4}c_2$ respectively. Prove that $\frac{1}{45}(c - 20c_1 + 64c_2)$ differs from x by a quantity of order x^7.

26. Sketch the graph of the function

$$y = e^{-(ax+bx^4)}$$

where a, b are both positive. Prove that there are always at least two points of inflexion.

Find the abscissæ of the points of inflexion when $a = \frac{10}{27}, b = 5$.

27. If $x = f(t)$, $y = g(t)$, express $\frac{dy}{dx}, \frac{dx}{dy}, \frac{d^2y}{dx^2}, \frac{d^2x}{dy^2}$ in terms of $\frac{dx}{dt}, \frac{dy}{dt}, \frac{d^2x}{dt^2}, \frac{d^2y}{dt^2}$.

Prove that

$$\left(\frac{dx}{dy}\right)^2 \frac{d^3y}{dx^3} + \left(\frac{dy}{dx}\right)^2 \frac{d^3x}{dy^3} + 3\frac{d^2y}{dx^2}\frac{d^2x}{dy^2} = 0.$$

28. The functions $\phi(x), \psi(x)$ are differentiable in the interval $a \leqslant x \leqslant b$, and $\psi'(x) > 0$ for $a \leqslant x \leqslant b$. Prove that there is at least one number ξ between a, b such that

$$\frac{\phi(\xi) - \phi(a)}{\psi(b) - \psi(\xi)} = \frac{\phi'(\xi)}{\psi'(\xi)}.$$

If $\phi(x) \equiv x^2$, $\psi(x) \equiv x$, find a value of ξ in terms of a and b.

29. Prove by differentiation, or otherwise, that

$$xy \leqslant e^{x-1} + y \log y$$

for all real x and positive y.

When does the sign of equality hold?

30. (i) Prove that, for positive values of x,

$$\log(1+x) < \frac{x(2+x)}{2(1+x)}.$$

(ii) Find whether $e^{-x^2} \sec^2 x$ has a maximum or a minimum value for $x = 0$.

31. Show that, if $f(0) = 0$ and if $f'(x)$ is an increasing function of x, then $y = \frac{f(x)}{x}$ is an increasing function of x for $x > 0$.

32. Prove that, if $x > 0$,

$$(1 - \tfrac{1}{2}x^2 + \tfrac{1}{24}x^4) \sin x > (x - \tfrac{1}{6}x^3 + \tfrac{1}{120}x^5) \cos x.$$

33. If $x>0$, prove that $(x-1)^2$ is not less than $x(\log x)^2$.

Discuss the general behaviour of the function

$$(\log x)^{-1}-(x-1)^{-1}$$

for positive values of x and with special reference to $x=1$.

Sketch the graph of the function.

34. Prove that, if x is positive,

$$\frac{2x}{2+x}<\log(1+x)<x.$$

Prove also that, if a, h are positive, then

$$\log(a+\theta h)-\log a-\theta\{\log(a+h)-\log a\},$$

considered as a function of θ, has a maximum for a value of θ between 0 and $\frac{1}{2}$.

35. If the sum of the lengths of the hypotenuse and one other side of a right-angled triangle is given, find the angle between the hypotenuse and that side when the area of the triangle has its maximum value.

36. A pyramid consists of a square base and four equal triangular faces meeting at its vertex. If the total surface area is kept fixed, show that the volume of the pyramid is greatest when each of the angles at its vertex is $36^\circ\,52'$.

37. Find the least volume of a right circular cone in which a sphere of unit radius can be placed.

38. Find the numerical values of

$$y=\sin(x+\tfrac{1}{4}\pi)+\tfrac{1}{4}\sin 4x$$

at its stationary values in the range $-\pi\leqslant x\leqslant\pi$. Distinguish between maxima, minima, and points of inflexion, and give a rough sketch of the curve.

39. Prove that, if

$$\theta=\cot^{-1}x \quad (0<\theta<\pi),$$

then
$$\frac{d^n\theta}{dx^n}=(-1)^n(n-1)!\sin^n\theta\sin n\theta,$$

when n is any positive integer.

Show that the absolute value of $d^n\theta/dx^n$ never exceeds $(n-1)!$ if n is odd, or $(n-1)!\cos^{n+1}\left(\frac{\pi}{2n+2}\right)$ if n is even.

40. Find the two nearest points on the curves

$$y^2-4x=0, \quad x^2+y^2-6y+8=0,$$

and evaluate their distance.

41. If

$$y=\frac{x(1-x)}{1+x^2},$$

(i) find the maximum and minimum values of y;

(ii) find the points of inflexion of the curve;

(iii) sketch a graph showing clearly the points determined in (i), (ii), and also the position of the curve relative to the line $y=x$.

42. Prove that the maxima of the curve $y=e^{-kx}\sin px$, where k, p are positive constants, all lie on a curve whose equation is $y=Ae^{-kx}$, and find A in terms of k, p.

Draw in the same diagram rough sketches of the curves $y=e^{-kx}$, $y=-e^{-kx}$ and $y=e^{-kx}\sin px$ for positive values of x.

43. The tangent and normal to a curve at a point O are taken as axes. P is a point on the curve at an arcual distance s from O. Prove that the coordinates of P are approximately

$$x=s-\tfrac{1}{6}\kappa^2s^3, \quad y=\tfrac{1}{2}\kappa s^2+\tfrac{1}{6}\kappa' s^3,$$

where κ is the curvature at O, and κ' the value of $d\kappa/ds$ at O.

The tangents at O and P meet at T. A circle is drawn through O to touch PT at T. Prove that the limiting value of its radius as P tends to O is $1/(4\kappa)$.

44. Obtain the coordinates of the centre of curvature at a point (x, y) of a curve in the form $(x-\rho\sin\psi, y+\rho\cos\psi)$, where ρ is the radius of curvature and $\tan\psi$ the slope of the tangent at the point. Prove also that, when s is the length of arc between (x, y) and a fixed point on the curve, $dx/ds=\cos\psi$, $dy/ds=\sin\psi$.

The centres of curvature of a certain curve lie on a fixed circle. Prove that $\rho\frac{d\rho}{ds}$ is constant.

45. Obtain the formula $\rho = r\,dr/dp$ for the radius of curvature of a curve in terms of the radius vector and the perpendicular from the origin on the tangent.

Prove that a curve for which $\rho = p$ satisfies the equation $r^2 = p^2 + a^2$, where a is constant, and deduce that the polar equation of the curve is

$$\sqrt{(r^2 - a^2)} = a\theta + a\cos^{-1}(a/r),$$

where the coordinate system is chosen so that the point $(a, 0)$ is on the curve.

46. (i) Prove that the (p, r) equation of the cardioid

$$r = a(1 + \cos\theta)$$

is

$$2ap^2 = r^3.$$

Hence, or otherwise, prove that the radius of curvature is

$$\tfrac{4}{3}a\cos\tfrac{1}{2}\theta.$$

(ii) Prove that the equation of the circle of curvature at the origin of the curve

$$x + y = x^2 + 2y^2 + 3x^3$$

is

$$3(x^2 + y^2) = 2(x + y).$$

47. If y is a function of x, and

$$x = \xi\cos\alpha - \eta\sin\alpha, \quad y = \xi\sin\alpha + \eta\cos\alpha,$$

where α is constant, express $\dfrac{dy}{dx}$ and $\dfrac{d^2y}{dx^2}$ in terms of $\dfrac{d\eta}{d\xi}$ and $\dfrac{d^2\eta}{d\xi^2}$.

Deduce that

$$\frac{\dfrac{d^2y}{dx^2}}{\left\{1 + \left(\dfrac{dy}{dx}\right)^2\right\}^{\frac{3}{2}}} = \frac{\dfrac{d^2\eta}{d\xi^2}}{\left\{1 + \left(\dfrac{d\eta}{d\xi}\right)^2\right\}^{\frac{3}{2}}},$$

and interpret this result.

48. A curve is such that its arc length s, measured from a certain point, and ordinate y are related by

$$y^2 = s^2 + c^2,$$

where c is a constant. Show that referred to suitably chosen rectangular axes the curve has the equation

$$y = c\cosh(x/c).$$

If C is the centre of curvature at a point P of the curve and G is the point in CP produced such that $CP = PG$, prove that the locus of G is a straight line.

49. Find the radius of curvature at the origin of the curve

$$y = 2x + 3x^2 - 2xy + y^2 + 2y^3,$$

and show that the circle of curvature at the origin has equation

$$3(x^2 + y^2) = 5(y - 2x).$$

50. Find the values of x for which $y = x^2(x-2)^3$ has maximum and minimum values, and evaluate for these values of x the curvature of the curve given by the equation above.

51. Sketch the curve defined by the parametric equations

$$x = a\cos^3 t, \quad y = a\sin^3 t.$$

Show that the intercept made on a variable tangent by the coordinate axes is of constant length.

Find the radius of curvature at the point 't'.

52. Find the points on the curve $y(x-1) = x$ at which the radius of curvature is least.

53. Sketch the curves

$$y = x^2 - x^3, \quad y^2 = x^2 - x^3,$$

and find their radii of curvature at the origin.

54. Prove that all the curves represented by the equation

$$\frac{x^{n+1}}{a} + \frac{y^{n+1}}{b} = \left(\frac{ab}{a+b}\right)^n,$$

for different positive values of n, touch each other at the point

$$\left(\frac{ab}{a+b}, \frac{ab}{a+b}\right).$$

Prove that the radius of curvature at the point of contact is equal to

$$\frac{(a^2+b^2)^{\frac{3}{2}}}{n(a+b)^2}.$$

55. Establish the (p, r) formula for the radius of curvature of a plane curve. For a certain curve it is known that the radius of curvature is a^n/r^{n-1}, where $n \neq -1$ and a is positive, and also that $p = a/(n+1)$ when $r = a$. Show that it is possible to express the curve by the polar equation $r^n = (n+1)\,a^n \cos n\theta$.

56. Show that the curvature of the curve $a/r = \cosh n\theta$ has a stationary value provided that $3n^2$ is not less than 1. Determine whether this value is a maximum or a minimum.

57. O is the middle point of a straight line AB of length $2a$, and a point P moves so that $AP.BP = c^2$. Show that the radius of curvature at P of the locus is $2c^2r^3/(3r^4+a^4-c^4)$, where $r = OP$.

58. Find the integrals:

$$\int \frac{dx}{x^4+4}, \quad \int e^{ax} \cos bx\, dx \quad (a \neq 0, b \neq 0), \quad \int \frac{dx}{x+(x^2-1)^{\frac{1}{2}}}.$$

59. Prove that the area enclosed by the curve traced by the foot of the perpendicular from the centre to a variable tangent of the ellipse $b^2x^2+a^2y^2 = a^2b^2$ is $\frac{1}{2}\pi(a^2+b^2)$.

60. (*a*) Prove that, if

$$F(x) = e^{2x}\int_0^x e^{-2t} f(t)\,dt - e^x \int_0^x e^{-t} f(t)\,dt,$$

then (i) $F(0) = 0$,

(ii) $F'(0) = 0$,

(iii) $F''(x) - 3F'(x) + 2F(x) = f(x)$.

(*b*) Find the most general function $\phi(x)$ which is such that

$$\int_0^x t\phi(t)\,dt = x^2\phi(x).$$

61. Sketch the curve whose polar equation is $r^2 = a^2(1+3\cos\theta)$, and find the area it encloses.

62. (i) By differentiating the expression $\dfrac{nx\cos x+\sin x}{\sin^{n+1} x}$, or otherwise, find

$$\int \frac{x\,dx}{\sin^6 x}.$$

(ii) Find

$$\int \frac{dx}{x^6(a+bx)}.$$

63. Sketch the general shape of the curve

$$x = at\cos t, \quad y = at\sin t$$

for positive values of the parameter t.

Find an expression for the length of the arc of the curve measured from the origin to the point t.

64. (i) Prove that, if n is a positive integer,

$$\int_0^{\frac{1}{2}\pi} e^x \cos nx\,dx = \frac{1}{n^2+1}\{\lambda e^{\frac{1}{2}\pi} - 1\},$$

where λ has one of the values $\pm 1, \pm n$; and classify the cases.

(ii) Find the area bounded by the parabola $y^2 = ax$ and the circle $x^2+y^2 = 2a^2$.

65. Find the indefinite integral

$$\int \log(1+x^2)\tan^{-1} x\,dx.$$

66. Show that the four figures bounded by the circle $r = 3a\cos\theta$ and the cardioid $r = a(1+\cos\theta)$ have areas

$$\tfrac{1}{8}\pi a^2, \quad \tfrac{1}{8}\pi a^2, \quad \pi a^2, \quad \tfrac{5}{4}\pi a^2.$$

67. The polar equation of a curve is

$$r^2 - 2r\cos\theta + \sin^2\theta = 0.$$

Sketch the curve and prove that the area enclosed by it is $\pi/\sqrt{2}$.

68. Determine the function $\phi(x)$ such that

$$1 + \int_0^x \phi(t)\,e^t\,dt \equiv (1+x)^2 e^x.$$

Prove also that it is possible to find a pair of quadratic functions $f(x)$, $g(x)$ such that

$$(1+x)f(x) \equiv 1 + \int_0^x g(t)\,dt,$$

$$(1+x)g(x) \equiv 3 + 9\int_0^x f(t)\,dt,$$

and determine these functions.

69. A plane curvilinear figure is bounded by the parabola $x^2 = \frac{169}{5}y$ from the origin to the point (13, 5); by the hyperbola $x^2 - y^2 = 144$ from the point (13, 5) to the point (15, 9); and by the parabola $y^2 = \frac{27}{5}x$ from the point (15, 9) to the origin. Prove that the area of the figure is $33\frac{1}{3} + 72\log\frac{4}{3}$.

70. Prove that

$$\frac{1}{2}\int_0^1 x^4(1-x)^4\,dx < \int_0^1 \frac{x^4(1-x)^4}{1+x^2}\,dx < \int_0^1 x^4(1-x)^4\,dx.$$

Verify the identity

$$x^4(1-x)^4 = (1+x^2)(4-4x^2+5x^4-4x^5+x^6)-4,$$

and, by using this identity in the second of these integrals, prove that

$$\frac{22}{7} - \frac{1}{630} < \pi < \frac{22}{7} - \frac{1}{1260}.$$

71. Determine constants A, B, C, D such that

$$\frac{x^4+1}{(x^2+1)^4} = \frac{d}{dx}\left(\frac{Ax^5+Bx^3+Cx}{(x^2+1)^3}\right) + \frac{D}{x^2+1}.$$

Hence or otherwise prove that

$$0{\cdot}502 < \int_0^1 \frac{x^4+1}{(x^2+1)^4}\,dx < 0{\cdot}503.$$

72. Determine A, B, C, D such that

$$\frac{x^2}{(x^2+1)^4} = \frac{d}{dx}\left(\frac{Ax^5+Bx^3+Cx}{(x^2+1)^3}\right) + \frac{D}{x^2+1},$$

and show that
$$\int_0^1 \frac{x^2}{(x^2+1)^4}\,dx = \frac{1}{48} + \frac{\pi}{64}.$$

73. If $y_r(x)$ satisfies the equation

$$\frac{d}{dx}\left\{(1-x^2)\frac{dy}{dx}\right\} + r(r+1)y = 0,$$

show that
$$\int_{-1}^1 y_m(x)\,y_n(x)\,dx = 0 \quad (m \neq n).$$

74. Polynomials $f_0(x), f_1(x), f_2(x), \ldots$ are defined by the relation

$$f_n(x) = \frac{d^n}{dx^n}(x^2-1)^n.$$

Prove that
$$\int_{-1}^{1} f_m(x) f_n(x)\,dx = 0 \quad (m \neq n),$$

$$\int_{-1}^{1} \{f_n(x)\}^2\,dx = \frac{(n!)^2}{2n+1}\,2^{2n+1}.$$

Show that, if $\phi(x)$ is any polynomial of degree m,

$$\phi(x) = \sum_{0}^{m} a_n f_n(x),$$

where
$$a_n = \frac{2n+1}{(n!)^2\,2^{2n+1}} \int_{-1}^{1} \phi(x)\, f_n(x)\,dx.$$

75. If m, n are positive integers greater than unity, prove that

$$I_{m,n} \equiv \int_0^{\frac{1}{2}\pi} \cos^m x \cos nx\,dx$$

$$= \frac{m}{m-n} I_{m-1,\,n+1} = \frac{m}{m+n} I_{m-1,\,n-1}.$$

Hence show, if p, q are positive integers, that

$$\int_0^{\frac{1}{2}\pi} \cos^{p+q} x \cos(p-q)x\,dx = \frac{\pi(p+q)!}{2^{p+q+1}\,p!\,q!}.$$

76. Find a reduction formula for

$$\int (1-x^2)^n \cosh ax\,dx.$$

Evaluate
$$\int_0^1 (1-x^2)^3 \cosh x\,dx.$$

77. If $y = \sin^p x \cos^q x \sqrt{(1-k^2\sin^2 x)}$ and p, q, k are constants, find $\sqrt{(1-k^2\sin^2 x)}\dfrac{dy}{dx}$.

Hence, by taking suitable values for p, q, express

$$I_m = \int_0^{\frac{1}{2}\pi} \frac{\sin^m x\,dx}{\sqrt{(1-k^2\sin^2 x)}}$$

in terms of I_{m-2}, I_{m-4}.

78. State and prove the formula for integration by parts, and show that

$$\int_0^1 x^n(1-x)^m\,dx = \frac{m!\,n!}{(m+n+1)!}.$$

79. Find the volume and area of the surface of the solid obtained by rotating the portion of the cycloid

$$x = a(\theta+\sin\theta), \quad y = a(1+\cos\theta)$$

between two consecutive cusps about the axis of x.

[Consider the range $-\pi \leqslant \theta \leqslant \pi$.]

80. Prove that the envelope of the system of lines

$$(t^3-3t)x+(t^3+2)y = t^3$$

is the curve $y^2(x+y-1) = x^3$.

81. Find the envelope of the line

$$x\sin 3t - y\cos 3t = 3a\sin t,$$

expressing the coordinates of a point of the envelope in terms of a and t.

82. Through a variable point $P\,(at^2, 2at)$ of the parabola $y^2 = 4ax$ a line is drawn perpendicular to SP, when S is the focus $(a, 0)$. Prove that the envelope of the line is the curve $27ay^2 = x(x-9a)^2$.

83. Sketch the locus (the cycloid) given by

$$x = a(t+\sin t), \quad y = a(1+\cos t)$$

for values of the parameter t between $0, 4\pi$.

Prove that the normals to this curve all touch an equal cycloid, and draw the second curve in your diagram.

84. Find the envelope, as t varies, of the straight line whose equation is

$$x\cos^3 t + y\sin^3 t = a.$$

Sketch this envelope, and find the radius of curvature at one of the points of the curve nearest to the origin.

85. As t varies, the line

$$x - t^2y + 2at^3 = 0$$

envelops a curve Γ. Show that for each value of t, other than $t = 0$, the line cuts Γ at a point P distinct from the point at which the line touches Γ.

Find the equation of the normal to Γ at P, and deduce that the centre of curvature at P is given by

$$x = -20at^3 - (3a/4t), \quad y = 48at^5 - 3at.$$

Discuss the case $t = 0$.

86. Find the equation of the normal and the centre and radius of curvature of the curve $ay^2 = x^3$ at the point (at^2, at^3).

Show that the length of the arc of the evolute between the points corresponding to $t = 0, t = 1$ is $(13^{3/2}/6)a$.

87. Show how to find the envelope of the line

$$y = tx + f(t).$$

Show that the length of an arc of the envelope is given by

$$\int_{t_1}^{t} f''(t)\sqrt{(1+t^2)}\,dt.$$

Hence obtain a formula for the radius of curvature in terms of t.

88. Prove that the equation of the normal to the curve

$$x^{2/3} + y^{2/3} = a^{2/3}$$

may be written in the form

$$x \sin t - y \cos t + a \cos 2t = 0,$$

and find parametrically the envelope of the normal.

89. Find the equations of the tangent and normal at any point of the curve

$$x = 3\sin t - 2\sin^3 t, \quad y = 3\cos t - 2\cos^3 t,$$

Prove that the evolute is

$$x^{2/3} + y^{2/3} = 2^{2/3}.$$

90. The coordinates of any point on the curve $x^3 + xy^2 = y^2$ can be expressed in the form

$$x = t^2/(1+t^2), \quad y = t^3/(1+t^2).$$

Find the equations of the tangent and normal at any point, and deduce that the coordinates of the corresponding centre of curvature are

$$(-t^2 - \tfrac{1}{6}t^4, \tfrac{4}{3}t).$$

91. Prove that the complete area of the curve traced out by the point

$$(2a\cos t + a\cos 2t, \quad 2a\sin t - a\sin 2t)$$

is $2\pi a^2$.

92. Find the area inside the curve given by

$$x = a\cos t + c\sin kt, \quad y = b\sin t + d\cos kt \quad (0 \leqslant t \leqslant 2\pi),$$

where k is an integer and a, b, c, d are positive.

[It may be assumed that c, d are so small in comparison with a, b that the curve has no double points.]

93. Sketch the curve

$$x = a\sin 2t, \quad y = b\cos^3 t,$$

where $a > 0, b > 0$, and find the area it encloses.

94. Show that the curve

$$x = (t-1)e^{-t}, \quad y = tx$$

has a loop, and find its area.

95. Trace the curve $x = \cos 2t, y = \sin 3t$ for real values of t, and find the area of the loop.

96. Show that, for the range $-a \leqslant x \leqslant a$, the area between the curve

$$\left(\frac{x}{a}\right)^{\frac{2}{3}} + \frac{y}{b} = 1$$

and the x-axis is $\frac{4}{5}ab$.

97. Sketch the curve given by

$$x = 2a(\sin^3 t + \cos^3 t), \quad y = 2b(\sin^3 t - \cos^3 t),$$

and prove that its area is $3\pi ab$.

98. Find the area of the loop $(-1 \leqslant t \leqslant 1)$ of the curve

$$x = \frac{1-t^2}{1+t^2}, \quad y = \frac{t(1-t^2)}{1+t^2}.$$

99. Prove that the area of the curve

$$x = a\cos t + b\sin t + c, \quad y = a'\cos t + b'\sin t + c'$$

is $\pi(ab' - a'b)$.

100. Explain the reasons for using the formula

$$\int 2\pi y \, ds$$

to calculate the area of a surface of revolution.

Apply the formula to show that the area of the surface obtained by revolving the curve

$$r = a(1+\cos\theta)$$

about the line $\theta = 0$ is equal to $\frac{32}{5}\pi a^2$.

101. The curve traced out by the point

$$x = a\log(\sec t + \tan t) - a\sin t,$$

$$y = a\cos t,$$

as t increases from $-\frac{1}{2}\pi$ to $+\frac{1}{2}\pi$, is rotated about the axis of x. Prove that the whole surface generated is equal to the surface of a sphere of radius a, and that the whole volume generated is half the volume of a sphere of radius a.

102. Find the length of the arc of the catenary

$$y = c\cosh(x/c)$$

between the points given by $x = \pm a$.

Find also the area of the curved surface generated by rotating this arc about the x-axis.

103. Evaluate the area of the surface generated by the revolution of the cycloid

$$x = a(t - \sin t), \quad y = a(1-\cos t)$$

about the line $y = 0$.

104. Two points $A(0, c), P(\xi, \eta)$ lie on the curve whose equation is

$$y = c\cosh(x/c),$$

and s is the length of the arc AP. If the curve makes a complete revolution about the x-axis, prove that the area S of the curved surface, bounded by planes through A and P perpendicular to the x-axis, and the corresponding volume V are given by

$$cS = 2V = \pi c(c\xi + s\eta).$$

105. A torus is the figure formed by rotating a circle of radius a about a line in its own plane at a distance $h(>a)$ from its centre. Find the volume and surface area of the torus.

106. AB, CD are two perpendicular diameters of a circle. Find the mean value of the distance of A from points on the semicircle BCD, and also the mean value of the reciprocal of that distance.

Prove that the product of these means is

$$8\pi^{-2}\sqrt{2}\log(1+\sqrt{2}).$$

107. Prove that the mean distance of points on a sphere of radius a from a point distant $f(\leqslant a)$ from the centre is

$$a+(f^2/3a).$$

What is the mean distance if $f>a$?

108. Calculate the average value, over the surface of a sphere of centre O and radius a, of the function $(1/r^3)$, where r is the distance of the point on the surface of the sphere from a fixed point C not on the surface and such that $OC=f$. Distinguish between the two cases $f>a$, $f<a$.

109. The centre of a disc of radius r is O, and P is a point on the line through O perpendicular to the plane of the disc. Prove that, if $OP=p$, the mean distance (with respect to area) of points of the disc from P is

$$\tfrac{2}{3}\{(p^2+r^2)^{\frac{3}{2}}-p^3\}/r^2.$$

Find the mean distance (with respect to volume) of the interior points of a sphere of radius a from a fixed point of its surface.

110. Prove that the mean value with respect to area over the surface of a sphere, of centre O and radius a, of the reciprocal of the distance from a fixed point C is equal to the reciprocal of OC if C is outside the sphere, but equal to the reciprocal of the radius a if C is inside the sphere.

111. A point P is taken on an ellipse whose foci are S, H. The distance SP is denoted by r, and the angle HSP by θ. Show that the mean value of r with respect to arc is the semi-major axis a, and that the mean value of r with respect to θ is the semi-minor axis b.

CHAPTER XI

COMPLEX NUMBERS

1. Introduction. It may be helpful if we begin this chapter by reminding the reader of the types of elements which he is already accustomed to use in arithmetic. These are

(i) the *integer* (positive, negative or zero);

(ii) the *rational number*, that is, the ratio of two integers;

(iii) the *irrational number* (such as $\sqrt{2}, \pi, e$) which cannot be expressed as the ratio of integers.

It is now necessary to extend our scope and to devise a system of numbers not included in any of these three classes.

In order to exhibit the need for the new numbers, we solve in succession a series of quadratic equations, similar to look at, but essentially distinct.

I. $$x^2-6x+5=0.$$

Following the usual 'completing the square' process, we have the solution

$$\begin{aligned} x^2-6x &= -5, \\ x^2-6x+9 &= -5+9, \\ &= 4, \\ (x-3)^2 &= 4. \end{aligned}$$

The important point at this stage is the existence of two integers, namely $+2$ and -2, whose square is equal to 4. Hence we can find *integral* values of x to satisfy the equation:

$$x-3=+2 \quad \textit{or} \quad x-3=-2,$$

so that $$x=5 \quad \textit{or} \quad x=1.$$

The equation $x^2-6x+5=0$ can therefore be solved by taking x as one or other of the *rational* numbers (integers, in fact) 5 or 1.

II. $$x^2-6x+7=0.$$

Proceeding as before, we have the solution

$$\begin{aligned}x^2-6x &= -7,\\ x^2-6x+9 &= -7+9,\\ &= 2,\\ (x-3)^2 &= 2.\end{aligned}$$

Now comes a break; for there is no rational number whose square is 2, and so we cannot find a rational number x to satisfy the given equation. We must therefore extend our idea of number beyond the elementary realm of integers and rational numbers. This is an advance which the reader absorbed, doubtless unconsciously, many years ago, but it represents a step of fundamental importance. The theory of the irrational numbers will be found in text-books of analysis; for our purpose, knowledge of its existence is sufficient. In particular, we regard as familiar the concept of the irrational number, written as $\sqrt{2}$, whose value is $1{\cdot}414\ldots$, with the property that $(\sqrt{2})^2=2$.

Once the irrational numbers are admitted, the solution of the equation follows:

$$x-3=+\sqrt{2} \quad \textit{or} \quad x-3=-\sqrt{2},$$

so that $$x=3+\sqrt{2} \quad \textit{or} \quad x=3-\sqrt{2}.$$

The equation $x^2-6x+7=0$ can therefore be solved by taking x as one or other of the *irrational* numbers $3+\sqrt{2}$ or $3-\sqrt{2}$.

III. $$x^2-6x+10=0.$$

As before, $$\begin{aligned}x^2-6x &= -10,\\ x^2-6x+9 &= -10+9,\\ &= -1,\\ (x-3)^2 &= -1.\end{aligned}$$

Again comes the break; for there is no number, rational or irrational, whose square is the NEGATIVE number -1. We have therefore two choices, to accept defeat and say that the equation has no solution, or to invent a new set of numbers from which a value of x can be selected. The second alternative is the purpose of this chapter.

To be strictly logical, we should now proceed to define these new numbers and then show how to frame the solution from among them. But their definition is, naturally, somewhat complicated, and it seems better to begin by merely postulating their 'existence'; we can then see what properties they will have to possess, and the reasons for the definition will become more apparent.

Just as, in earlier days, we learned to write the symbol '$\sqrt{2}$' for a number whose square is 2, so now we use the symbol '$\sqrt{(-1)}$' for a 'number' (in some sense of the word) whose square is -1; but, in practice, it is more convenient to have a single mark instead of the five marks $\sqrt{}$, (, $-$, 1,), and so we write

$$i \equiv \sqrt{(-1)}$$

as a convenient abbreviation.

We shall subject this symbol i to all the usual laws of algebra, but first endow it further with the property that

$$i^2 = -1.$$

This will be its sole distinguishing feature—though, of course, that feature is itself so overwhelming as to introduce us into an entirely new number-world.

It may be noticed at once that the 'number' $-i$ also has -1 for its square, since, in accordance with the normal rules,

$$\begin{aligned}(-i)^2 &= (-1 \times i)^2 = (-1)^2 \times (i)^2 = (+1) \times (-1) \\ &= -1.\end{aligned}$$

We round off this introduction by displaying the solution of the above equation:

$$x - 3 = i \quad \textit{or} \quad x - 3 = -i,$$

so that

$$x = 3 + i \quad \textit{or} \quad x = 3 - i.$$

By selecting x from this extended range of 'numbers', we are therefore able to find two solutions of the equation.

It may, perhaps, make this statement clearer if we verify how $3 + i$, say, does exactly suffice. By saying that $3 + i$ is a solution of the equation $x^2 - 6x + 10 = 0$, we mean that

$$(3 + i)^2 - 6(3 + i) + 10 = 0.$$

The left-hand side is

$$\begin{aligned} 9+6i+i^2-18-6i+10 & \\ &= 1+i^2 \\ &= 1+(-1) \\ &= 0 \\ &= \text{right-hand side,} \end{aligned}$$

so that the solution is verified.

We now investigate the elementary properties of our extended number-system, and then, when the ideas are a little more familiar, return to put the basis on a surer foundation.

EXAMPLES I

Solve after the manner of the text the following sets of equations:

1. $x^2-4x+3 = 0, \quad x^2-4x+1 = 0, \quad x^2-4x+5 = 0.$
2. $x^2+8x+15 = 0, \quad x^2+8x+11 = 0, \quad x^2+8x+20 = 0.$
3. $x^2-2x-3 = 0, \quad x^2-2x-4 = 0, \quad x^2-2x+10 = 0.$

2. Definitions. The numbers of ordinary arithmetic (positive or negative integers, rational numbers and irrational numbers) are called *real* numbers.

If a, b are two real numbers, then numbers of the form

$$a+ib \quad (i^2 = -1)$$

are called *complex* numbers. When $b = 0$, such a number is real. When $a = 0$, the number

$$ib \quad (b \text{ real})$$

is called a *pure imaginary* number.

Two numbers such as

$$a+ib, \quad a-ib \quad (a, b \text{ real})$$

are called *conjugate* complex numbers.

A single symbol, such as c, is often used for a complex number.

If

$$c \equiv a + ib,$$

then a is called the *real part* of c and b the *imaginary part*. The notations

$$a = \mathscr{R}c, \quad b = \mathscr{I}c$$

are often used.

The conjugate of a complex number c is often denoted by $\bar{c}$, so that, if $c = a + ib$, then $\bar{c} = a - ib$. Clearly the conjugate of $\bar{c}$ is c itself.

If c is real, then $c = \bar{c}$.

The magnitude

$$+\sqrt{(a^2 + b^2)} \quad (a, b \text{ real})$$

is called the *modulus* of the complex number $c \equiv a + ib$; it is often denoted by the notations

$$|c|, \quad |a + ib|.$$

It follows that

$$|\bar{c}| = |c|.$$

Moreover, we have the relation

$$c\bar{c} = |c|^2,$$

for

$$\begin{aligned} c\bar{c} &= (a + ib)(a - ib) \\ &= a^2 - i^2 b^2 = a^2 - (-1) b^2 \\ &= a^2 + b^2. \end{aligned}$$

Finally, since

$$c = a + ib, \quad \bar{c} = a - ib,$$

we have the relations

$$\mathscr{R}c = a = \tfrac{1}{2}(c + \bar{c}),$$

$$\mathscr{I}c = b = \frac{1}{2i}(c - \bar{c}).$$

Note. All the numbers now to be considered are actually complex, of the form $a + ib$, even when $b = 0$. Correctly speaking we should use the phrase 'real complex number' for a number such as 2, and 'pure imaginary complex number' for $2i$. But this becomes tedious once it has been firmly grasped that *all* numbers are complex anyway.

EXAMPLES II

The following examples are designed to make the reader familiar with the use of the symbol i. Use normal algebra, except that $i^2 = -1$.

Express the following complex numbers in the form $a+ib$, where a, b are both real:

1. $(3+5i)+(7-2i)$.
2. $(-2+3i)-(6-5i)$.
3. $(4+5i)(6-2i)$.
4. $(2+7i)(-1+2i)$.
5. $3+2i+i(5+i)$.
6. $4-3i+i(3+4i)$.
7. $(1+i)^2$.
8. $(2+i)^3$.
9. $(2-i)^2-(3+2i)^2$.
10. $(1+i)(1+2i)(1+3i)$.

Prove the following identities:

11. $\dfrac{1}{i} = -i$.
12. $\dfrac{1}{1+i} = \tfrac{1}{2}(1-i)$.
13. $\dfrac{1}{3-4i} = \dfrac{1}{25}(3+4i)$.
14. $\dfrac{1}{a+ib} = \dfrac{a-ib}{a^2+b^2}$.
15. $\dfrac{p+iq}{a+ib} = \dfrac{(ap+bq)+i(aq-bp)}{a^2+b^2}$.

3. Addition, subtraction, multiplication.

Let

$$c \equiv a+ib, \quad w \equiv u+iv \quad (a, b, u, v \text{ real})$$

be two given complex numbers. In virtue of the meaning which we have given to i, we obtain expressions for their sum, difference and product as follows:

(i) SUM $\qquad c+w = (a+u)+i(b+v)$.

(ii) DIFFERENCE $\qquad c-w = (a-u)+i(b-v)$.

(iii) PRODUCT

$$\begin{aligned} cw &= (a+ib)(u+iv) \\ &= au+ibu+iav+i^2bv \\ &= (au-bv)+i(bu+av). \end{aligned}$$

The awkward one of these is the product. At present it is best not to remember the formula, but to be able to derive it when required.

EXAMPLES III

Express the following products in the form $a+ib$ (a, b real):

1. $(2+3i)(4+5i)$.
2. $(1+4i)(2-6i)$.
3. $(-1-i)(-3+4i)$.
4. $(a+ib)(a-ib)$.
5. $i(4+3i)$.
6. $(\cos A+i\sin A)(\cos B+i\sin B)$.
7. $(3-i)(3+i)$.
8. $(2+3i)^3$.

4. Division. As before, let

$$c \equiv a+ib, \quad w \equiv u+iv \quad (a, b, u, v \text{ real}).$$

Then the QUOTIENT is, by definition, the fraction

$$\frac{c}{w} \equiv \frac{a+ib}{u+iv}.$$

It is customary to express such fractions in a form in which the denominator is real. For this, multiply numerator and denominator by $\bar{w} \equiv u-iv$. Then

$$\frac{c}{w} = \frac{(a+ib)(u-iv)}{(u+iv)(u-iv)}$$

$$= \frac{(au+bv)+i(bu-av)}{u^2+v^2}$$

$$= \frac{au+bv}{u^2+v^2}+i\frac{bu-av}{u^2+v^2}.$$

It is implicit that u, v are not both zero.

For example, $(2+i) \div (3-i)$

$$= \frac{2+i}{3-i}$$

$$= \frac{(2+i)(3+i)}{(3-i)(3+i)}$$

$$= \frac{6+5i+i^2}{9-i^2} = \frac{6+5i-1}{9+1}$$

$$= \frac{5+5i}{10}$$

$$= \tfrac{1}{2}+\tfrac{1}{2}i.$$

EXAMPLES IV

Express the following quotients in the form $a+ib$ (a, b real):

1. $\dfrac{1+i}{1-i}$. 2. $\dfrac{3+4i}{4+3i}$. 3. $\dfrac{5i+6}{7i}$.

4. $\dfrac{2-3i}{4+i}$. 5. $\dfrac{3+5i}{1-6i}$. 6. $\dfrac{\cos 2\theta + i\sin 2\theta}{\cos\theta + i\sin\theta}$.

5. Equal complex numbers. To verify that, *if two complex numbers are equal, then their real parts are equal and their imaginary parts are equal.*

Let
$$c \equiv a+ib, \quad r \equiv p+iq$$
be two given complex numbers (a, b, p, q all real), such that $c = r$. Then
$$a+ib = p+iq,$$
or
$$a-p = i(q-b).$$
Squaring each side, we have
$$(a-p)^2 = i^2(q-b)^2$$
$$= -(q-b)^2.$$
But $a-p, q-b$ are real, so that their squares are positive or zero; hence this relation cannot hold unless each side is zero. That is,
$$a = p, \quad b = q,$$
as required.

COROLLARY. *If* a, b *are real and*
$$a+ib = 0,$$
then
$$a = 0, \quad b = 0.$$

EXAMPLES V

Find the sum, difference, product and quotient of each of the following pairs of complex numbers:

1. $2+3i, \quad 3-5i$. 2. $-4+2i, \quad -3+7i$.

3. $4, \quad 2i$. 4. $3+2i, \quad -i$.

5. $2+3i, \quad 2-3i$. 6. $-3+4i, \quad -3-4i$.

Find, in the form $a+ib$ (a, b real) the reciprocal of each of the following complex numbers:

7. $3+4i$.
8. $-5+12i$.
9. $6i$.
10. $-6-8i$.

Solve the following quadratic equations, expressing your answers in the form $a \pm ib$:

11. $x^2+4x+13=0$.
12. $x^2-2x+2=0$.
13. $x^2+6x+10=0$.
14. $4x^2+9=0$.
15. $x^2-8x+25=0$.
16. $x^2+4x+5=0$.

6. The complex number as a number-pair.

A complex number

$$c \equiv a+ib$$

consists essentially of the pair of real numbers a, b linked by the symbol i to which we have given a specific property (i^2 negative) not enjoyed by any real number. Hence c is of the nature of an *ordered pair* of real numbers, the ordering being an important feature since, for example, the two complex numbers $4+5i$ and $5+4i$ are quite distinct.

The concept of number-pair forms the foundation for the more logical development promised at the end of § 1, for it is precisely this concept which enables us to extend the range of numbers required for the solution of the third quadratic equation ($x^2-6x+10=0$). We therefore make a fresh start, with a series of definitions designed to exhibit the properties hitherto obtained by the use of the fictitious 'number' i.

We define a *complex number* to be an ordered number-pair (of *real* numbers), denoted by the symbol $[a, b]$, subject to the following rules of operation (chosen, of course, to fit in to the treatment given at the start of this chapter):

(i) The symbols $[a, b]$, $[u, v]$ are equal if, and only if, the two relations $a=u, b=v$ are satisfied;

(ii) The sum

$$[a, b]+[u, v]$$

is the number-pair $[a+u, b+v]$;

(iii) The product

$$[a, b] \times [u, v]$$

is the number-pair $[au - bv, bu + av]$.

Compare the formulæ in § 3.

A number $[a, 0]$, whose second component is zero, is called a *real complex number*; a number $[0, b]$, whose first component is zero, is called a *pure imaginary complex number*. These terms are usually abbreviated to *real* and *pure imaginary* numbers. (Compare p. 162.)

In order to achieve correlation with the normal notation for real numbers, and with the customary notation based on the letter i for pure imaginaries, we make the following conventions:

When we are working in a system of algebra requiring the use of complex numbers,

(i) the symbol a will be used as an abbreviation for the number-pair $[a, 0]$,

(ii) the symbol ib will be used as an abbreviation for the number-pair $[0, b]$.

It follows that the symbol $a + ib$ may be used for the number-pair $[a, 0] + [0, b]$, or $[a, b]$.

In particular, we write 1 for the unit $[1, 0]$ of real numbers, and $i \times 1$ for the unit $[0, 1]$ of pure imaginary numbers; but no confusion arises if we abbreviate $i \times 1$ to the symbol i itself.

We do not propose to go into much greater detail, but the reader may easily check that the complex numbers defined in this way by number-pairs have all the properties tentatively proposed for them in the earlier paragraphs of this chapter. We ought, however, to verify explicitly that the square of the number-pair i is the real number -1. For this, we appeal directly to the definition of multiplication:

$$[a, b] \times [u, v] = [au - bv, bu + av].$$

Write $a = u = 0$, $b = v = 1$. Then

$$[0, 1] \times [0, 1] = [-1, 0]$$

or, in terms of the abbreviated symbols,

$$i \times i = -1.$$

Finally, we remark that, although the ordinary language of real numbers is retained without apparent change, the symbols in complex algebra carry an entirely different meaning. For example, the statement

$$2 \times 2 = 4$$

remains true; but what we really mean is that

$$[2,0] \times [2,0] = [2 \times 2 - 0 \times 0, 0 \times 2 + 2 \times 0]$$
$$= [4,0].$$

From now on, however, we shall revert to the normal symbolism without square brackets, writing a complex number in the form $a+ib$ as before. It must always be remembered that number-pairs are intended.

7. The Argand diagram. Abstractly viewed, a complex number $a+ib$ is a 'number-pair' $[a,b]$ in which the first component of the pair corresponds to the real part and the second component to the imaginary. A number-pair is, however, also capable of a familiar *geometrical* interpretation, when the two numbers a, b in assigned order are taken to be the Cartesian coordinates of a point in a plane. We therefore seek next to link these two conceptions of number-pair so that we can incorporate the ideas of analytical geometry into the development of complex algebra.

With a change of notation, denote a complex number by the letter z, where

$$z \equiv x+iy.$$

We do this to strengthen the implication of the real and imaginary parts x and y as the Cartesian coordinates of a point $P(x,y)$ (Fig. 93).

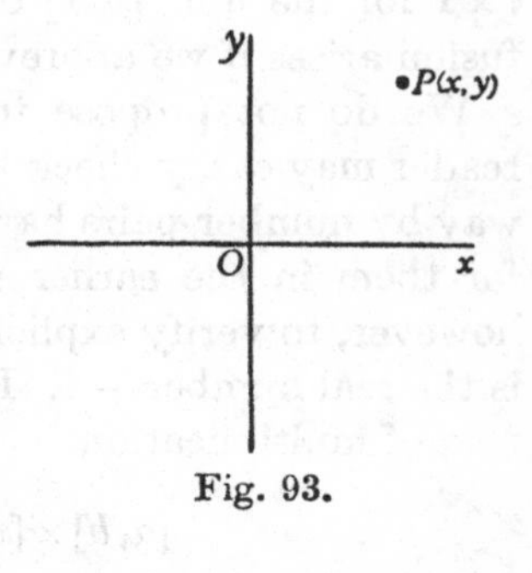

Fig. 93.

There is then an exact correspondence between the complex numbers $z \equiv x+iy$ and the points $P(x,y)$ of the plane:

(i) If z is given, then its real and imaginary parts are determined and so P is known;

(ii) If P is given, then its coordinates are determined, and so z is known.

We say that P *represents* the complex number z. The diagram in which the complex numbers are represented is called an *Argand diagram*, and the plane in which it is drawn is called the *complex plane*, or, when precision is required, the *z-plane.*

EXAMPLES VI

Mark on an Argand diagram the points which represent the following complex numbers:

1. $3+4i$. 2. $5-i$. 3. $6i$.

4. -3. 5. $1+i$. 6. $-2-3i$.

7. $\cos\theta+i\sin\theta$ for $\theta = 0°, 30°, 60°, \ldots, 330°, 360°$.

8. $2\cos^2\frac{1}{2}\theta+i\sin\theta$ for $\theta = 0, 45°, 90°, \ldots, 315°, 360°$.

8. Modulus and argument. Given the point $P(x, y)$ representing the complex number

$$z \equiv x+iy,$$

we may describe its position alternatively by means of polar coordinates r, θ (Fig. 94), where r is the distance OP, ESSENTIALLY POSITIVE and θ is the angle $\angle xOP$ which OP makes with the x-axis, measured in the counter-clockwise sense from Ox. Thus r is defined uniquely, but θ is ambiguous by multiples of 2π.

We know from elementary trigonometry that

$$x = r\cos\theta, \quad y = r\sin\theta,$$

whatever the quadrant in which P lies.

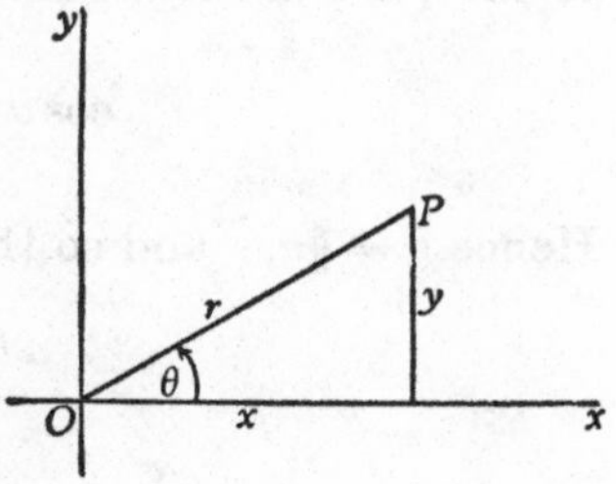

Squaring and adding these relations, we have

$$r^2 = x^2+y^2,$$

so that, since r is positive,

$$r = +\sqrt{(x^2+y^2)}.$$

Fig. 94.

Dividing the relations, we have

$$\tan\theta = y/x,$$

or

$$\theta = \tan^{-1}(y/x).$$

The choice of quadrant for θ depends on the signs of BOTH y AND x; if x, y are $+, +$, the quadrant is the first; if x, y are $-, +$, the second; if x, y are $-, -$, the third; if x, y are $+, -$, the fourth.

The two numbers
$$r = \sqrt{(x^2+y^2)},$$
$$\theta = \tan^{-1}(y/x),$$
with appropriate choice of the quadrant for θ, are called the *modulus* and *argument* respectively of the complex number z. If r, θ are given, then
$$z = r(\cos\theta + i\sin\theta).$$

We have already (p. 162) explained the notation $|z| = r$, and proved the formula $z\bar{z} = |z|^2$.

ILLUSTRATION 1. *To find the modulus and argument of the complex number* $-1+i\sqrt{3}$.

If the modulus is r, then
$$r^2 = (-1)^2 + (\sqrt{3})^2 = 1+3 = 4,$$
so that $r = 2$. The number is therefore
$$2\left\{-\frac{1}{2}+i\frac{\sqrt{3}}{2}\right\},$$
so that, if θ is the argument,
$$\cos\theta = -\tfrac{1}{2}, \quad \sin\theta = \frac{\sqrt{3}}{2}.$$
Hence $\theta = \tfrac{2}{3}\pi$, and so the number is
$$2\{\cos\tfrac{2}{3}\pi + i\sin\tfrac{2}{3}\pi\}.$$

EXAMPLES VII

1. Plot in an Argand diagram the points which represent the numbers $4+3i$, $-3+4i$, $-4-3i$, $3-4i$, and verify that the points are at the vertices of a square.

2. Plot in an Argand diagram the points which represent the numbers $7+3i$, $6i$, $-3-i$, $4-4i$, and verify that the points are at the vertices of a square.

3. Find the modulus and argument in degrees and minutes of each of the complex numbers

$$\sqrt{3}-i, \quad 3+4i, \quad -5+12i, \quad 3, \quad 6-8i, \quad -2i.$$

4. Prove that, if the point z in the complex plane lies on the circle whose centre is the (complex) point a and whose radius is the real number k, then

$$|z-a| = k.$$

5. Find the locus in an Argand diagram of the point representing the complex number z subject to the relation

$$|z+2| = |z-5i|.$$

9. The representation in an Argand diagram of the sum of two numbers.

Suppose that

$$z_1 \equiv x_1 + iy_1, \quad z_2 \equiv x_2 + iy_2$$

are two complex numbers represented in an Argand diagram by the points

$$P_1(x_1, y_1), \quad P_2(x_2, y_2)$$

respectively (Fig. 95).

Their sum is the number

$$\begin{aligned} z &= z_1 + z_2 \\ &= (x_1 + x_2) + i(y_1 + y_2) \end{aligned}$$

represented by the point $P(x, y)$, where

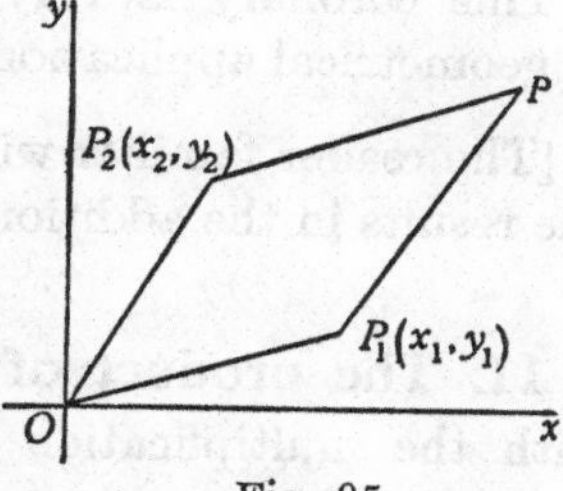

Fig. 95.

$$x = x_1 + x_2, \quad y = y_1 + y_2.$$

By elementary analytical geometry, the lines P_1P_2 and OP have the same middle point $\left(\frac{x_1+x_2}{2}, \frac{y_1+y_2}{2}\right)$, so that OP_1PP_2 is a parallelogram. Hence P (*representing* z_1+z_2) *is the fourth vertex of the parallelogram of which* OP_1, OP_2 *are adjacent sides.*

In particular, *OP is the modulus* $|z_1+z_2|$ *of the sum of the two numbers* z_1, z_2.

10. The representation in an Argand diagram of the difference of two numbers. To find the point representing the difference

$$z \equiv z_2 - z_1,$$

we proceed as follows:

The relation is equivalent to

$$z_2 = z + z_1,$$

so that OP_2 (Fig. 96) is the diagonal of the parallelogram of which OP_1, OP (where P represents z) are adjacent sides.

Hence $\overrightarrow{OP}$ *is the line through* O *parallel to* $\overrightarrow{P_1P_2}$, *the sense along the lines being indicated by the arrows, where* OP, P_1P_2 *are equal in magnitude.*

COROLLARY. *If* $P_1(x_1, y_1)$, $P_2(x_2, y_2)$ *represent the two complex numbers* $z_1 \equiv x_1 + iy_1, z_2 \equiv x_2 + iy_2$, *then the line* $\overrightarrow{P_1P_2}$ *represents the difference* $z_2 - z_1$, *in the sense that the length* P_1P_2 *is the modulus* $|z_2 - z_1|$ *and the angle which* $\overrightarrow{P_1P_2}$ *makes with the* x*-axis is the argument of* $z_2 - z_1$.

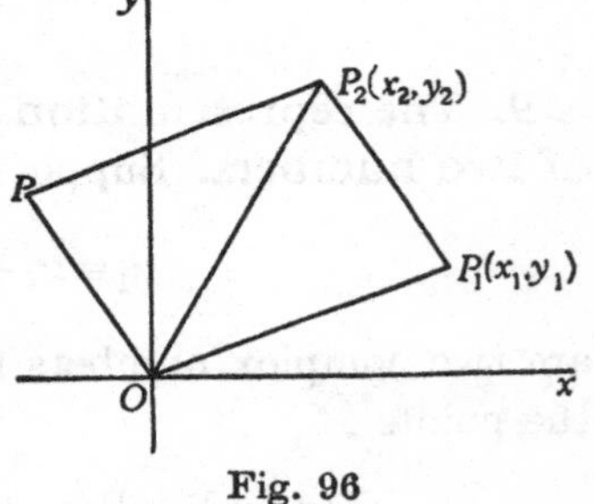

Fig. 96

This corollary is very important in geometrical applications.

[The reader familiar with the theory of vectors should compare the results in the addition and subtraction of two vectors.]

11. The product of two complex numbers. In dealing with the multiplication of complex numbers, it is often more convenient to express them in modulus-argument form. We therefore write

$$z_1 \equiv r_1(\cos\theta_1 + i\sin\theta_1), \quad z_2 \equiv r_2(\cos\theta_2 + i\sin\theta_2).$$

Then
$$\begin{aligned} z_1z_2 &= r_1r_2(\cos\theta_1 + i\sin\theta_1)(\cos\theta_2 + i\sin\theta_2) \\ &= r_1r_2\{(\cos\theta_1\cos\theta_2 - \sin\theta_1\sin\theta_2) \\ &\qquad + i(\sin\theta_1\cos\theta_2 + \cos\theta_1\sin\theta_2)\} \\ &= r_1r_2\{\cos(\theta_1+\theta_2) + i\sin(\theta_1+\theta_2)\}, \end{aligned}$$

which is a complex number of modulus r_1r_2 and argument $\theta_1 + \theta_2$.

Hence

(i) *the modulus of a product is the product of the moduli;*

(ii) *the argument of a product is the sum of the arguments.*

We may obtain similarly the results for division:

(iii) *the modulus of a quotient is the quotient of the moduli;*

(iv) *the argument of a quotient is the difference of the arguments.*

12. The product in an Argand diagram. In the diagram (Fig. 97), let P_1, P_2, P represent the two complex numbers z_1, z_2 and their product $z \equiv z_1 z_2$ respectively, and let U be the 'unit' point $z = 1$.

Then (§ 10)

$$\angle xOP = \theta_1 + \theta_2,$$

where θ_1, θ_2 are the arguments of z_1, z_2, and so

$$\begin{aligned} \angle P_2OP &= \angle xOP - \angle xOP_2 \\ &= (\theta_1 + \theta_2) - \theta_2 \\ &= \theta_1 \\ &= \angle UOP_1. \end{aligned}$$

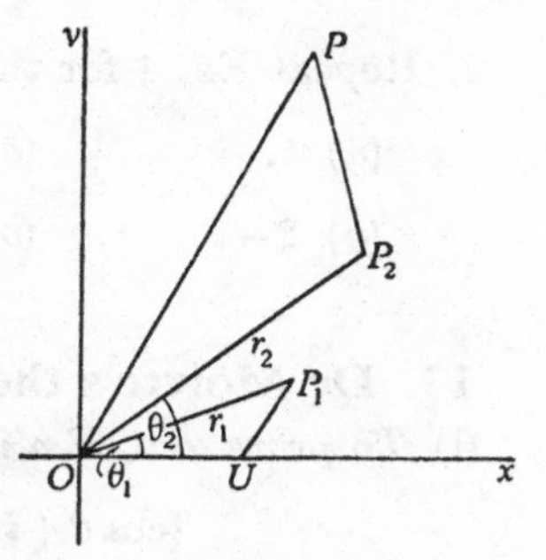

Fig. 97.

Moreover, if r_1, r_2 are the moduli of z_1, z_2,

$$\begin{aligned} OP/OP_2 &= (r_1 r_2)/r_2 = r_1/1 \\ &= OP_1/OU. \end{aligned}$$

Now in measuring the angle $\angle P_2OP$ or $\angle UOP_1$, we proceed, by convention, from the radius vector $\overrightarrow{OP_2}$ or $\overrightarrow{OU}$, through an angle θ_1, in the counter-clockwise sense, to the vector $\overrightarrow{OP}$ or $\overrightarrow{OP_1}$. It may be that θ_1 itself does not lie between $0, \pi$, but nevertheless the fact that the angles $\angle P_2OP, \angle UOP_1$ so described are equal ensures that the Euclidean angles $\angle P_2OP$, $\angle UOP_1$ of the triangles ΔP_2OP, ΔUOP_1 are also equal. Moreover, the sides about these equal angles have been proved proportional, and so the triangles are similar.

Hence follows the construction for P:

Let P_1, P_2 represent the two given numbers z_1, z_2; let O be the origin and U the unit point $z = 1$. Describe the triangle POP_2 similar to the triangle P_1OU, in such a sense that $\angle P_2OP = \angle UOP_1$. Then the point P represents the product $z_1 z_2$.

COROLLARY. *The effect of multiplying a complex number z by i is to rotate the vector OP representing z through an angle $\frac{1}{2}\pi$ in the counter-clockwise sense.* This follows at once if we put $z_1 = i$ in the preceding work, so that P_1 in the diagram becomes the point $(0, 1)$ or, in polar coordinates, $(1, \frac{1}{2}\pi)$.

Thus i acts in the Argand diagram like an *operator*, turning the radius vector through a right angle.

EXAMPLES VIII

1. Mark in an Argand diagram the points which represent (*a*) $3+2i$, (*b*) $2+i$, (*c*) their product. Verify from your diagram the theorem of the text.

2. Repeat Ex. 1 for the points

(*a*) 1, (*b*) $3-5i$, (*c*) their product;

(*a*) $2-i$, (*b*) $2+i$, (*c*) their product.

13. De Moivre's theorem.

(i) *To prove that, if n is a positive integer, then*

$$(\cos\theta + i\sin\theta)^n = \cos n\theta + i\sin n\theta.$$

We use a proof by induction. Suppose that the theorem is true for a particular number N, a positive integer, so that

$$(\cos\theta + i\sin\theta)^N = \cos N\theta + i\sin N\theta.$$

Then

$$\begin{aligned}(\cos\theta + i\sin\theta)^{N+1} &= (\cos N\theta + i\sin N\theta)(\cos\theta + i\sin\theta)\\ &= (\cos N\theta\cos\theta - \sin N\theta\sin\theta)\\ &\qquad + i(\cos N\theta\sin\theta + \sin N\theta\cos\theta)\\ &= \cos(N+1)\theta + i\sin(N+1)\theta.\end{aligned}$$

Hence if the theorem is true for any particular value N, it is true for $N+1, N+2, \ldots$ and so on. But it is clearly true for $N = 0$, and so the theorem is established.

(ii) *To prove that, if n is a negative integer, then*

$$(\cos\theta + i\sin\theta)^n = \cos n\theta + i\sin n\theta.$$

Write $$n = -m.$$

Then m is a positive integer, and so

$$
\begin{aligned}
(\cos\theta + i\sin\theta)^n &= (\cos\theta + i\sin\theta)^{-m} \\
&= \frac{1}{(\cos\theta + i\sin\theta)^m} \\
&= \frac{1}{\cos m\theta + i\sin m\theta} \quad \text{(as above)} \\
&= \frac{\cos m\theta - i\sin m\theta}{(\cos m\theta + i\sin m\theta)(\cos m\theta - i\sin m\theta)} \\
&= \frac{\cos m\theta - i\sin m\theta}{\cos^2 m\theta - i^2\sin^2 m\theta} \\
&= \frac{\cos m\theta - i\sin m\theta}{\cos^2 m\theta + \sin^2 m\theta} \\
&= \cos m\theta - i\sin m\theta \\
&= \cos(-n\theta) - i\sin(-n\theta) \\
&= \cos n\theta + i\sin n\theta,
\end{aligned}
$$

as required.

(iii) *To prove that, if n is a rational fraction, then*

$$(\cos\theta + i\sin\theta)^n = \cos n(\theta + 2k\pi) + i\sin n(\theta + 2k\pi),$$

where k is an integer, positive or negative.

Suppose that

$$n = p/q,$$

where p, q are integers (q positive) without a common factor.

Consider the expression

$$(\cos\theta + i\sin\theta)^{1/q},$$

and suppose that it CAN be expressed in the form

$$\cos\phi + i\sin\phi.$$

If so, then

$$
\begin{aligned}
\cos\theta + i\sin\theta &= (\cos\phi + i\sin\phi)^q \\
&= \cos q\phi + i\sin q\phi \quad \text{[by (i)]}.
\end{aligned}
$$

Equating real and imaginary parts, we have

$$\cos\theta = \cos q\phi, \quad \sin\theta = \sin q\phi.$$

These equalities hold simultaneously if, and only if,

$$q\phi = \theta + 2k\pi,$$

where k is any integer, positive or negative. Hence

$$\phi = \frac{1}{q}\theta + \frac{2k}{q}\pi,$$

so that

$$(\cos\theta + i\sin\theta)^{1/q} = \cos\left(\frac{1}{q}\theta + \frac{2k}{q}\pi\right) + i\sin\left(\frac{1}{q}\theta + \frac{2k}{q}\pi\right).$$

Raising each side to the power p, by (i) or (ii) above, and then writing $p/q = n$, we obtain the relation

$$(\cos\theta + i\sin\theta)^n = \cos n(\theta + 2k\pi) + i\sin n(\theta + 2k\pi).$$

The expression on the right-hand side takes various values according to the choice of k. That is to say, there are several values for $(\cos\theta + i\sin\theta)^n$ when n is a rational fraction; the case $n = \frac{1}{2}$ is familiar, leading to two distinct square roots. We see by elementary trigonometry that distinct values are obtained when k assumes in succession the values $0, 1, 2, \ldots, q-1$, where q is the denominator in the expression of n as a proper fraction in its lowest terms. Thereafter they repeat themselves, any block of q consecutive values of k giving the whole set.

SUMMARY. *The formula*

$$(\cos\theta + i\sin\theta)^n = \cos n(\theta + 2k\pi) + i\sin n(\theta + 2k\pi)$$

is true for all values of n, where n may be a positive or negative integer or rational fraction. When n is an integer, k may be taken as zero; when n is fractional, distinct values are obtained on the right-hand side for $k = 0, 1, 2, \ldots, q-1$, where q is the denominator in the expression of n as a proper fraction in its lowest terms.

EXAMPLES IX

1. Express $\frac{1}{2} + i\frac{\sqrt{3}}{2}$ in modulus-argument form $r(\cos\theta + i\sin\theta)$ and hence, with the help of your tables, find (i) its square, (ii) its square roots, (iii) its cube roots, (iv) its fourth roots.

2. Repeat Ex. 1 for the number $4 + 3i$.

3. Repeat Ex. 1 for the number $12 - 5i$.

14. The nth roots of unity. In virtue of the formula

$$1 = \cos 0 + i \sin 0,$$

it follows from De Moivre's theorem that, if n is any integer (assumed positive here) then an nth root of unity, that is

$$\sqrt[n]{1} \equiv 1^{1/n},$$

can be expressed in the form

$$\cos\left(\frac{1}{n}.2k\pi\right) + i \sin\left(\frac{1}{n}.2k\pi\right)$$

for $k = 0, 1, 2, \ldots, n-1$. Hence *for any given positive integer n, there are n distinct roots of unity, namely*

$$\cos\frac{2k\pi}{n} + i \sin\frac{2k\pi}{n},$$

where $k = 0, 1, 2, \ldots, n-1$.

For example:

When $n = 2$, there are two square roots, namely

$$\cos\frac{2k\pi}{2} + i \sin\frac{2k\pi}{2},$$

where $k = 0, 1$. When $k = 0$, this gives $\cos 0$, or 1; when $k = 1$, it gives $\cos\pi$, or -1.

When $n = 3$, there are three cube roots, namely

$$\cos\frac{2k\pi}{3} + i \sin\frac{2k\pi}{3},$$

where $k = 0, 1, 2$. When $k = 0$, this gives 1 itself. The values $k = 1, 2$ give

$$\cos\frac{2\pi}{3} + i \sin\frac{2\pi}{3} \equiv \tfrac{1}{2}(-1 + i\sqrt{3}),$$

$$\cos\frac{4\pi}{3} + i \sin\frac{4\pi}{3} \equiv \tfrac{1}{2}(-1 - i\sqrt{3}).$$

These complex cube roots of unity are usually denoted by the symbols ω, ω^2, either being the square of the other. They are also connected by the relation

$$1 + \omega + \omega^2 = 0.$$

EXAMPLES X

Use De Moivre's theorem to express the following powers in 'modulus-argument' form $r(\cos\theta + i\sin\theta)$:

1. $(1+i)^4$.
2. $\dfrac{1}{(1-i)^5}$.
3. $(1+i\sqrt{3})^6$.
4. $(\sqrt{3}-i)^{\frac{1}{3}}$.
5. $(1-i)^{\frac{1}{5}}$.
6. $(\sqrt{3}+i)^{\frac{2}{3}}$.
7. Find expressions for

$$\text{(i)}\quad 1^{\frac{1}{5}}, \quad \text{(ii)}\quad 1^{\frac{1}{6}}$$

analogous to the roots evaluated in the text.

15. Complex powers. The reader will remember that, when seeking an interpretation for symbols such as

$$a^0, \quad a^{-n}, \quad a^{1/n}, \quad a^{-1/n}$$

for real positive a and positive integral n, he was guided by the principle that the formula of multiplication

$$a^p \times a^q = a^{p+q}$$

should hold for all values of p and q. This led to the interpretations

$$a^0 = 1, \quad a^{-n} = 1/a^n, \quad a^{1/n} = \sqrt[n]{a}, \quad a^{-1/n} = 1/\sqrt[n]{a}.$$

We now consider the meanings which should be given to a^n when a, n are complex numbers, still retaining the validity for the formula of multiplication.

(i) *When n is real and rational*, the treatment is comparatively simple, for we know that any complex number a can be expressed in the form

$$a \equiv r(\cos\theta + i\sin\theta),$$

and so, by De Moivre's theorem,

$$a^n = r^n\{\cos n(\theta + 2k\pi) + i\sin n(\theta + 2k\pi)\}$$

where the values $k = 0, 1, \ldots, q-1$ (q being the denominator in the expression of n as a proper fraction in its lowest terms) give distinct values for a^n. Thus a^n is, in general, a q-valued function. (For example, if $n = \frac{1}{2}$, there are two square roots.)

(ii) *When n is complex*, say

$$n = x + iy, \quad (x, y \text{ real})$$

then we must interpret a^n in such a way that

$$a^{x+iy} = a^x \times a^{iy}.$$

We have already dealt with the factor a^x for real and rational values of x, so that we are left to consider what meaning may be given to the expression a^{iy} when the power iy is pure imaginary.

16. Pure imaginary powers. In the preceding paragraph we reduced the interpretation of a^n to the case, which we now consider, when n is pure imaginary. With a slight change of notation, we write

$$n = ix,$$

where x is real.

(i) ***When a is real and positive.***

We note first that any real positive number a (other than zero) can be expressed in the form

$$a \equiv e^{\log_e a},$$

so that, if the laws of indices are to be preserved,

$$a^n \equiv (e^{\log_e a})^n \equiv e^{n \log_e a}.$$

Writing $\quad n \log_e a \equiv ix \log_e a \equiv ix' \quad (x' \text{ real}),$

we obtain the expression, as an imaginary power of e, in the form

$$a^n \equiv e^{ix'}$$

which is found to be more amenable to treatment. Dropping the dash, then, we consider what interpretation should be given to the number

$$e^{ix} \quad (x \text{ real}).$$

We take the hint from the expansion of p. 54, which, if valid for pure imaginary numbers, would yield the relation

$$e^{ix} = 1 + (ix) + \frac{(ix)^2}{2!} + \frac{(ix)^3}{3!} + \frac{(ix)^4}{4!} + \dots$$

$$= \left(1 - \frac{x^2}{2!} + \frac{x^4}{4!} - \dots\right) + i\left(x - \frac{x^3}{3!} + \frac{x^5}{5!} - \dots\right).$$

Now (p. 50) we have the relations, valid for all real x,

$$\cos x = 1 - \frac{x^2}{2!} + \frac{x^4}{4!} - \ldots,$$

$$\sin x = x - \frac{x^3}{3!} + \frac{x^5}{5!} - \ldots,$$

and so we are led to propose for consideration the relation

$$e^{ix} = \cos x + i \sin x.$$

Before we can accept this as a valid interpretation for e^{ix}, however, we must satisfy ourselves that it obeys the index law

$$e^{ix} \times e^{iy} = e^{i(x+y)} \quad (x, y \text{ real}).$$

Under the proposed interpretation, the left-hand side is

$$(\cos x + i \sin x) \times (\cos y + i \sin y)$$

$$= (\cos x \cos y - \sin x \sin y) + i(\sin x \cos y + \cos x \sin y)$$

$$= \cos(x+y) + i \sin(x+y),$$

and this is precisely the interpretation to be given to the right-hand side also.

The interpretation is therefore consistent with the series expansions and with the index law, and so we DEFINE the interpretation of e^{ix} (x real) to be given by the relation

$$e^{ix} = \cos x + i \sin x.$$

Note. The mere writing of ix in the expansion in series does not of itself allow us to use the notation e^{ix} under the index laws. The step that the product of the expressions proposed for e^{ix}, e^{iy} is indeed $e^{i(x+y)}$ is vital to the interpretation.

(ii) *When a is any complex number.*

We are now in a position to give an interpretation to the expression a^{ix} for any complex number a. We have seen that a may be written in the form

$$a \equiv r(\cos\theta + i \sin\theta) \quad (r \text{ positive})$$

and, by what we have just done, we have the two relations

$$r = e^{\log_e r} \quad (r \neq 0),$$

$$\cos\theta + i \sin\theta = e^{i\theta}.$$

Hence
$$a = e^{\log_e r} \times e^{i\theta}$$
$$= e^u \times e^{i\theta},$$

say, where u, or $\log_e r$, is a real number, positive or negative. For interpretations consistent with the laws of indices, we must therefore take
$$a^{ix} = (e^u)^{ix} \times (e^{i\theta})^{ix}$$
$$= e^{iux} \times e^{-\theta x}$$
$$= e^{-\theta x}(\cos ux + i \sin ux)$$
$$= e^{-\theta x}\{\cos(x \log_e r) + i \sin(x \log_e r)\}.$$

COROLLARY. *Any (non-zero) complex number can be expressed in the form e^{u+iv}, where u, v are real.*

17. Multiplicity of values. There is one difficulty which we have slurred over, as it should not be over-emphasized at the present stage. An example will illustrate the point.

Let us consider what interpretation we are to give to the symbol
$$i^i.$$

Our first task is to express the number to be raised to the power i (in this case, i itself) in the form $re^{i\theta}$, and this we do easily by noticing that
$$i = \cos(2k + \tfrac{1}{2})\pi + i \sin(2k + \tfrac{1}{2})\pi$$
$$= e^{(2k+\frac{1}{2})\pi i}$$

for any integer k, positive or negative. Hence the interpretation
$$i^i \equiv \{e^{(2k+\frac{1}{2})\pi i}\}^i$$
$$= e^{(2k+\frac{1}{2})\pi i \times i}$$
$$= e^{-(2k+\frac{1}{2})\pi}$$

gives an *infinite succession* of values, all of which, oddly enough, are real.

In a full discussion, we should therefore be on our guard to include many-valued powers. A safe way of doing this is to include the factor $e^{2k\pi i}$ (which is just $\cos 2k\pi + i \sin 2k\pi$, or 1, for any integer k) in the expression in exponential form of the number whose power we seek. That is to say, in order to evaluate a^n, we write a in the form
$$a \equiv e^{u+i(v+2k\pi)}$$

and allow k to take integral values.

It should be noticed that we established the interpretation

$$e^{ix} \equiv \cos x + i \sin x$$

under the conditions of De Moivre's theorem, namely that x is a real number, being a positive or negative integer or rational fraction. When x is not integral, there is really an ambiguity of interpretation, since

$$\begin{aligned} e^{ix} &= (e^i)^x \\ &= e^{i(1+2k\pi)x} \\ &= \cos(1+2k\pi)x + i \sin(1+2k\pi)x, \end{aligned}$$

where distinct values are obtained for $k = 0, 1, \ldots, q-1$ as usual, q being the denominator when the fraction x is expressed in its lowest terms. Our interpretation is therefore subject to the convention $k = 0$.

ILLUSTRATION 2. *To find an expression for*

$$(1+i\sqrt{3})^{4+\frac{1}{2}i}.$$

We first write $1+i\sqrt{3}$ in the form

$$\begin{aligned} 1+i\sqrt{3} &\equiv 2\left(\frac{1}{2}+i\frac{\sqrt{3}}{2}\right) \\ &= 2\left(\cos\frac{\pi}{3}+i\sin\frac{\pi}{3}\right) \\ &= 2e^{i(\frac{1}{3}\pi+2k\pi)} \quad (k \text{ integral}). \end{aligned}$$

We next use the fact that

$$(1+i\sqrt{3})^{4+\frac{1}{2}i} = (1+i\sqrt{3})^4 \times (1+i\sqrt{3})^{\frac{1}{2}i}.$$

Now
$$\begin{aligned} (1+i\sqrt{3})^4 &= 2^4 e^{4i(\frac{1}{3}\pi+2k\pi)} \\ &= 16e^{4\pi i/3}. \end{aligned}$$

Also
$$\begin{aligned} (1+i\sqrt{3})^{\frac{1}{2}i} &= 2^{\frac{1}{2}i} \times \{e^{i(\frac{1}{3}\pi+2k\pi)}\}^{\frac{1}{2}i} \\ &= 2^{\frac{1}{2}i} \times e^{-(\frac{1}{6}\pi+k\pi)}. \end{aligned}$$

Finally,
$$2 = e^{\log_e 2},$$

giving
$$2^{\frac{1}{2}i} = e^{\frac{1}{2}i\log_e 2},$$

so that, in all,

$$(1+i\sqrt{3})^{4+\frac{1}{2}i} = 16e^{4\pi i/3} \times e^{\frac{1}{2}i\log_e 2} \times e^{-(\frac{1}{6}\pi+k\pi)}$$
$$= 16e^{-(\frac{1}{6}\pi+k\pi)} \times e^{(\frac{4}{3}\pi+\frac{1}{2}\log_e 2)i}$$
$$= 16e^{-(\frac{1}{6}\pi+k\pi)} \{\cos(\tfrac{4}{3}\pi + \tfrac{1}{2}\log_e 2) + i\sin(\tfrac{4}{3}\pi + \tfrac{1}{2}\log_e 2)\},$$

where k is any integer, positive, negative, or zero.

EXAMPLES XI

Find expressions for

1. $(1+i)^i$.
2. $(1-i)^{2-i}$.
3. $(\sqrt{3}+i)^{\frac{1}{2}i}$.
4. $(1-i\sqrt{3})^{3+2i}$.
5. $(\sqrt{3}-i)^{1+\frac{1}{2}i}$.
6. $(1+i)^{4+\frac{1}{2}i}$.

18. The logarithm of a complex number, to the base *e*. We have seen (p. 181) that any (non-zero) complex number can be expressed in the form

$$a \equiv e^{u+iv}.$$

We DEFINE the logarithm of a to be the complex number

$$u+iv \quad (u, v \text{ real}).$$

This is consistent with the treatment already given for real numbers ($v = 0$) and, in virtue of the interpretation already given to indices, it retains the property

$$\log_e a + \log_e b = \log_e (ab).$$

We have now, of course, released the restriction (p. 5) that a must be positive in order to have a logarithm. It need not even be real.

The ambiguous determination of v, as seen by the equivalent formula

$$a \equiv e^{u+i(v+2k\pi)} \quad (k \text{ integral})$$

means that *the logarithm is ambiguous to the extent of additive multiples of* $2\pi i$.

Note. The relation (pp. 170, 180)

$$z = re^{i\theta},$$

where r is the modulus and θ the argument of z, gives the formula

$$\log z = \log|z| + i\arg z.$$

NOTATION. The notation

$$\mathrm{Log}_e a$$

is often used to denote a value of the logarithm when ambiguity may be present. Then the notation

$$\log_e a$$

denotes that determination of the logarithm whose imaginary part lies between $-\pi$ and π, and $\log_e a$ is called the *principal value* of $\mathrm{Log}_e\, a$.

This distinction, which is often of importance, will not, in fact, be used much in this book, and we shall usually take the principal value without further comment.

19. The sine and cosine. We have made no statements so far about the formula

$$e^{ix} = \cos x + i \sin x$$

except when x is real; naturally, because $\cos x, \sin x$ are otherwise undefined. Our next task is to define the trigonometric functions of a complex variable $z \equiv x + iy$ (x, y real), and this we do by what is in some ways a reversal of the process hitherto adopted. We first observe that, for real x,

$$e^{ix} = \cos x + i \sin x,$$

$$e^{-ix} = \cos x - i \sin x,$$

so that

$$\cos x = \tfrac{1}{2}(e^{ix} + e^{-ix}),$$

$$\sin x = \frac{1}{2i}(e^{ix} - e^{-ix}).$$

We now adopt these formulæ, and make them the basis of the following more general definitions:

If z is a complex number, then

$$\cos z = \tfrac{1}{2}(e^{iz} + e^{-iz}),$$

$$\sin z = \frac{1}{2i}(e^{iz} - e^{-iz}).$$

It follows at once that these definitions give the normal functions when z is real, and also that

$$e^{iz} = \cos z + i \sin z.$$

We must, however, now make sure that they retain the NORMAL properties of the trigonometric functions.

Firstly, we have

$$4(\cos^2 z+\sin^2 z) = (e^{iz}+e^{-iz})^2-(e^{iz}-e^{-iz})^2$$
$$= (e^{2iz}+2+e^{-2iz})-(e^{2iz}-2+e^{-2iz})$$
$$= 4,$$

so that $\qquad \cos^2 z+\sin^2 z = 1.$

Again,

$$4(\cos z_1 \cos z_2-\sin z_1 \sin z_2)$$
$$= (e^{iz_1}+e^{-iz_1})(e^{iz_2}+e^{-iz_2})+(e^{iz_1}-e^{-iz_1})(e^{iz_2}-e^{-iz_2})$$
$$= 2(e^{i(z_1+z_2)}+e^{-i(z_1+z_2)})$$

on reduction, so that

$$\cos z_1 \cos z_2-\sin z_1 \sin z_2 = \cos(z_1+z_2).$$

Also

$$4i(\sin z_1 \cos z_2+\cos z_1 \sin z_2)$$
$$= (e^{iz_1}-e^{-iz_1})(e^{iz_2}+e^{-iz_2})+(e^{iz_1}+e^{-iz_1})(e^{iz_2}-e^{-iz_2})$$
$$= 2(e^{i(z_1+z_2)}-e^{-i(z_1+z_2)}),$$

so that $\qquad \sin z_1 \cos z_2+\cos z_1 \sin z_2 = \sin(z_1+z_2).$

These are the formulæ on which the theory of real angles is based, and so the structure will stand for complex 'angles' also.

Further details are left to the reader.

Note. The functions $|\cos z|, |\sin z|$ are not now subject to the restriction of being less than unity. For example, the equation

$$\cos z = 2$$

is satisfied if $\qquad e^{iz}+e^{-iz} = 4,$

or $\qquad e^{2iz}-4e^{iz}+1 = 0,$

so that $\qquad e^{iz} = 2 \pm \sqrt{3}.$

Thus $\qquad iz = \log(2 \pm \sqrt{3})$

or $\qquad z = -i \log(2 \pm \sqrt{3}).$

For reference we record the following formulæ:

(i) $\cosh iz = \frac{1}{2}(e^{iz}+e^{-iz})$
$= \cos z;$

(ii) $\sinh iz = \frac{1}{2}(e^{iz}-e^{-iz})$
$= i \sin z;$

(iii) $e^{2k\pi i} = \cos 2k\pi + i \sin 2k\pi = 1$ (k integral);

(iv) $e^{(2k+1)\pi i} = \cos(2k+1)\pi + i\sin(2k+1)\pi = -1;$

(v) $e^{(2k+\frac{1}{2})\pi i} = \cos(2k+\frac{1}{2})\pi + i\sin(2k+\frac{1}{2})\pi = i.$

20. The modulus of e^z. If z is complex, so that

$$z \equiv x + iy \quad (x, y \text{ real}),$$

then

$$e^z = e^{x+iy} = e^x e^{iy}$$
$$= e^x(\cos y + i \sin y).$$

Hence *the modulus of e^z is e^x and its argument is y*, the argument being undetermined to the extent of multiples of 2π.

COROLLARY. An important corollary, found by putting $x = 0$, is that

$$|e^{iy}| = 1,$$

when iy is a pure imaginary.

21. The differentiation and integration of complex numbers. We confine our attention to the only case which we shall use, the complex functions of a REAL variable x. The general theory for a complex variable is much beyond the scope of this book.

By a complex function of a real variable x, we shall mean a function $w(x)$ which is *either* given in the form

$$w(x) \equiv u(x) + iv(x),$$

where $u(x), v(x)$ are real functions of x, *or* can be reduced to that form. For example, the function e^{ix} can be reduced to

$$\cos x + i \sin x.$$

We then use the following DEFINITIONS:

(i)
$$\frac{dw}{dx} = \frac{du}{dx} + i\frac{dv}{dx},$$

(ii)
$$\int w\,dx = \int u\,dx + i\int v\,dx.$$

Particular interest is attached to the function e^{cx}, where c may be complex of the form $a+ib$. We have

$$\frac{d}{dx}(e^{cx}) = \frac{d}{dx}(e^{(a+ib)x})$$

$$= \frac{d}{dx}\{e^{ax}(\cos bx + i\sin bx)\}$$

$$= \frac{d}{dx}[\{e^{ax}\cos bx\} + i\{e^{ax}\sin bx\}].$$

Hence, by definition,

$$\frac{d}{dx}(e^{cx}) = (a\,e^{ax}\cos bx - b\,e^{ax}\sin bx) + i(a\,e^{ax}\sin bx + b\,e^{ax}\cos bx)$$

$$= a\,e^{ax}(\cos bx + i\sin bx) + ib\,e^{ax}(\cos bx + i\sin bx)$$

$$= (a+ib)\,e^{ax}.e^{ibx}$$

$$= (a+ib)\,e^{(a+ib)x}$$

$$= ce^{cx}.$$

Hence *the rule*
$$\frac{d}{dx}(e^{cx}) = ce^{cx}$$
holds whether c is real or complex.

Similarly we may prove that the formula

$$\int e^{cx}\,dx = \frac{1}{c}\,e^{cx}$$

holds whether c is real or complex.

Because of its importance, we ought perhaps to mention also the differentiation and integration of x^c, where x is assumed to be real but where c may be complex. Since

$$x^c = e^{c \log x},$$

we have

$$\frac{d}{dx}(x^c) = e^{c \log x} . c . \frac{d}{dx}(\log x) \quad \text{(as above)}$$

$$= e^{c \log x} . \frac{c}{x}$$

$$= x^c . \frac{c}{x}$$

$$= cx^{c-1},$$

in accordance with the similar rule for real functions.

It follows that

$$\int x^c \, dx = \frac{1}{c+1} x^{c+1} \quad (c \neq -1)$$

whether c is real or complex.

ILLUSTRATION 3. *To prove the rule*

$$\frac{d}{dx}\{f(x)\,g(x)\} = g(x)\frac{d}{dx}\{f(x)\} + f(x)\frac{d}{dx}\{g(x)\}$$

when $f(x), g(x)$ are complex functions of the real variable x.

Write

$$f(x) \equiv u(x) + iv(x) \equiv u + iv,$$

$$g(x) \equiv p(x) + iq(x) \equiv p + iq.$$

Then

$$f(x)g(x) = (up - vq) + i(uq + vp),$$

so that

$$\frac{d}{dx}\{f(x)g(x)\}$$

$$= [(u'p - v'q) + (up' - vq')] + i[(u'q + v'p) + (uq' + vp')]$$

$$= [u'g + iv'g] + [p'f + iq'f]$$

$$= f'g + g'f.$$

Reversal of this formula leads to the rule for *integration by parts*. Hence this method is also at our disposal for these functions.

The two illustrations which follow show how complex numbers may be used to sum series and to evaluate integrals.

ILLUSTRATION 4. *To find the sum of the first n terms of the series*

$$1+\cos\theta.\cos\theta+\cos^2\theta.\cos 2\theta+\ldots+\cos^{n-1}\theta.\cos(n-1)\theta,$$

where θ is a real angle.

Write

$$C\equiv 1+\cos\theta.\cos\theta+\cos^2\theta.\cos 2\theta+\ldots+\cos^{n-1}\theta.\cos(n-1)\theta,$$

$$S\equiv \quad \cos\theta.\sin\theta+\cos^2\theta.\sin 2\theta+\ldots+\cos^{n-1}\theta.\sin(n-1)\theta.$$

Then

$$C+iS = 1+\cos\theta\, e^{i\theta}+\cos^2\theta\, e^{2i\theta}+\ldots+\cos^{n-1}\theta\, e^{(n-1)i\theta}.$$

This is a geometric progression, of n terms, with first term 1 and ratio $\cos\theta\, e^{i\theta}$. It may be proved, exactly as for real numbers, that the sum for first term a and ratio r is

$$\frac{a(1-r^n)}{1-r}$$

so that

$$C+iS = \frac{1-\cos^n\theta\, e^{ni\theta}}{1-\cos\theta\, e^{i\theta}}.$$

In order to find 'real and imaginary parts', multiply numerator and denominator by $1-\cos\theta\, e^{-i\theta}$. The new denominator is

$$\begin{aligned}(1-\cos\theta\, e^{i\theta})(1-\cos\theta\, e^{-i\theta}) &= 1-\cos\theta(e^{i\theta}+e^{-i\theta})+\cos^2\theta\\ &= 1-\cos\theta(2\cos\theta)+\cos^2\theta = 1-\cos^2\theta\\ &= \sin^2\theta.\end{aligned}$$

Thus

$$\begin{aligned}C+iS &= \frac{1}{\sin^2\theta}\{(1-\cos^n\theta\, e^{ni\theta})(1-\cos\theta\, e^{-i\theta})\}\\ &= \frac{1}{\sin^2\theta}\{1-\cos^n\theta\, e^{ni\theta}-\cos\theta\, e^{-i\theta}+\cos^{n+1}\theta\, e^{(n-1)i\theta}\}.\end{aligned}$$

Equating real parts, we have

$$\begin{aligned}C &= \frac{1}{\sin^2\theta}\{1-\cos^n\theta\cos n\theta-\cos^2\theta+\cos^{n+1}\theta\cos(n-1)\theta\}\\ &= \frac{1}{\sin^2\theta}\{\sin^2\theta-\cos^n\theta(\cos n\theta-\cos\theta\cos(n-1)\theta)\}\\ &= \frac{1}{\sin^2\theta}\{\sin^2\theta-\cos^n\theta(-\sin\theta\sin(n-1)\theta)\}\\ &= 1+\frac{\cos^n\theta\sin(n-1)\theta}{\sin\theta}.\end{aligned}$$

It is implicit in the working that $\sin\theta$ is not zero. That case can easily be given separate treatment.

ILLUSTRATION 5. *To find the real integral*

$$\int x e^x \cos x \, dx.$$

Write
$$C \equiv \int x e^x \cos x \, dx,$$
$$S \equiv \int x e^x \sin x \, dx,$$

so that
$$C + iS = \int x e^x e^{ix} \, dx$$
$$= \int x e^{(1+i)x} \, dx.$$

Integrating by parts (p. 188), we have

$$C + iS = \frac{1}{1+i} e^{(1+i)x} . x - \frac{1}{1+i} \int e^{(1+i)x} . 1 . dx$$
$$= \frac{1}{1+i} x e^{(1+i)x} - \frac{1}{(1+i)^2} e^{(1+i)x}.$$

Now
$$\frac{1}{1+i} = \frac{1-i}{1-i^2} = \tfrac{1}{2}(1-i),$$
$$e^{(1+i)x} = e^x (\cos x + i \sin x).$$

Hence
$$C + iS = [\tfrac{1}{2}(1-i)x - \tfrac{1}{4}(1 - 2i + i^2)] e^x (\cos x + i \sin x)$$
$$= [\tfrac{1}{2}x - \tfrac{1}{2}ix + \tfrac{1}{2}i] e^x (\cos x + i \sin x).$$

Equating real parts, we have
$$C = \tfrac{1}{2} e^x \{x \cos x + (x-1) \sin x\}.$$

EXAMPLES XII

Use the method of the two illustrations to find:

1. $1 + \cos\theta + \cos 2\theta + \ldots + \cos n\theta$.

2. $\sin\theta - x \sin 2\theta + x^2 \sin 3\theta - x^3 \sin 4\theta + \ldots$ (to n terms).

Find the integrals:

3. $\int x \cos x \, dx$. 4. $\int e^x \sin x \, dx$. 5. $\int e^{4x} \sin 3x \, dx$.

6. $\int x e^{2x} \sin x \, dx$. 7. $\int x \sin 3x \, dx$. 8. $\int x e^{-4x} \cos 3x \, dx$.

Finally, we give an illustration to show how the use of complex numbers can help in dealing with the geometry of a plane curve. We use for the curve the notation of the preceding chapter.

ILLUSTRATION 6. Let $P(x, y)$ be a point of a plane curve, and let s, ψ denote as usual the length of the arc measured from a fixed point and the angle between the tangent at P and the x-axis. (It is assumed that x, y are real.)

Write
$$z = x + iy.$$

Then
$$\frac{dz}{ds} = \frac{dx}{ds} + i\frac{dy}{ds}$$
$$= \cos\psi + i\sin\psi \quad \text{(p. 108)}$$
$$= e^{i\psi} \quad \text{(p. 180).}$$

In particular,
$$\left|\frac{dz}{ds}\right| = 1.$$

We also have the relation
$$\frac{d^2z}{ds^2} = i\,e^{i\psi}\frac{d\psi}{ds}$$
$$= i\kappa e^{i\psi} \quad \text{(p. 113).}$$

This is equivalent to
$$\frac{d^2x}{ds^2} + i\frac{d^2y}{ds^2} = i\kappa e^{i\psi},$$

so that, taking moduli on each side,
$$|\kappa| = \sqrt{\left\{\left(\frac{d^2x}{ds^2}\right)^2 + \left(\frac{d^2y}{ds^2}\right)^2\right\}}.$$

Again,
$$\frac{dx}{ds} - i\frac{dy}{ds} = \cos\psi - i\sin\psi$$
$$= e^{-i\psi},$$

so that
$$\left(\frac{dx}{ds} - i\frac{dy}{ds}\right)\left(\frac{d^2x}{ds^2} + i\frac{d^2y}{ds^2}\right) = i\kappa.$$

Equating imaginary parts, we obtain the formula
$$\kappa = \frac{dx}{ds}\frac{d^2y}{ds^2} - \frac{dy}{ds}\frac{d^2x}{ds^2}.$$

REVISION EXAMPLES VII

'Scholarship' Level

1. Two complex numbers z, z' are connected by the relation $z' = (2+z)/(2-z)$. Show that, as the point which represents z on the Argand diagram describes the axis of y, from the negative end to the positive end, the point which represents z' describes completely the circle $x^2+y^2 = 1$ in the counter-clockwise sense.

2. Explain what is meant by the principal value of the logarithm of a complex number $x+iy$ as distinct from the general value.

Show that, considering only principal values, the real part of

$$(1+i)^{\log_e(1+i)}$$

is
$$2^{\frac{1}{4}\log_e 2}e^{-\frac{1}{16}\pi^2}\cos\left(\tfrac{1}{4}\pi\log_e 2\right).$$

3. Express $\tan(a+ib)$ in the form $A+iB$, where A, B are real when a, b are real.

Show that, if $x+iy = \tan\frac{1}{2}(\xi+i\eta)$, then

$$\tfrac{1}{2}x = e^{-\eta}\sin\xi - e^{-2\eta}\sin 2\xi + e^{-3\eta}\sin 3\xi - \ldots,$$

when η is positive; and that there is a like expansion with the sign of η changed, valid when η is negative.

4. Prove that, if

$$\sin(x+iy) = \operatorname{cosec}(u+iv),$$

where x, y, u, v are real, then

$$\cosh^2 v\tanh^2 y = \cos^2 u, \quad \cosh^2 y\tanh^2 v = \cos^2 x.$$

5. Solve the equation

$$(1-xi)^n + i(1+xi)^n = 0,$$

giving the roots as trigonometrical functions of angles between 0 and π.

6. Prove that, if $x+iy = c\cot(u+iv)$, then

$$\frac{x}{\sin 2u} = \frac{-y}{\sinh 2v} = \frac{c}{\cosh 2v - \cos 2u},$$

and show that, if x, y are coordinates of a point in a plane, then for a given value of v the point lies on the circle

$$x^2+y^2+2cy\coth 2v+c^2 = 0.$$

Also verify that, if a_1, a_2 denote the radii of the circles for the values v_1, v_2 of v, and d denotes the distance between their centres, then

$$\frac{a_1^2+a_2^2-d^2}{2a_1 a_2} = \cosh 2(v_1-v_2).$$

7. Obtain an expression for $|\cos(x+iy)|^2$ in terms of trigonometric and hyperbolic functions of the real variables x, y.

Show that $|\cos(x+ix)|$ increases with x for all positive values of x.

8. Express
$$\frac{z_1+z_2}{z_1 z_2-1}$$
in the form $X+iY$, where $z_1 = x_1+iy_1$, $z_2 = x_2+iy_2$.

Deduce that the points which represent three complex numbers z_1, z_2, z_3 in the Argand diagram cannot all lie on the same side of the real axis if

$$z_1+z_2+z_3 = z_1z_2z_3.$$

9. Prove that the roots of the equation

$$(x-1)^n = -(x+1)^n,$$

where n is a positive integer, are

$$i\cot\frac{(2k+1)\pi}{2n} \quad (k = 0, 1, 2, \ldots, n-1).$$

Deduce, or prove otherwise, that

$$\sum_{k=0}^{n-1}\cot\frac{(2k+1)\pi}{2n} = 0, \quad \sum_{k=0}^{n-1}\operatorname{cosec}^2\frac{(2k+1)\pi}{2n} = n^2.$$

10. Write down the complex numbers conjugate to $x+iy$ and to $r(\cos\theta+i\sin\theta)$.

Prove that, if a quadratic equation with real coefficients has one complex root, the other root is the conjugate complex number. Deduce a similar result for a cubic equation with real coefficients.

Given that $x = 1+3i$ is one solution of the equation

$$x^4+16x^2+100 = 0,$$

find all the solutions.

11. Express each of the complex numbers

$$z_1 = (1+i)\sqrt{2}, \quad z_2 = 4(-1+i)\sqrt{2}$$

in the form $r(\cos\theta + i\sin\theta)$, where r is positive.

Prove that $z_1^3 = z_2$, and find the other cube roots of z_2 in the form $r(\cos\theta + i\sin\theta)$.

12. The complex number $z = x+iy = r(\cos\theta + i\sin\theta)$ is represented in the Argand diagram by the point (x, y). Prove that, if three variable points z_1, z_2, z_3 are such that $z_3 = \lambda z_2 + (1-\lambda)z_1$, where λ is a complex constant, then the triangle whose vertices are z_1, z_2, z_3 is similar to the triangle with vertices at the points $0, 1, \lambda$.

ABC is a triangle. On the sides BC, CA, AB triangles BCA', CAB', ABC' are described similar to a given triangle DEF. Prove that the median points of the triangles ABC, $A'B'C'$ are coincident.

13. The complex numbers $a, b, (1-k)a+kb$ are represented in the Argand diagram by the points A, B, C. Prove that the vector $\overrightarrow{AB}$ represents $b-a$, and that

(i) when k is real, C lies on AB and divides AB in the ratio $k : 1-k$;

(ii) when k is not real, ABC is a triangle in which the sides AB, AC are in the ratio $1 : |k|$ and the angle of turn from $\overrightarrow{AB}$ to $\overrightarrow{AC}$ is θ, where $k = |k|(\cos\theta + i\sin\theta)$.

The vertices of two triangles ABC and XYZ represent the complex numbers a, b, c and x, y, z. Prove that a necessary and sufficient condition for the triangles to be similar and similarly situated is

$$\begin{vmatrix} 1 & 1 & 1 \\ a & b & c \\ x & y & z \end{vmatrix} = 0.$$

14. [In this problem small letters stand for complex numbers and capital letters for the corresponding points in the Argand diagram.]

The circumcentre, centroid, and orthocentre of a triangle ABC are S, G, H respectively. Prove that

$$g = \tfrac{1}{3}(a+b+c), \quad 2s+h = a+b+c.$$

ABC is a triangle with its circumcentre at the origin. The internal bisectors of the angles A, B, C of the triangle meet the circumcircle again at L, M, N respectively. Prove (by pure geometry if you wish) that the incentre P of ABC is the orthocentre of LMN, and deduce that

$$p = l+m+n.$$

Prove also that the excentres P_1, P_2, P_3 of the triangle ABC are given by

$$p_1 = l-m-n, \quad p_2 = -l+m-n, \quad p_3 = -l-m+n,$$

and that the circumcentre K of $P_1P_2P_3$ is given by

$$k = -(l+m+n).$$

Prove, finally, that KP_1 is perpendicular to BC.

15. Points A, B in the Argand diagram represent complex numbers a, b respectively; O is the origin, and P represents one of the values of $\sqrt{(ab)}$. Prove that, if $OA = OB = r$, then also $OP = r$, and OP is perpendicular to AB.

A, B, C lie on a circle with centre O, and represent complex numbers a, b, c respectively. Prove that the point D which represents $-bc/a$ also lies on the circle, and that AD is perpendicular to BC.

The perpendiculars from B, C to CA, AB meet the circle again at E, F respectively. Prove that OA is perpendicular to EF.

16. Prove by means of an Argand diagram, that

$$|z_1+z_2| \leqslant |z_1|+|z_2|.$$

Find all the points in the complex plane which represent numbers satisfying the equations

(i) $z^2-2z = 3$, (ii) $|z-2| = 3$, (iii) $|z-1|+|z-2| = 3$.

17. Find the fifth roots of unity, and hence solve the equation

$$(2x-1)^5 = 32x^5.$$

18. Prove that, if α, β are the roots of the equation

$$t^2-2t+2 = 0,$$

then
$$\frac{(x+\alpha)^n-(x+\beta)^n}{\alpha-\beta} = \frac{\sin n\theta}{\sin^n \theta},$$

where
$$\cot\theta = x+1.$$

19. A point $z = x+iy$ in the Argand diagram is such that $|z| = 2, x = 1$, and $y > 0$. Determine the point and find the distance between the point and $\frac{1}{2}z^2$.

Show also that the points $2, z, \frac{1}{2}z^2, \frac{1}{4}z^3, \frac{1}{8}z^4, \frac{1}{16}z^5$ are the vertices of a regular hexagon.

20. The points z_1, z_2, z_3 form an equilateral triangle in the Argand diagram, and $z_1 = 4+6i, z_2 = (1-i)z_1$. Show that z_3 must have one of two values, and determine these values.

What are the vertices of the regular hexagon of which z_1 is the centre, and z_2 is one vertex?

21. A regular pentagon $ABCDE$ is inscribed in the circle $x^2+y^2 = 1$, the vertex A being the point $(1, 0)$. Obtain the complex numbers $x+iy$ of which the points A, B, C, D, E form a representation in an Argand diagram.

22. Use De Moivre's theorem to find all the roots of the equation

$$(2x-1)^5 = (x-2)^5$$

in the form $a+ib$, where a, b are real numbers.

23. Mark on an Argand diagram the points $\sqrt{3}+i, 2+2i\sqrt{3}$, and their product. What is the general relation between the positions of the points $a+ib$ and $(\sqrt{3}+i)(a+ib)$?

A triangle ABC has its vertices at the points $0, 2+2i\sqrt{3}$, $-1+i\sqrt{3}$ respectively. A similar triangle $A'B'C'$ has its vertices A', B' at the points $0, 8i$ respectively. Find the position of C'.

24. Find all values of $(5-12i)^{\frac{1}{2}}, (2i-2)^{\frac{1}{3}}$, expressing the answers in the form $a+ib$, where a, b are fractions or surds.

25. Prove that the triangle formed by the three points representing the complex numbers f, g, h is similar to the triangle formed by the points representing the numbers u, v, w if

$$fv+gw+hu = fw+gu+hv.$$

26. By expressing $3+4i$, $1+2i$ in the form $r(\cos\theta+i\sin\theta)$, or otherwise, evaluate

$$(3+4i)^{30}/(1+2i)^{50}$$

in the form $a+ib$, obtaining each of the real numbers a, b correct to two significant figures.

27. Identify all the points in the z plane which satisfy the following relations:

(i) $z^2+2=2z$, (ii) $\left|\frac{z-2}{z+2}\right| \leqslant 1$,

(iii) $\left|\frac{z-1}{z+1}\right| \geqslant 2$, (iv) $|z-1|+|z+1| \leqslant 4$.

28. Prove that three points z_1, z_2, z_3 in the complex plane are collinear if, and only if, the ratio

$$(z_3-z_1)/(z_2-z_1)$$

is real.

29. Find the real and imaginary parts and the modulus for each of the expressions

(i) $\frac{1}{1+2i}$, (ii) $\frac{a+ib}{b+ia}$, (iii) $\frac{1+\cos\alpha+i\sin\alpha}{1+\cos\beta+i\sin\beta}$.

30. Find the logarithms of $-1, 2-i, 10^{x+iy}$.

31. The three points in an Argand diagram which correspond to the roots of the equation

$$z^3-3pz^2+3qz-r=0$$

are the vertices of a triangle ABC. Prove that the centroid of the triangle is the point corresponding to p.

If the triangle ABC is equilateral, prove that $p^2=q$.

32. Find the cube roots of $4-3i$.

33. If the vertices A, B, C of an equilateral triangle represent the numbers z_1, z_2, z_3 respectively in an Argand diagram, state the number represented by the mid-point M of AB, and show that

$$z_3 = \tfrac{1}{2}(z_1+z_2) \pm \frac{\sqrt{3}}{2} i(z_2-z_1).$$

If $$z_1 = 2+2i, \quad z_2 = 4+(2+2\sqrt{3})i,$$

and if C is on the same side of AB as the origin, find the number z_3.

34. If $z=\cos\theta+i\sin\theta$, express $\cos\theta, \sin\theta, \cos n\theta, \sin n\theta$ in terms of z, where n is an integer.

Hence express $\sin^7\theta$ in the form

$$A\sin\theta+B\sin 3\theta+C\sin 5\theta+D\sin 7\theta.$$

35. Prove that

$$\left(\frac{1+\sin\theta-i\cos\theta}{1+\sin\theta+i\cos\theta}\right)^6 = -\cos 6\theta - i\sin 6\theta.$$

36. Prove that

$$\left(\frac{1+\sin\theta+i\cos\theta}{1+\sin\theta-i\cos\theta}\right)^n = \cos\left(\tfrac{1}{2}n\pi-n\theta\right)+i\sin\left(\tfrac{1}{2}n\pi-n\theta\right).$$

37. Show that the equation

$$|z-1-2i| = 3$$

represents a circle in an Argand diagram.

If z lies on $|z-1-2i| = 3$, what is the locus of the point $u = z+4-3i$?

Find the greatest and least values of $|z-4-6i|$ if z is subject to the inequality $|z-1-2i| \leqslant 3$.

38. Find the modulus of

$$\frac{3z+i}{(3z-i)^2}$$

when $|z| = 1$.

39. Two complex numbers z, z_1 are connected by the relation

$$z_1 = \frac{2+z}{2-z}.$$

Prove that, if $z = iq$, where q is real, then the locus of z_1 in an Argand diagram is a circle. Describe the variation in the amplitude of z_1 as q increases from $-\infty$ to $+\infty$.

40. If a, b are real and n is an integer, prove that

$$\sqrt[n]{(a+ib)}+\sqrt[n]{(a-ib)}$$

has n real values, and find those of

$$\sqrt[3]{(1+i\sqrt{3})}+\sqrt[3]{(1-i\sqrt{3})}.$$

41. If x, y are real, separate

$$\sec(x+iy)$$

into its real and imaginary parts.

42. If a, b, c, d are real, express each of a^2+b^2 and c^2+d^2 as a product of linear factors.

Deduce that the product of two factors, each of which is a sum of two squares, is itself a sum of two squares.

43. The intrinsic coordinates of a point on a plane curve are s, ψ, and the cartesian coordinates are x, y. The complex coordinate $x+iy$ is denoted by z and the curvature by κ. Prove that

$$\frac{dz}{ds} = e^{i\psi}, \quad \frac{d^2z}{ds^2} = i\kappa e^{i\psi}.$$

Hence, or otherwise, prove that

$$\kappa = (x''^2+y''^2)^{\frac{1}{2}} = (x'y''-x''y'),$$

where dashes denote differentiations with respect to s.

Prove further that

$$\begin{vmatrix} x' & y' & 1 \\ x'' & y'' & \kappa \\ x''' & y''' & \kappa^2 \end{vmatrix} = \kappa(2\kappa^2-\kappa').$$

44. Any point P is taken on a given curve. The tangent at P is drawn in the direction of increasing s, and a point Q is taken at a constant distance l along this tangent. In this way Q describes a curve specified by X, Y, S, Z analogous to the specification x, y, s, z for P. [Notation of previous problem.] Show that

(i) $\dfrac{dS}{ds} = (1+l^2\kappa^2)^{\frac{1}{2}}$;

(ii) the curvature K of the derived curve at Q is given by the formula

$$K(1+l^2\kappa^2)^{\frac{3}{2}} = \kappa(1+l^2\kappa^2)+l'\kappa,$$

where $\kappa' = d\kappa/ds$.

CHAPTER XII

SYSTEMATIC INTEGRATION

AT first sight the integration of functions seems to depend as much upon luck as upon skill. This is largely because the teacher or author must, in the early stages, select examples which are known to 'come out'. Nor is it easy to be sure, even with years of experience, that any particular integral is capable of evaluation; for example, $x \sin x$ can be integrated easily, whereas $\sin x/x$ cannot be integrated at all in finite terms by means of functions studied hitherto.

The purpose of this chapter is to explain how to set about the processes of integration in an orderly way. This naturally involves the recognition of a number of 'types', followed by a set of rules for each of them. But first we make two general remarks.

(i) The rules will ensure that an integral of given type MUST come out; but it is always wise to examine any particular example carefully to make sure that an easier method (such as substitution) cannot be used instead.

(ii) It is probably true to say that more integrals remain unsolved through faulty manipulation of algebra and trigonometry than through difficulties inherent in the integration itself. The reader is urged to acquire facility in the normal technique of these subjects. For details a text-book should be consulted.

1. Polynomials. The first type presents no difficulty. If $f(x)$ is the polynomial

$$f(x) \equiv a_0 x^n + a_1 x^{n-1} + \ldots + a_n,$$

then
$$\int f(x)\,dx = \frac{a_0 x^{n+1}}{n+1} + \frac{a_1 x^n}{n} + \ldots + a_n x.$$

2. Rational functions. (Compare also p. 11.) A rational function $f(x)$ of the variable x is defined to be the ratio of two *polynomials*, so that

$$f(x) = \frac{P(x)}{Q(x)}$$

for polynomials $P(x), Q(x)$. If the degree of $P(x)$ is not less than that of $Q(x)$, we can divide out, getting a polynomial (easily integrated) together with a rational fraction in which the degree of the numerator is less than that of the denominator.

We therefore confine our attention to the case in which the degree of $P(x)$ is less than that of $Q(x)$. It is assumed, too, that all coefficients are real.

It is a theorem of algebra that any (real) polynomial, and, in particular, $Q(x)$, can be expressed as a product of factors, of which typical terms are

$$(\alpha x+\beta)^m,$$

$$(ax^2+2hx+b)^n,$$

where α, β, a, h, b are real constants, but where

$$h^2 < ab$$

so that the quadratic $ax^2+2hx+b$ cannot be further resolved into real factors.

[We ought to add that, for a given polynomial, the difficulty of factorization may be very great indeed.]

It is a further theorem of algebra that the rational function may then be expressed in the form

$$\Sigma\left\{\frac{A_1}{(\alpha x+\beta)^m}+\frac{A_2}{(\alpha x+\beta)^{m-1}}+\ldots+\frac{A_m}{\alpha x+\beta}\right\}$$

$$+\Sigma\left\{\frac{B_1x+C_1}{(ax^2+2hx+b)^n}+\frac{B_2x+C_2}{(ax^2+2hx+b)^{n-1}}+\ldots+\frac{B_nx+C_n}{ax^2+2hx+b}\right\},$$

where the first summation extends over all linear factors $\alpha x+\beta$, and the second over all quadratic factors $ax^2+2hx+b$.

The integrals from the first summation are of the type

$$\int\frac{A\,dx}{(\alpha x+\beta)^k},$$

which give

$$\begin{cases}\dfrac{A}{(1-k)\,\alpha(\alpha x+\beta)^{k-1}} & (k\neq 1)\\[2ex] \dfrac{A}{\alpha}\log|\,\alpha x+\beta\,| & (k=1).\end{cases}$$

The integrals from the second summation are of the type

$$\int \frac{Bx+C}{(ax^2+2hx+b)^p}\,dx,$$

and require more detailed consideration.

Note first that

$$\int \frac{ax+h}{(ax^2+2hx+b)^p}\,dx$$

$$= \begin{cases} \dfrac{1}{2(1-p)(ax^2+2hx+b)^{p-1}} & (p \neq 1) \\ \tfrac{1}{2}\log|ax^2+2hx+b| & (p=1), \end{cases}$$

and so, by writing

$$Bx+C \equiv (B/a)(ax+h)+(aC-hB)/a,$$

we can reduce our problem to the evaluation of integrals such as

$$\int \frac{dx}{(ax^2+2hx+b)^p}.$$

Write $$a(ax^2+2hx+b) \equiv (ax+h)^2+ab-h^2,$$

and make the substitution (remembering that $ab-h^2$ is positive by assumption)

$$ax+h = t\sqrt{(ab-h^2)}.$$

Then $$\int \frac{dx}{(ax^2+2hx+b)^p} = \int \frac{a^{p-1}\,dt\sqrt{(ab-h^2)}}{\{t^2(ab-h^2)+(ab-h^2)\}^p}$$

$$= \frac{a^{p-1}}{(ab-h^2)^{p-\frac{1}{2}}}\int \frac{dt}{(t^2+1)^p}.$$

Our final problem is therefore to evaluate

$$I_p \equiv \int \frac{dt}{(t^2+1)^p},$$

and for this we need a formula of reduction. On integration by parts, we have

$$I_p = \frac{t}{(t^2+1)^p} - \int \frac{t(-2pt)}{(t^2+1)^{p+1}}\,dt$$

$$= \frac{t}{(t^2+1)^p} + 2p\int \frac{(t^2+1)-1}{(t^2+1)^{p+1}}\,dt$$

$$= \frac{t}{(t^2+1)^p} + 2p(I_p - I_{p+1}).$$

Hence $$2pI_{p+1}-(2p-1)I_p=\frac{t}{(t^2+1)^p},$$
or, replacing p by $p-1$,
$$2(p-1)I_p-(2p-3)I_{p-1}=\frac{t}{(t^2+1)^{p-1}}.$$

By applying this formula successively, we make the evaluation of I_p depend on that of $I_{p-1}, I_{p-2}, \ldots$, and, ultimately, on I_1. But
$$I_1=\int\frac{dt}{t^2+1}$$
$$=\tan^{-1}t,$$
and so the whole integration is effected.

3. Integrals involving $\sqrt{(ax+b)}$. Rational functions of x and $\sqrt{(ax+b)}$ may be integrated readily by means of the substitution
$$t=\sqrt{(ax+b)}$$
or
$$x=\frac{1}{a}(t^2-b).$$
The result is the integration of a rational function of t.

Note. The reader is unlikely to remember all the details to be given in § 4 following. The methods should be thoroughly understood, but it may well be found necessary to refer to the book for details in actual examples. If desired, §§ 5, 6 and 7, which will be of more immediate practical value, may be read next.

4. Integrals involving $\sqrt{(ax^2+2hx+b)}$. We first take steps to simplify the quadratic expression under the square root sign.

(i) *Suppose that a is positive.* Then
$$a(ax^2+2hx+b)=(ax+h)^2+ab-h^2.$$
Write
$$ax+h=x';$$
then
$$a(ax^2+2hx+b)=\begin{cases}x'^2+p^2 & (ab>h^2)\\ x'^2-q^2 & (ab<h^2)\end{cases}$$
where $p^2=ab-h^2, q^2=h^2-ab$ respectively.

(ii) *Suppose that a is negative.* Then $-a$ is positive, so we write
$$-a(ax^2+2hx+b)=h^2-ab-(ax+h)^2.$$

If $h^2 - ab$ is negative, the right side is necessarily negative, so that $ax^2 + 2hx + b$ is also negative and (in real algebra) has no square root. We therefore take $h^2 - ab$ to be positive, and write

$$r^2 = h^2 - ab;$$

thus, if $$ax + h = x',$$
as before, we have

$$-a(ax^2 + 2hx + b) = r^2 - x'^2.$$

If, then, we have to evaluate a rational function of x and $\sqrt{(ax^2 + 2hx + b)}$, we may first apply the transformation

$$ax + h = x'.$$

The integrand becomes a rational function of x' and of a surd which may assume one or other of the three forms

$$\text{(i)}\ \sqrt{(x'^2 + p^2)} \quad a > 0,\ ab - h^2 > 0,$$

$$\text{(ii)}\ \sqrt{(x'^2 - q^2)} \quad a > 0,\ ab - h^2 < 0,$$

$$\text{(iii)}\ \sqrt{(r^2 - x'^2)} \quad a < 0,\ ab - h^2 < 0.$$

We may now drop the dashes and treat x as the variable.

(i) *The surd* $\sqrt{(x^2 + p^2)}$.

Consider the transformation

$$x = \frac{2pt}{1 - t^2}.$$

The graph (Fig. 98) indicates (what can also be proved algebraically) that all values of x are obtained by allowing t to vary continuously from -1 to $+1$. We therefore impose the restriction

$$-1 < t < 1$$

on the values of t which we select.

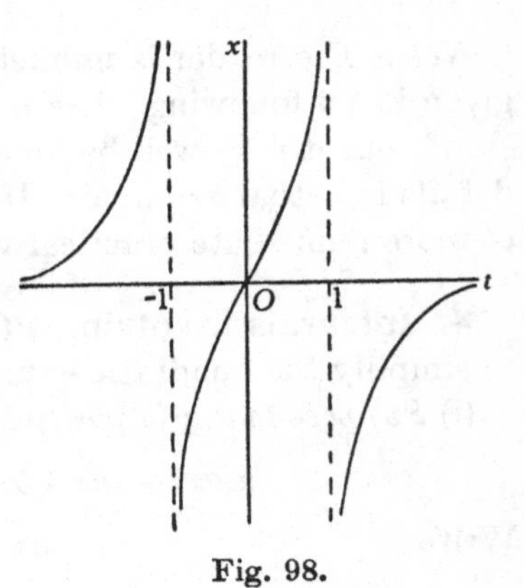

Fig. 98.

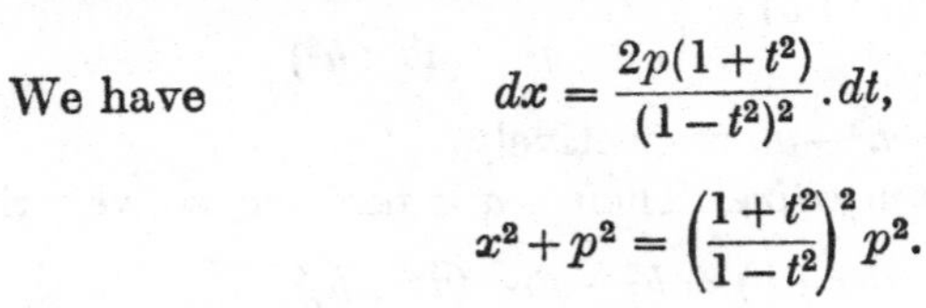

We have $$dx = \frac{2p(1 + t^2)}{(1 - t^2)^2} . dt,$$

$$x^2 + p^2 = \left(\frac{1 + t^2}{1 - t^2}\right)^2 p^2.$$

Since $t^2<1$, we have, without ambiguity (assuming, as we may, that p is positive),

$$\sqrt{(x^2+p^2)} = +\frac{1+t^2}{1-t^2}p.$$

Hence a rational function of x, $\sqrt{(x^2+p^2)}$ is transformed into a rational function of t, and may be integrated accordingly.

(ii) *The surd* $\sqrt{(x^2-q^2)}$.

Consider the transformation

$$x = \frac{t^2+1}{t^2-1}q.$$

The graph (Fig. 99) indicates (what can also be proved algebraically) that all values of x for which $x>q$ are obtained by allowing t to vary continuously from $+1$ to $+\infty$.

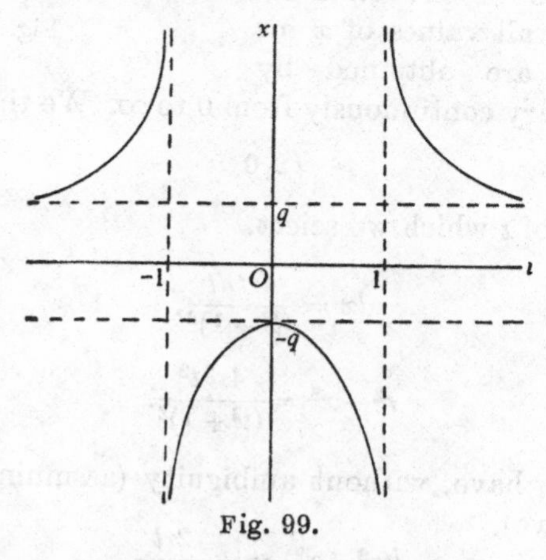

Fig. 99.

[We can restrict ourselves to positive values of x, since a range of integration which involved the two signs for x would have to pass through the region $-q<x<q$ where $\sqrt{(x^2-q^2)}$ is undefined. We could, of course, equally restrict ourselves to negative values of x if necessary.]

We may therefore impose the restriction

$$t>1.$$

Now
$$dx = \frac{-4qt\,dt}{(t^2-1)^2},$$

$$x^2-q^2 = \frac{4t^2q^2}{(t^2-1)^2}.$$

Since $t > 1$, we have, without ambiguity (assuming, as we may, that q is positive),

$$\sqrt{(x^2 - q^2)} = +\frac{2tq}{t^2 - 1}.$$

Hence a rational function of x, $\sqrt{(x^2 - q^2)}$ is transformed into a rational function of t, and may be integrated accordingly.

(iii) *The surd* $\sqrt{(r^2 - x^2)}$.

Consider the transformation

$$x = \frac{t^2 - 1}{t^2 + 1} r.$$

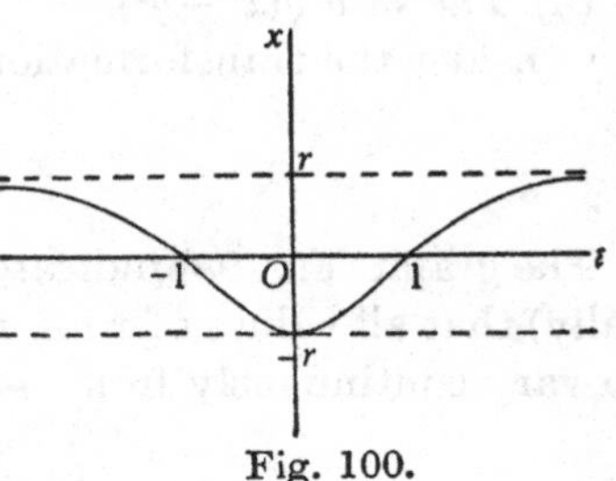

Fig. 100.

The graph (Fig. 100) indicates (what can also be proved algebraically) that all values of x for which $x^2 < r^2$ are obtained by allowing t to vary continuously from 0 to ∞. We therefore impose the restriction

$$t > 0$$

on the values of t which we select.

We have
$$dx = \frac{4rt\,dt}{(t^2 + 1)^2},$$

$$r^2 - x^2 = \frac{4r^2 t^2}{(t^2 + 1)^2}.$$

Since $t > 0$, we have, without ambiguity (assuming, as we may, that r is positive),

$$\sqrt{(r^2 - x^2)} = +\frac{2rt}{t^2 + 1}.$$

Hence a rational function of x, $\sqrt{(r^2 - x^2)}$ is transformed into a rational function of t, and may be integrated accordingly.

Note. The work just completed proves that the integrations are POSSIBLE; it does not necessarily give the easiest method. For example, the quadratic surds may also be subjected to the following substitutions:

(i) For $\sqrt{(x^2 + p^2)}$, let $x = p \tan \theta$, or $x = p \sinh \theta$;

(ii) For $\sqrt{(x^2 - q^2)}$, let $x = q \sec \theta$, or $x = q \cosh \theta$;

(iii) For $\sqrt{(r^2 - x^2)}$, let $x = r \sin \theta$.

5. Integrals involving $\sqrt{(ax^2+2hx+b)}$: alternative treatment.

Integrals of rational functions of x, $\sqrt{(ax^2+2hx+b)}$ may also be treated by methods which, by leaving transformation until the last stages, retain the identity of the surd. We begin by reducing such a rational function to a more amenable form.

Since even powers of $\sqrt{(ax^2+2hx+b)}$ are polynomials and odd powers are polynomials multiplying the square root, we may express any polynomial in x, $\sqrt{(ax^2+2hx+b)}$ in the form

$$P+Q\sqrt{(ax^2+2hx+b)},$$

where P, Q are polynomials in x. Hence any rational function, being by definition the quotient of two polynomials, is

$$\frac{P+Q\sqrt{(ax^2+2hx+b)}}{R+S\sqrt{(ax^2+2hx+b)}},$$

where P, Q, R, S are polynomials in x. Multiply numerator and denominator by $R-S\sqrt{(ax^2+2hx+b)}$, so that the new denominator is the polynomial $R^2-S^2(ax^2+2hx+b)$ and the numerator $PR-QS(ax^2+2hx+b)+(QR-PS)\sqrt{(ax^2+2hx+b)}$, and we obtain the form

$$\frac{A+B\sqrt{(ax^2+2hx+b)}}{C},$$

where A, B, C are polynomials in x. We already know how to deal with the rational function A/C, so our problem reduces to the integration of

$$\frac{B\sqrt{(ax^2+2hx+b)}}{C}.$$

It is found (surprisingly, perhaps) more convenient to have the surd on the denominator, so we multiply numerator and denominator by $\sqrt{(ax^2+2hx+b)}$, obtaining

$$\frac{U}{V\sqrt{(ax^2+2hx+b)}},$$

or

$$\frac{F}{\sqrt{(ax^2+2hx+b)}},$$

where F is a rational function of x.

By § 2 above, the problems to which F gives rise involve as typical terms

$$x^m \quad (m \geqslant 0),$$

$$\frac{1}{(x-k)^n} \quad (n \geqslant 1),$$

$$\frac{Ax+B}{(px^2+2qx+r)^n} \quad (n \geqslant 1,\ q^2 < pr).$$

We have therefore to consider the three types of integral:

$$\text{(i)} \int \frac{x^m\,dx}{\sqrt{(ax^2+2hx+b)}} \quad (m \geqslant 0),$$

$$\text{(ii)} \int \frac{dx}{(x-k)^n \sqrt{(ax^2+2hx+b)}} \quad (n \geqslant 1),$$

$$\text{(iii)} \int \frac{(Ax+B)\,dx}{(px^2+2qx+r)^n \sqrt{(ax^2+2hx+b)}} \quad (n \geqslant 1,\ q^2 < pr).$$

(i) *The evaluation of the integral*

$$I_m \equiv \int \frac{x^m\,dx}{\sqrt{(ax^2+2hx+b)}} \quad (m \geqslant 0).$$

Preparing the ground for an integration by parts, we observe the identity

$$aI_{m+2} + 2hI_{m+1} + bI_m = \int \frac{(ax^2+2hx+b)\,x^m\,dx}{\sqrt{(ax^2+2hx+b)}}$$

$$= \int x^m \sqrt{(ax^2+2hx+b)}\,dx.$$

Now performing the integration, we have, on the right-hand side,

$$\frac{x^{m+1}}{m+1}\sqrt{(ax^2+2hx+b)} - \frac{1}{m+1}\int \frac{x^{m+1}(ax+h)}{\sqrt{(ax^2+2hx+b)}}\,dx$$

$$= \frac{x^{m+1}}{m+1}\sqrt{(ax^2+2hx+b)} - \frac{1}{m+1}(aI_{m+2} + hI_{m+1}).$$

Equating the two sides, we have the *recurrence relation*

$$(m+2)\,aI_{m+2} + (2m+3)\,hI_{m+1} + (m+1)\,bI_m$$
$$= x^{m+1}\sqrt{(ax^2+2hx+b)}.$$

This enables us, once I_0, I_1 have been determined, to calculate $I_2, I_3, I_4, \ldots$ successively, and so to evaluate I_m for any positive integral value of m.

For I_0, we must reduce the surd to one or other of the forms enumerated earlier in this section, giving

$$\int \frac{dx}{\sqrt{(x^2+p^2)}} = \log\{x+\sqrt{(x^2+p^2)}\}$$

or

$$\int \frac{dx}{\sqrt{(x^2-q^2)}} = \log\{x+\sqrt{(x^2-q^2)}\}$$

or

$$\int \frac{dx}{\sqrt{(r^2-x^2)}} = \sin^{-1}(x/r).$$

For I_1, we have to consider the integral

$$I_1 \equiv \int \frac{x\,dx}{\sqrt{(ax^2+2hx+b)}}.$$

Now

$$aI_1 + hI_0 = \int \frac{(ax+h)\,dx}{\sqrt{(ax^2+2hx+b)}} = \sqrt{(ax^2+2hx+b)},$$

so that

$$I_1 = \frac{1}{a}\{\sqrt{(ax^2+2hx+b)} - hI_0\}.$$

(ii) *The evaluation of the integral*

$$\frac{dx}{(x-k)^n\sqrt{(ax^2+2hx+b)}}.$$

We can reduce this type to the form of type (i) and evaluate it at once by the substitution.

$$x-k = \frac{1}{t},$$

$$dx = -\frac{1}{t^2}\,dt.$$

Then

$$ax^2+2hx+b = a\left(k+\frac{1}{t}\right)^2 + 2h\left(k+\frac{1}{t}\right)+b = \frac{1}{t^2}(\alpha t^2+2\beta t+\gamma),$$

where

$$\alpha = ak^2+2hk+b, \quad \beta = ak+h, \quad \gamma = a.$$

The integral is therefore

$$\int \frac{-\dfrac{dt}{t^2}}{\dfrac{1}{t^n}\cdot\dfrac{1}{t}\sqrt{(\alpha t^2+2\beta t+\gamma)}}$$

$$= -\int \frac{t^{n-1}\,dt}{\sqrt{(\alpha t^2+2\beta t+\gamma)}} \quad (n-1 \geqslant 0),$$

which is of the first form

$$\int \frac{x^m\,dx}{\sqrt{(ax^2+2hx+b)}} \quad (m \geqslant 0).$$

(iii) *The evaluation of the integral*

$$I_n \equiv \int \frac{(Ax+B)\,dx}{(px^2+2qx+r)^n\sqrt{(ax^2+2hx+b)}},$$

where $n \geqslant 1$, $q^2 < pr$.

The general case is very difficult. It is, of course, possible to avoid it by expressing the rational function as a sum of *complex* partial fractions of the type $1/(x-k)^n$, where k is complex. This reduces the integration to type (ii)

$$\int \frac{dx}{(x-k)^n\sqrt{(ax^2+2hx+b)}} \quad (k \text{ complex})$$

already discussed. The final return to real form is an added point of difficulty.

We begin with an algebraic lemma, designed to reduce two quadratic expressions *simultaneously* to simpler form.

LEMMA. *To establish the existence of a (real) transformation of the type*

$$x = \frac{\alpha t+\beta}{t+1}$$

which reduces the quadratic expressions

$$ax^2+2hx+b,$$

$$px^2+2qx+r$$

to the forms

$$\frac{ut^2+v}{(t+1)^2}$$

$$\frac{u't^2+v'}{(t+1)^2}$$

simultaneously.

Since $(t+1)^2(ax^2+2hx+b)$

$$= a(\alpha t+\beta)^2+2h(\alpha t+\beta)(1+t)+b(1+t)^2$$

under the transformation, we see that the coefficient of t vanishes on the right-hand side if α, β are chosen so that

$$a\alpha\beta+h(\alpha+\beta)+b = 0.$$

Similarly the coefficient of t from $px^2+2qx+r$ vanishes if

$$p\alpha\beta+q(\alpha+\beta)+r = 0.$$

Also α, β necessarily satisfy the equation in θ

$$\alpha\beta-\theta(\alpha+\beta)+\theta^2 = 0.$$

On eliminating the ratios $\alpha\beta:\alpha+\beta:1$ between these three relations, we obtain the equation

$$\begin{vmatrix} a & h & b \\ p & q & r \\ 1 & -\theta & \theta^2 \end{vmatrix} = 0,$$

which is a quadratic in θ with (by definition) roots α, β. Hence α, β may be found and the transformation determined. (If, exceptionally, $a/h = p/q$, the two quadratics differ only by a constant. The substitution $ax+h = at$ reduces the integral I_n to one or other of the types discussed on p. 213.)

Moreover α, β are *real*. If not, they must be conjugate complex numbers, since all coefficients are real. Suppose that

$$\alpha = \lambda+i\mu, \quad \beta = \lambda-i\mu \quad (\lambda, \mu \text{ real}).$$

Since

$$p\alpha\beta+q(\alpha+\beta)+r = 0,$$

we have the relation

$$p(\lambda^2+\mu^2)+2q\lambda+r = 0.$$

Multiply by p (which is not zero, by the nature of the problem). Then

$$p^2\lambda^2+2pq\lambda+pr+p^2\mu^2 = 0,$$

or

$$(p\lambda+q)^2+(pr-q^2)+(p\mu)^2 = 0.$$

Since p, q, λ, μ are all real, the two squares are positive; and, by hypothesis, $pr-q^2 > 0$. Hence the left-hand side is positive. We are therefore led to a contradiction, and so the supposition that α, β are not real is untenable.

Moreover, the condition $pr-q^2 > 0$ shows that α, β are DIFFERENT numbers.

SUMMARY. We can find two distinct real numbers, α, β, the roots of the quadratic equation

$$\begin{vmatrix} a & h & b \\ p & q & r \\ 1 & -\theta & \theta^2 \end{vmatrix} = 0,$$

which enable us, by means of the transformation

$$x = \frac{\alpha t + \beta}{t+1},$$

to reduce the two quadratic forms

$$ax^2 + 2hx + b = 0,$$

$$px^2 + 2qx + r = 0 \quad (q^2 < pr)$$

to the forms

$$\frac{ut^2 + v}{(t+1)^2},$$

$$\frac{u't^2 + v'}{(t+1)^2}$$

simultaneously.

Let us now return to the integral. On substituting for x in the expression $Ax + B$, we obtain an expression of the form

$$\frac{Ct + D}{t+1},$$

where $C = A\alpha + B$, $D = A\beta + B$; also

$$dx = \frac{\alpha - \beta}{(t+1)^2}\, dt.$$

Hence
$$I_n = \int \frac{\dfrac{Ct+D}{t+1} \cdot \dfrac{\alpha - \beta}{(t+1)^2} dt}{\dfrac{(u't^2 + v')^n}{(t+1)^{2n}} \cdot \dfrac{\sqrt{(ut^2 + v)}}{t+1}}$$

$$= \int \frac{(\alpha - \beta)(Ct + D)(t+1)^{2n-2}\, dt}{(u't^2 + v')^n \sqrt{(ut^2 + v)}}.$$

The numerator, which is a polynomial in t, can be expressed in the form $P + Qt$, where P, Q are polynomials in t^2; and so, writing $u't^2 + v' = s$, or

$$t^2 = \frac{s - v'}{u'},$$

we obtain P and Q as polynomials in s. Hence the numerator can be expressed as a polynomial in $s \equiv (u't^2+v')$, together with t times another polynomial in $(u't^2+v')$, the order of the whole numerator being $2n-1$ in t, which is less than that of the denominator. Hence I_n is found as a sum of integrals of the form

$$U_k \equiv \int \frac{dt}{(u't^2+v')^k \sqrt{(ut^2+v)}}, \quad V_k \equiv \int \frac{t\,dt}{(u't^2+v')^k \sqrt{(ut^2+v)}}.$$

For the latter, put
$$y^2 = ut^2+v,$$
so that
$$y\,dy = ut\,dt;$$
hence
$$V_k = \frac{1}{u}\int \frac{y\,dy}{\left\{\frac{u'}{u}(y^2-v)+v'\right\}^k y}$$
$$= u^{k-1}\int \frac{dy}{\{u'y^2+(uv'-u'v)\}^k},$$
which reduces the problem to the integration of a rational function.

Finally, consider
$$U_k \equiv \int \frac{dt}{(u't^2+v')^k \sqrt{(ut^2+v)}}.$$

Write
$$u't^2+v' = \frac{1}{z},$$
$$2u't\,dt = -\frac{1}{z^2}dz,$$
so that
$$t = \sqrt{\left(\frac{1-v'z}{u'z}\right)},$$
$$ut^2+v = \frac{u-(uv'-u'v)z}{u'z}.$$

Then
$$U_k \equiv \int \left\{-\frac{dz}{2u'z^2}\sqrt{\left(\frac{u'z}{1-v'z}\right)}\right\}.z^k.\sqrt{\left\{\frac{u'z}{u-(uv'-u'v)z}\right\}}$$
$$= -\frac{1}{2}\int \frac{z^{k-1}\,dz}{\sqrt{[(1-v'z)\{u-(uv'-u'v)z\}]}}.$$

But the integral is of the form
$$\int \frac{z^{k-1}\,dz}{\sqrt{(a+2hz+bz^2)}},$$
whose evaluation we considered in the preceding section.

The whole integration may therefore be effected.

6. Trigonometric functions. Since

$$\cos x = \frac{1-\tan^2 \frac{1}{2}x}{1+\tan^2 \frac{1}{2}x}, \quad \sin x = \frac{2\tan \frac{1}{2}x}{1+\tan^2 \frac{1}{2}x},$$

trigonometric integrals may be reduced by means of the substitution

$$t = \tan \tfrac{1}{2}x.$$

Then

$$\cos x = \frac{1-t^2}{1+t^2}, \quad \sin x = \frac{2t}{1+t^2},$$

$$dx = \frac{2dt}{1+t^2}.$$

Hence a rational function of $\sin x$ and $\cos x$ is transformed into a rational function of t, and may be integrated by the methods of § 2.

For reduction formulæ for

$$\int \sin^m x \cos^n x \, dx$$

and similar integrals, see Vol. I, pp. 106–10.

The substitution $t = \tan \frac{1}{2}x$ should not be applied blindly. For example, if the integrand has period π, the alternative

$$t = \tan x$$

may be better.

Thus, consider the integral

$$I \equiv \int \frac{dx}{(\sec x + \cos x)^2}.$$

Since

$$\frac{1}{\{\sec(x+\pi) + \cos(x+\pi)\}^2} = \frac{1}{(\sec x + \cos x)^2},$$

we may use the substitution

$$t = \tan x.$$

Now

$$I = \int \frac{\sec^2 x \, dx}{(\sec^2 x + 1)^2}$$

$$= \int \frac{dt}{(t^2+2)^2}.$$

This is the integral of a rational function, and may be evaluated according to the usual rules. Alternatively, the substitution $t = \sqrt{2}\tan\theta$ makes the integral trigonometric again, but leads to an easy solution.

7. Exponential times polynomial in x. A polynomial is a sum of multiples of powers of x, and so, in order to evaluate the integral

$$\int e^{px} f(x)\, dx,$$

where $f(x)$ is a polynomial, we need only consider integrals of the type

$$u_n \equiv \int e^{px} x^n\, dx \quad (n \geqslant 0).$$

Integrating by parts, we have the relation

$$u_n = \frac{1}{p} e^{px} x^n - \frac{n}{p} \int e^{px} x^{n-1}\, dx,$$

leading to the recurrence formula

$$u_n = \frac{1}{p} e^{px} x^n - \frac{n}{p} u_{n-1} \quad (n \geqslant 1)$$

which enables us to express u_n in terms of

$$u_0 \equiv \int e^{px}\, dx = \frac{1}{p} e^{px}.$$

8. Exponential times polynomial in sines and cosines of multiples of x. Consider the integration of a sum of terms of the type

$$e^{px} \sin a_1 x \sin a_2 x \ldots \sin a_m x \cos b_1 x \cos b_2 x \ldots \cos b_n x,$$

involving $m+n$ factors in sines and cosines. The use of the formulæ such as

$$2 \sin ux \cos vx = \sin (u+v) x + \sin (u-v) x,$$

$$2 \sin ux \sin vx = \cos (u-v) x - \cos (u+v) x,$$

and so on, enables us to reduce the number of terms in a typical product to $m+n-1, m+n-2, m+n-3, \ldots$ successively, while retaining the same type of expression. Ultimately we reach one or other of the forms

$$C \equiv \int e^{px} \cos qx dx,$$

$$S \equiv \int e^{px} \sin qx dx,$$

whose solution should now be familiar. We find

$$C = \frac{1}{p^2+q^2} e^{px}(p\cos qx + q\sin qx),$$

$$S = \frac{1}{p^2+q^2} e^{px}(p\sin qx - q\cos qx).$$

WARNING. The work of this chapter will enable the reader to evaluate (ultimately) any integral that comes within its scope. But there are many functions which cannot be integrated in terms yet known to him. Some of these are very innocent to look at; for example,

$$\int \frac{\sin x}{x}\,dx, \quad \int e^{x^2}dx,$$

$$\int \frac{dx}{\sqrt{\{(1-x^2)(1-k^2x^2)\}}}.$$

Thorough familiarity with the FORMS of the integrals that can be evaluated should help in the avoidance of these pitfalls.

[For examples on the work of this chapter, see Revision Examples VIII and IX, pp. 226, 227.]

CHAPTER XIII

INTEGRALS INVOLVING 'INFINITY'

1. 'Infinite' limits of integration. It is often necessary to evaluate an integral such as

$$\int_a^b f(x)\,dx$$

under conditions where b tends to infinity; we then speak about 'the integral from a to infinity'

$$\int_a^\infty f(x)\,dx.$$

Similarly we meet the integral

$$\int_{-\infty}^b f(x)\,dx,$$

or, combining both possibilities,

$$\int_{-\infty}^\infty f(x)\,dx.$$

Our problem is to discuss what is meant by such integrals, and to show how to evaluate them. The general theory is difficult, so we confine ourselves to the simplest cases. We also assume that the function $f(x)$ is continuous throughout the range of integration.

We define the *integral to infinity*

$$\int_a^\infty f(x)\,dx$$

by means of the relation

$$\int_a^\infty f(x)\,dx = \lim_{N\to\infty}\int_a^N f(x)\,dx.$$

In the cases with which we shall be concerned, the indefinite integral $F(x)$ of $f(x)$ is considered to be known, so that

$$\int_a^N f(x)\,dx = F(N) - F(a).$$

We can therefore evaluate the integral to infinity in the form

$$\int_a^\infty f(x)\,dx = \lim_{N\to\infty} F(N) - F(a).$$

It may be added that it is often possible to evaluate the definite integral $\int_a^\infty f(x)\,dx$ even when the indefinite integral $\int f(x)\,dx$ cannot be found. A well-known example is $\int_0^\infty \frac{\sin x}{x}\,dx$ whose value is $\frac{1}{2}\pi$.

ILLUSTRATION 1. *To evaluate*

$$\int_1^\infty x^n\,dx.$$

Consider the integral

$$\int_1^N x^n\,dx \quad (n \neq -1)$$

which, by elementary theory, is

$$\left[\frac{1}{n+1}x^{n+1}\right]_1^N$$

or

$$\frac{N^{n+1}}{n+1} - \frac{1}{n+1}.$$

If $n+1$ is POSITIVE, then N^{n+1} increases indefinitely with N, so that

$$\lim_{N\to\infty}\left\{\frac{N^{n+1}}{n+1} - \frac{1}{n+1}\right\}$$

does not exist.

If $n+1$ is NEGATIVE, then N^{n+1} tends to zero as N increases indefinitely, so that

$$\lim_{N\to\infty}\left\{\frac{N^{n+1}}{n+1} - \frac{1}{n+1}\right\} = -\frac{1}{n+1}.$$

Hence

$$\int_1^\infty x^n\,dx = -\frac{1}{n+1} \quad (n < -1).$$

ILLUSTRATION 2. *To evaluate*

$$\int_0^\infty xe^{-x}\,dx.$$

We know that, on integrating by parts,

$$\int xe^{-x}\,dx = -xe^{-x} + \int 1.e^{-x}\,dx$$
$$= -xe^{-x} - e^{-x},$$

so that

$$\int_0^N xe^{-x}\,dx = \Big[-(x+1)e^{-x}\Big]_0^N$$
$$= -(N+1)e^{-N} + 1.$$

Now

$$(N+1)e^{-N} = \frac{N+1}{e^N} = \frac{N+1}{1+N+\dfrac{N^2}{2!}+\dots},$$

and, for positive values of N, the denominator is certainly greater than $\frac{1}{2}N^2$, so that

$$(N+1)\,e^{-N} < \frac{2(N+1)}{N^2}.$$

Thus

$$(N+1)e^{-N} \to 0$$

as N increases indefinitely. Hence

$$\int_0^\infty xe^{-x}\,dx = \lim_{N\to\infty}\{-(N+1)e^{-N}+1\}$$
$$= 0+1$$
$$= 1.$$

ILLUSTRATION 3. *To evaluate*

$$\int_{-\infty}^{\infty} \frac{dx}{1+x^2}.$$

Consider the integral

$$\int_{-M}^{N} \frac{dx}{1+x^2},$$

where M, N are large positive numbers. The value of this integral is

$$\Big[\tan^{-1} x\Big]_{-M}^{N}$$

or $\tan^{-1} N - \tan^{-1}(-M)$.

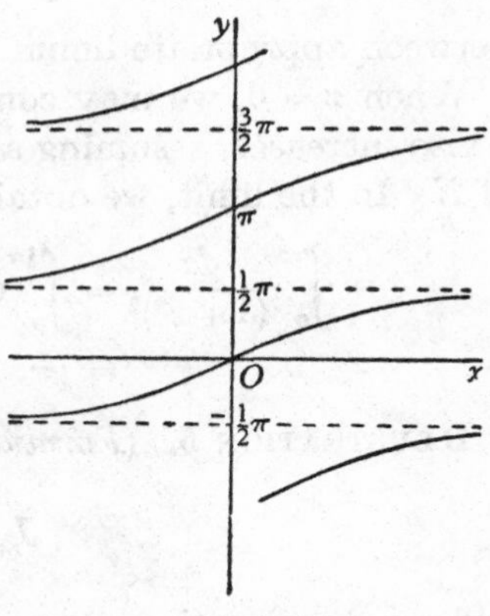

Fig. 101.

Care must be exercised in selecting the correct angles from the many-valued inverse tangents. The graph

$$y = \tan^{-1} x,$$

shown in the diagram (Fig. 101), illustrates the 'parallel' curves along which y must run, and the values selected must be confined to one of them. The most natural choice is to work with the curve through the origin O. Then for large negative values of x, the value $\tan^{-1}(-M)$ just exceeds $-\frac{1}{2}\pi$; as x increases, $\tan^{-1} x$ rises continuously through, say, $-\frac{1}{3}\pi, -\frac{1}{4}\pi, -\frac{1}{6}\pi, 0, \frac{1}{3}\pi, \frac{1}{6}\pi$ and so on, approaching the value $\frac{1}{2}\pi$ for the limit of $\tan^{-1} N$.

Hence
$$\begin{aligned}\int_{-\infty}^{\infty} \frac{dx}{1+x^2} &= \lim_{N\to\infty} \{\tan^{-1} N\} - \lim_{M\to\infty} \{\tan^{-1}(-M)\} \\ &= \tfrac{1}{2}\pi - (-\tfrac{1}{2}\pi) \\ &= \pi.\end{aligned}$$

ILLUSTRATION 4. (*Change of variable.*) *To evaluate*

$$\int_0^{\infty} \frac{dx}{(1+x^2)^2}.$$

Consider the integral
$$u_N \equiv \int_0^N \frac{dx}{(1+x^2)^2}.$$

Make the substitution
$$x = \tan\theta,$$
$$dx = \sec^2\theta\, d\theta.$$

Then
$$u_N = \int \frac{\sec^2\theta\, d\theta}{\sec^4\theta} = \int \cos^2\theta\, d\theta$$

between appropriate limits.

When $x = 0$, we may conveniently take $\theta = 0$. As x increases, θ also increases, assuming a value very near to $\frac{1}{2}\pi$ for large values of N. In the limit, we obtain the value $\frac{1}{2}\pi$. Hence

$$\begin{aligned}\int_0^{\infty} \frac{dx}{(1+x^2)^2} &= \int_0^{\frac{1}{2}\pi} \cos^2\theta\, d\theta = \frac{1}{2}\int_0^{\frac{1}{2}\pi} (1+\cos 2\theta)\, d\theta \\ &= \tfrac{1}{4}\pi.\end{aligned}$$

ILLUSTRATION 5. (*Formula of reduction.*) *To evaluate*

$$J_m \equiv \int_0^{\infty} x^m e^{-x}\, dx,$$

where m is a positive integer, greater than zero.

Let
$$I_m = \int_0^N x^m e^{-x}\,dx$$
$$= \Big[-x^m e^{-x}\Big]_0^N + m\int_0^N x^{m-1} e^{-x}\,dx.$$

Now
$$N^m e^{-N} = \frac{N^m}{e^N} = \frac{N^m}{1+N+\frac{N^2}{2!}+\frac{N^3}{3!}+\dots};$$

and the denominator, being certainly greater than
$$N^{m+1}/(m+1)!,$$
greatly exceeds the numerator for large values of N, whatever m may be, so that
$$\lim_{N\to\infty} N^m e^{-N} = 0.$$

Also
$$x^m e^{-x} = 0$$
when $x = 0$, since $m > 0$.

Hence
$$\lim_{N\to\infty}\Big[-x^m e^{-x}\Big]_0^N = 0,$$
and so, as $N\to\infty$,
$$J_m = mJ_{m-1}.$$

The formula of reduction enables us to make the value of J_m depend on that of J_0; for

$$J_m = mJ_{m-1}$$
$$= m(m-1)J_{m-2}$$
$$\dots\dots\dots\dots\dots\dots$$
$$= m(m-1)\dots 2.1J_0$$
$$= m!\,J_0.$$

Moreover,
$$J_0 = \int_0^\infty e^{-x}\,dx$$
$$= \lim_{N\to\infty}\Big[-e^{-x}\Big]_0^N$$
$$= \lim_{N\to\infty}\Big[-e^{-N}+1\Big]$$
$$= 1.$$

Hence
$$\int_0^\infty x^m e^{-x}\,dx = m!$$

EXAMPLES I

Evaluate:

1. $\displaystyle\int_1^\infty \frac{dx}{x^2}$.
2. $\displaystyle\int_2^\infty \frac{dx}{x\sqrt{x}}$.
3. $\displaystyle\int_0^\infty x^2 e^{-x}\,dx$.
4. $\displaystyle\int_1^\infty \frac{dx}{1+x^2}$.
5. $\displaystyle\int_{-\infty}^1 \frac{dx}{1+x^2}$.
6. $\displaystyle\int_0^\infty e^{-x}\sin x\,dx$.
7. $\displaystyle\int_0^\infty e^{ax}$, stating for what values of a the integration is possible.

2. 'Infinite' integrand. It may happen that, in evaluating the function
$$\int_a^b f(x)\,dx,$$

we find that the integrand $f(x)$ tends to infinity—or, indeed, has some other discontinuity—in the range of integration. Suppose, for example, that $f(x)$ increases without bound near $x = c$. If c is inside the interval (that is, not equal to a or b) we 'cut it out' by considering the sum
$$\int_a^{c-\epsilon} f(x)\,dx + \int_{c+\eta}^b f(x)\,dx,$$

where ϵ, η are small positive constants, over ranges which just miss c on either side. *We then define the value of the integral to be*
$$\lim_{\epsilon\to 0}\int_a^{c-\epsilon} f(x)\,dx + \lim_{\eta\to 0}\int_{c+\eta}^b f(x)\,dx,$$

supposing that these limits exist.

When c is at a or b the modification is obvious; only one limiting value is then required.

ILLUSTRATION 6. *To determine the values of n for which the integral*
$$\int_0^1 \frac{dx}{x^n} \quad (n>0)$$

exists.

Consider the integral

$$\int_\epsilon^1 \frac{dx}{x^n}$$

$$= \left[\frac{1}{1-n}x^{1-n}\right]_\epsilon^1$$

$$= \frac{1}{1-n}\{1-\epsilon^{1-n}\}.$$

As ϵ tends to zero, the term ϵ^{1-n} also tends to zero if the exponent $1-n$ is positive; otherwise ϵ^{1-n} increases indefinitely. Hence the integral exists provided that

$$n < 1,$$

and its value is then $\quad \dfrac{1}{1-n}\{1-0\}$

$$= \frac{1}{1-n}.$$

The case $n = 1$ requires separate treatment. We then have

$$\int_\epsilon^1 \frac{dx}{x} = \Big[\log x\Big]_\epsilon^1 = \log 1 - \log \epsilon,$$

so that (compare p. 4) the integral does not exist.

The required condition is therefore

$$n < 1.$$

ILLUSTRATION 7. *To evaluate*

$$\int_1^\infty \frac{dx}{(x+3)\sqrt{(x-1)}}.$$

This integral involves two infinities; the upper limit of integration is infinite, and the integrand is infinite when $x = 1$. The two phenomena must be kept separate.

We consider the integral

$$I \equiv \int_{1+\epsilon}^N \frac{dx}{(x+3)\sqrt{(x-1)}}$$

Write $$x-1 = t^2,$$
$$dx = 2t\,dt.$$

Then $$I = \int \frac{2t\,dt}{(t^2+4)t}$$

between appropriate limits.

Now we have cut out the value $x = 1, t = 0$ by making the lower limit $1+\epsilon$. Thus t is never zero, so that the factor t may be cancelled from numerator and denominator of the integrand. Hence

$$I = \int \frac{2\,dt}{t^2+4}$$
$$= \left[\tan^{-1}\left(\frac{t}{2}\right)\right]$$

between appropriate limits.

Consider the range of variation of t as x increases continuously from $1+\epsilon$ to N. We have

$$t = +\sqrt{(x-1)},$$

having committed ourselves to the positive square root by putting $\sqrt{(x-1)} = t$ in the integrand during the substitution. Hence t increases continuously from the small value $\sqrt{\epsilon}$ to the large value $\sqrt{(N-1)}$; and as it does so, $\tan^{-1}(\frac{1}{2}t)$ increases continuously from just above zero to just short of $\frac{1}{2}\pi$. Thus, in the limit,

$$I = \tfrac{1}{2}\pi - 0$$
$$= \tfrac{1}{2}\pi.$$

ILLUSTRATION 8. *To evaluate*

$$\int_a^b \frac{dx}{\sqrt{\{(x-a)(b-x)\}}} \quad (0<a<b).$$

The denominator in the integrand becomes zero at $x = a$ and at $x = b$, and so we must consider the integral

$$I \equiv \int_{a+\epsilon}^{b-\eta} \frac{dx}{\sqrt{\{(x-a)(b-x)\}}}.$$

Make the substitution

$$x = a\cos^2 t + b\sin^2 t,$$
$$dx = 2(b-a)\sin t\cos t\,dt.$$

Then

$$x - a = (b-a)\sin^2 t,$$
$$b - x = (b-a)\cos^2 t.$$

Consider the range of integration in the variable t. When $x = a$, we have $a\sin^2 t = b\sin^2 t$, so that $\sin t = 0$; and when $x = b$, we have $b\cos^2 t = a\cos^2 t$, so that $\cos t = 0$. We must select as starting-point for t a value for which $\sin t = 0$ (corresponding to $x = a$), and the obvious value is $t = 0$. The relation

$$\frac{dx}{dt} = 2(b-a)\sin t\cos t$$

shows that, at any rate for small values of t, the variables x, t increase together; and this process continues until $t = \frac{1}{2}\pi$, at which point x has the value b. The range is therefore $0, \frac{1}{2}\pi$. But we have had to exclude these points themselves because of trouble with the integrand, and so the range must run from just above zero to just short of $\frac{1}{2}\pi$; say from ϵ' to $\frac{1}{2}\pi - \eta'$. Thus

$$I = \int_{\epsilon'}^{\frac{1}{2}\pi-\eta'} \frac{2(b-a)\sin t\cos t\,dt}{\sqrt{\{(b-a)^2\sin^2 t\cos^2 t\}}}.$$

Moreover the POSITIVE value of $\sqrt{\{(b-a)^2\sin^2 t\cos^2 t\}}$ in the interval $0, \frac{1}{2}\pi$ is $(b-a)\sin t\cos t$, so that

$$I = \int_{\epsilon'}^{\frac{1}{2}\pi-\eta'} \frac{2(b-a)\sin t\cos t\,dt}{(b-a)\sin t\cos t}.$$

We have excluded the end-points, so that $\sin t$ and $\cos t$ are not zero in the range of integration; we may therefore cancel factors in the numerator and denominator of the integrand, giving

$$I = \int_{\epsilon'}^{\frac{1}{2}\pi-\eta'} 2\,dt$$
$$= 2\{(\tfrac{1}{2}\pi - \eta') - \epsilon'\}.$$

Proceeding to the limit as ϵ', η' tend to zero independently, we have

$$I = 2(\tfrac{1}{2}\pi)$$
$$= \pi.$$

REVISION EXAMPLES VIII

'Advanced' Level

1. Find the following integrals:

$$\int \frac{dx}{(x-1)(x-3)}, \quad \int \frac{x^2\,dx}{(x-1)(x-3)}, \quad \int \frac{dx}{\sqrt{\{(x-1)(3-x)\}}}.$$

2. Integrate $\dfrac{(x^2+1)}{x^2-4x+3}$, $\dfrac{x}{\sqrt{(x^2+2x+2)}}$.

Show that $\displaystyle\int_0^{\frac{1}{2}\pi} \frac{dx}{a+b\sin x} = \frac{1}{\sqrt{(a^2-b^2)}}\cos^{-1}\left(\frac{b}{a}\right)$ $(b<a)$.

3. Evaluate the integral

$$\int_a^b \frac{x\,dx}{\sqrt{\{(x-a)(b-x)\}}}$$

by means of the substitution $x = a\cos^2\theta + b\sin^2\theta$.

Prove that the integrals

$$\int_0^1 x^7(1-x)^8\,dx, \quad \int_0^1 x^8(1-x)^7\,dx$$

are equal, and show that their common value is

$$\frac{7!\,8!}{16!}.$$

4. Integrate

$$\frac{x^3}{(x^2+1)(x-2)}, \quad (x+a)(x+b)^{\frac{1}{2}}, \quad \sin^5 x.$$

5. Integrate

$$\frac{x}{(1+x^2)(1-x)}, \quad \frac{x}{\sqrt{(x^2+4x+5)}}, \quad \frac{x^5}{(a^2+x^2)^2}.$$

Prove that, when a, b are positive,

$$\int_0^\pi \frac{\cos^2 x\,dx}{a^2\cos^2 x + b^2\sin^2 x} = \frac{\pi}{a(a+b)}.$$

6. Integrate

$$\frac{(2x+3)}{x^3+x^2-2x}, \quad \frac{1}{x\sqrt{(x^2+2x-1)}}, \quad \frac{xe^x}{(x+1)^2}.$$

7. Integrate

$$\frac{(x+1)}{x^3-x^2-6x}, \quad \sqrt{(x^2+x+1)}, \quad (\log x)^2.$$

8. Find a reduction formula for

$$\int_0^1 (1+x^2)^{n+\frac{1}{2}}\,dx,$$

and evaluate the integral when $n=2$.

9. By integration by parts, show that, if $0<m<n$, and

$$I=\int_0^1 x^m \frac{d^n}{dx^n}\{x^n(1-x)^n\}\,dx,$$

then
$$I=-m\int_0^1 x^{m-1}\frac{d^{n-1}}{dx^{n-1}}\{x^n(1-x)^n\}\,dx.$$

Deduce that
$$I=0.$$

10. Taking
$$y=\frac{\sqrt{(2x^2+6x+5)}}{x+2},$$

verify that the result of changing the variable from y to x in the integral

$$\int\frac{dy}{\sqrt{(y^2-1)}}$$

is
$$\int\frac{dx}{(x+2)\sqrt{(2x^2+6x+5)}}.$$

Deduce that
$$\int_{-1}^{2}\frac{dx}{(x+2)\sqrt{(2x^2+6x+5)}}=\log_e 2.$$

REVISION EXAMPLES IX

'Scholarship' Level

1. Show that, if
$$u_n=\int_0^\pi \frac{\cos nx\,dx}{5-4\cos x},$$

then, provided that $n \neq 1$,

$$2u_n-5u_{n-1}+2u_{n-2}=0.$$

Hence, or otherwise, show that, if n is a positive integer,

$$u_n=\pi/(3.2^n).$$

2. Prove that

$$\int_0^{\frac{1}{2}\pi} \frac{d\theta}{a^2\cos^2\theta + b^2\sin^2\theta} = \frac{\pi}{2ab}, \quad \int_0^{\frac{1}{2}\pi} \log\tan\theta\, d\theta = 0,$$

$$\int_0^{\frac{1}{2}\pi} \sin^2\theta \log\tan\theta\, d\theta = \tfrac{1}{4}\pi.$$

3. (i) Prove that values of a, b can be chosen so that the substitution

$$x = \frac{at+b}{t+1}$$

yields the result

$$\int \frac{dx}{(2x^2-6x+5)\sqrt{(5x^2-12x+8)}} = -\int \frac{(t+1)\,dt}{(t^2+1)\sqrt{(t^2+4)}}.$$

(ii) Find the indefinite integral of

$$\frac{t+1}{(t^2+1)\sqrt{(t^2+4)}}.$$

4. Evaluate $\displaystyle\int_0^{\frac{1}{2}} (\sin^{-1}x)^2\,dx, \quad \int_1^2 \sqrt{\left(\frac{x-1}{x+1}\right)}\frac{dx}{x}.$

5. Prove that

$$\int_0^{\frac{1}{2}\pi} \sqrt{(\cos 2x - \cos 4x)}\,dx = \tfrac{1}{2}\sqrt{6} - \tfrac{1}{4}\sqrt{2}\log(2+\sqrt{3}).$$

6. Evaluate $\displaystyle\int_0^1 \frac{dx}{x+\sqrt{(1-x)}}, \quad \int_{-1}^1 \frac{x^4\,dx}{\sqrt{(1-x^2)}}.$

7. Evaluate

$$\int_0^\infty \frac{x\tan^{-1}x}{(1+x^2)^2}\,dx, \quad \int_0^{\frac{1}{2}\pi} \frac{\cos^2\theta\,d\theta}{a^2\cos^2\theta + b^2\sin^2\theta}.$$

Find

$$\int \frac{dx}{(1-x)\sqrt{(1+x)}}.$$

8. Evaluate

$$\int_0^\infty \frac{x\,dx}{x^3+x^2+x+1}.$$

Find

$$\int \frac{dx}{(x^2-1)^{\frac{3}{2}}}, \quad \int x^3\sin^2 x\,dx.$$

9. If

$$I_n \equiv \int_0^{2\pi} \frac{\cos(n-1)x - \cos nx}{1-\cos x}\,dx,$$

show that I_n is independent of n, where n is a positive integer.

Hence evaluate I_n and prove that

$$\int_0^{2\pi} \left(\frac{\sin \frac{1}{2}nx}{\sin \frac{1}{2}x}\right)^2 dx = 2n\pi.$$

10. Evaluate

$$\int_0^a \frac{x\,dx}{x+\sqrt{(a^2-x^2)}}, \quad \int_0^{\frac{1}{2}\pi} \frac{3+2\cos x}{(3+\cos x)^2}\,dx.$$

11. Evaluate the integrals:

$$\int_0^1 \frac{\sin^{-1} x}{(1+x)^2}\,dx, \quad \int_0^{\frac{3}{4}\pi} \frac{dx}{2\cos^2 x + 2\cos x \sin x + \sin^2 x},$$

$$\int_{-\infty}^{\infty} \frac{dx}{(e^{x-a}+1)(1+e^{-x})}.$$

12. Evaluate $$\int_2^{\infty} \frac{x^2\,dx}{(1-x^2)(1+x^2)^2}.$$

13. Evaluate $$\int_0^{\infty} \frac{dx}{(x^2+a^2)(x^2+b^2)(x^2+c^2)},$$

where a, b, c are positive.

14. Prove that

$$\int_0^{\infty} \frac{dx}{x^2+2x\cos\alpha+1} = \frac{\alpha}{\sin\alpha} \quad (0<\alpha<\pi).$$

Evaluate

$$\int_0^{\infty} \frac{dx}{x^4+2x^2\cos\alpha+1}, \quad \int_0^{\infty} \frac{(x^2+1)\,dx}{x^4+2x^2\cos\alpha+1}.$$

15. Prove that, when $b>a>0$,

$$\int_0^{\pi} \frac{\sin\theta\,d\theta}{\sqrt{(a^2+b^2-2ab\cos\theta)}} = \frac{2}{b},$$

$$\int_0^{\pi} \frac{\sin\theta\cos\theta\,d\theta}{\sqrt{(a^2+b^2-2ab\cos\theta)}} = \frac{2a}{3b^2}.$$

Make it clear at what points of your proofs you use the condition $b>a>0$.

16. Evaluate the definite integrals:

$$\int_{\sqrt{2}}^{\infty} \frac{dx}{(x^2-1)\sqrt{(1+x^2)}}, \quad \int_0^1 (1+x^2)\tan^{-1} x\,dx.$$

17. Evaluate

$$I = \int_{-\delta}^{\delta} \frac{1 - r\cos\theta}{1 - 2r\cos\theta + r^2}\, d\theta \quad (0 < \delta < \pi)$$

when (i) $0 < r < 1$, (ii) $r > 1$.

Prove that, δ being fixed, I tends to one limit as $r \to 1$ through values less than 1, and to a different limit as $r \to 1$ through values greater than 1.

Show, also, that neither limit is equal to the value of I when

$$r = 1.$$

18. Show that, by proper choice of a new variable, the integration of any rational function of $\sin x$ and $\cos x$ can be reduced to the integration of a rational algebraic function of that variable.

Integrate

$$\frac{\sin x}{\sin(x-\alpha)}, \quad \frac{1}{\sin x \cos x + \sin x + \cos x - 1}.$$

19. Integrate $$\frac{(x^2-1)}{x\sqrt{(x^4+x^2+1)}}.$$

Prove that

$$\int_1^{\infty} \frac{dx}{(x+\cos\alpha)\sqrt{(x^2-1)}} = \frac{\alpha}{\sin\alpha} \quad (0 < \alpha < \pi),$$

$$\int_0^{\pi} \frac{x\,dx}{1+\cos\alpha\sin x} = \frac{\pi\alpha}{\sin\alpha} \quad (0 < \alpha < \tfrac{1}{2}\pi).$$

20. Integrate $$\frac{1}{(x^2+1)\sqrt{(x^2+b^2)}} \; (b < a).$$

Evaluate

$$\int_a^b (x-a)^{\frac{3}{2}}(b-x)^{\frac{3}{2}}\, dx, \quad \int_0^{\pi} \frac{x\sin x\, dx}{\sqrt{(1-a^2\sin^2 x)}} \quad |a| < 1.$$

Prove that

$$\int_0^{\frac{1}{2}\pi} f(\sin 2x)\sin x\, dx = \sqrt{2}\int_0^{\frac{1}{4}\pi} f(\cos 2x)\cos x\, dx.$$

21. Evaluate the integrals:

$$\int_6^7 \sqrt{\left(\frac{x-3}{x-6}\right)}\, dx, \quad \int_{\frac{1}{2}}^1 \frac{\sqrt{(1-x^2)}}{x}\, dx, \quad \int_1^2 \frac{\sqrt{(x^2-1)}}{x}\, dx.$$

22. Find the integrals:

$$\int \frac{x^2\,dx}{x^4-x^2-12}, \quad \int e^x(1+x)\log x\,dx.$$

Evaluate
$$\int_0^1 \frac{dx}{(1+x)\sqrt{(1+2x-x^2)}}.$$

23. Prove that, if a is positive, the value of the integral

$$\int_0^{\frac{1}{2}\pi} \frac{1-a\cos\theta}{1-2a\cos\theta+a^2}\,d\theta$$

is $\pi-\cot^{-1}a$ or $-\cot^{-1}a$, according as $a<1$ or $a>1$, where the value of $\cot^{-1}$ is taken between $0, \frac{1}{2}\pi$.

What is the value of the integral when $a=1$?

24. Show that, if P, Q are polynomials in s, c (where $s=\sin\theta$, $c=\cos\theta$), of which Q contains only even powers of both s and c, then

$$\int \frac{P}{Q}\,d\theta$$

can be expressed as a sum of integrals of the form $\int R_1(s)\,ds$, $\int R_2(c)\,dc$, $\int R_3(t)\,dt$, where $t=\tan\theta$ and each of the functions R_1, R_2, R_3 is a rational function of its argument.

Apply this method to obtain

$$\text{(i)} \int \frac{(1+\sin\theta)(1+\cos\theta)}{1+\cos^2\theta}\,d\theta, \quad \text{(ii)} \int \frac{1}{1+\sin\theta}\,d\theta.$$

In case (ii) obtain the integral also by the substitution

$$x=\tan\tfrac{1}{2}\theta,$$

and reconcile the results obtained by the two methods.

25. Show that the substitution

$$4x = -\frac{b}{a}\left(t+\frac{1}{t}\right)^2+\frac{d}{c}\left(t-\frac{1}{t}\right)^2$$

reduces the integral

$$\int F\{x, \sqrt{(ax+b)}, \sqrt{(cx+d)}\}\,dx,$$

where $F(x,y,z)$ is a rational function of x, y, z, to the integral of a rational function of t.

Hence, or otherwise, find

$$\int \frac{(x-1)^{\frac{1}{2}}}{(x+1)^{\frac{3}{2}}}\,dx.$$

26. Find a reduction formula for

$$I_n = \int \frac{dx}{(5+4\cos x)^n}$$

in terms of I_{n-1}, I_{n-2} $(n \geqslant 2)$, and use it to show that

$$\int_0^{\frac{2}{3}\pi} \frac{dx}{(5+4\cos x)^2} = \frac{1}{81}(5\pi - 6\sqrt{3}).$$

27. Find a reduction formula for

$$\int_1^{\infty} \frac{dx}{x^n\sqrt{(x^2-1)}}.$$

Evaluate the integral when $n = 1$ and when $n = 2$.

28. If $$I_{p,q} = \int_0^{\frac{1}{2}\pi} \sin^p x \cos^q x\,dx,$$

show that $$(p+q)I_{p,q} = \begin{cases} (q-1)\,I_{p,q-2} & (q \geqslant 2), \\ (p-1)\,I_{p-2,q} & (p \geqslant 2), \end{cases}$$

and evaluate $I_{\alpha-1,5}$ where α is any positive real number.

29. If $$I_n = \int_0^{\infty} x^n e^{-ax} \cos bx\,dx, \quad J_n = \int_0^{\infty} x^n e^{-ax} \sin bx,$$

where n is a positive integer and a, b are positive, prove that

$$I_n(a^2+b^2) = n(aI_{n-1} - bJ_{n-1}),$$
$$J_n(a^2+b^2) = n(bI_{n-1} + aJ_{n-1}).$$

Show that $$(a^2+b^2)^{\frac{1}{2}(n+1)} I_n = n!\cos(n+1)\alpha,$$
$$(a^2+b^2)^{\frac{1}{2}(n+1)} J_n = n!\sin(n+1)\alpha,$$

where $\tan\alpha = b/a$ and $0 < \alpha < \frac{1}{2}\pi$.

30. (i) Prove that, if

$$u_n = \int_0^{\frac{1}{2}\pi} \frac{dx}{(a+b\tan x)^n} \quad (n > 1),$$

then $$(a^2+b^2)\,u_n - 2au_{n-1} + u_{n-2} = \frac{b}{(n-1)\,a^{n-1}}.$$

(ii) Prove that, if n is an integer and $m > 1$, then

$$(m^2-4n^2)\int_0^{\frac{1}{2}\pi}\sin^m x\cos 2nx\,dx = m(m-1)\int_0^{\frac{1}{2}\pi}\sin^{m-2}x\cos 2nx\,dx.$$

31. Find the reduction formula for

$$\int\frac{dx}{(ax^2+2hx+b)^{n+1}}.$$

Prove that, if a and $ab-h^2$ are positive and n is a positive integer, then

$$\int_{-\infty}^{\infty}\frac{dx}{(ax^2+2hx+b)^{n+1}} = \frac{2n-1}{2n}\cdot\frac{a}{ab-h^2}\int_{-\infty}^{\infty}\frac{dx}{(ax^2+2hx+b)^n}$$
$$= \frac{1.3\ldots(2n-1)}{2.4\ldots 2n}\cdot\frac{\pi a^n}{(ab-h^2)^{n+\frac{1}{2}}}.$$

32. Show that

$$\int\frac{(x-p)^{2n+1}\,dx}{(ax^2+2hx+b)^{n+\frac{3}{2}}} = \frac{1}{(h^2-ab)^{n+1}}\int(y^2-q^2)^n\,dy,$$

where
$$q^2 = ap^2+2hp+b,$$
$$y\sqrt{(ax^2+2hx+b)} = (ap+h)x+(hp+b).$$

Deduce that, when $a, b, h+\sqrt{(ab)}$ are positive,

$$\int_0^{\infty}\frac{dx}{(ax^2+2hx+b)^{\frac{3}{2}}} = \frac{1}{\{h+\sqrt{(ab)}\}\sqrt{b}},$$

$$\int_0^{\infty}\frac{x\,dx}{(ax^2+2hx+b)^{\frac{3}{2}}} = \frac{1}{\{h+\sqrt{(ab)}\}\sqrt{a}}.$$

33. Find a reduction formula for the integral

$$\int\frac{dx}{(1+x)^n\sqrt{(1+x^2)}}.$$

Prove that

$$\int_0^{\infty}\frac{dx}{(1+x)^2\sqrt{(1+x^2)}} = \tfrac{1}{2}\sqrt{2}\log(1+\sqrt{2}),$$

$$\int_0^{\infty}\frac{dx}{(1+x)^3\sqrt{(1+x^2)}} = \tfrac{1}{8}\{2+\sqrt{2}\log(1+\sqrt{2})\}.$$

34. (i) Integrate $$\frac{1}{(x^2+4)\sqrt{(x^2+1)}}.$$

(ii) Find a formula of reduction for

$$\int (1+x^2)^n e^{ax}\,dx.$$

35. (i) Evaluate

$$\int_{-1}^{1} \frac{dx}{(a-x)\sqrt{(1-x^2)}} \quad (a>1).$$

(ii) If $$u_n = \int_0^1 \frac{x^n\,dx}{\sqrt{(1-2x\cos\alpha+x^2)}},$$

prove that

$$nu_n-(2n-1)\,u_{n-1}\cos\alpha+(n-1)\,u_{n-2} = 2\sin\tfrac{1}{2}\alpha$$

when $n \geqslant 2$, and evaluate u_0, u_1.

36. Obtain reduction formulæ for the integrals

$$\int_1^\infty \frac{(\log x)^n}{x^2}\,dx, \quad \int_0^\pi x^n \sin^2 x\,dx,$$

and evaluate the first integral for any positive integer n.

37. If $$I_p = \int_0^1 \frac{x^p\,dx}{\sqrt{(x^q+1)}} \quad (p, q \text{ real}, p>q-1),$$

prove that

$$(2p-q+2)\,I_p+(2p-2q+2)\,I_{p-q} = 2\sqrt{2}.$$

Hence, or otherwise, prove that

$$\int_0^1 \frac{x^8\,dx}{\sqrt{(x^3+1)}} = \frac{1}{45}(14\sqrt{2}-16).$$

38. If $$I_n = \int_a^b \frac{dx}{(x^2+k)^n\sqrt{(x^2+l)}},$$

verify by differentiation that, when n is a positive integer or zero,

$$2(n+1)\,k(k-l)\,I_{n+2}-(2n+1)(2k-l)\,I_{n+1}+2nI_n = \left[\frac{-x\sqrt{(x^2+l)}}{(x^2+k)^{n+1}}\right]_a^b.$$

Show further that I_1 can be integrated by the substitution

$$\sqrt{(x^2+l)} = xt.$$

Hence, or otherwise, find

$$\int_0^1 \frac{dx}{(x^2+2)^2\sqrt{(x^2+3)}}.$$

39. Show that, if

$$I_n = \frac{1}{n!}\int_0^\infty x^n e^{-x}\cos x\,dx, \quad J_n = \frac{1}{n!}\int_0^\infty x^n e^{-x}\sin x\,dx,$$

then $$2J_n = I_{n-2}, \quad 2I_n = -J_{n-2}.$$

Hence, or otherwise, prove that

$$I_n = \frac{\cos\frac{1}{4}(n+1)\pi}{2^{\frac{1}{2}(n+1)}}.$$

40. Find a reduction formula for the integral

$$\int_0^{\frac{1}{4}\pi} \frac{dx}{\cos^n x},$$

and evaluate the integral for the cases $n = 1, 2$.

41. Prove that, if

$$I_n = \int \frac{dx}{(2x^2+1)^n\sqrt{(x^2+1)}} \quad (n \geqslant 0),$$

then $$(n+1)I_{n+2} - nI_n = \frac{x\sqrt{(x^2+1)}}{(2x^2+1)^{n+1}}.$$

Hence evaluate the integral

$$\int \frac{dx}{(2x^2+1)^4\sqrt{(x^2+1)}}.$$

42. Find a reduction formula for the integral

$$I_n = \int \frac{x^n\,dx}{\sqrt{(ax^2+2hx+b)}},$$

and use it to evaluate

$$\int \frac{x^3\,dx}{\sqrt{(x^2+2x+2)}}.$$

APPENDIX

First Steps in Partial Differentiation

The functions which we have considered in this volume have always involved a single variable. The work on functions of several variables belongs to a later stage, but it may be convenient to set down one or two of the most elementary properties—mainly definitions and first applications.

Consider, as an illustration, the expression

$$u \equiv x^4 y^3 z^2.$$

As x, y, z take various values, so also does u. For example,

if $$x = 1, \quad y = -2, \quad z = 3,$$

then $$u = -72;$$

if $$x = -1, \quad y = 0, \quad z = 3,$$

then $$u = 0;$$

if $$x = 2, \quad y = -2, \quad z = 1,$$

then $$u = -128;$$

and so on.

We say that u is then a *function of the three independent variables* x, y, z. To denote this functional dependence, we may use the notation

$$u(x, y, z) \equiv x^4 y^3 z^2.$$

The function u no longer has a unique differential coefficient. Each of the variables x, y, z is capable of its own independent variation, and each of these variations produces a differential coefficient of its own. More precisely, we use the notation

$\dfrac{\partial u}{\partial x} \equiv$ the differential coefficient of u with respect to x only, calculated on the assumption that y, z are constant;

$\dfrac{\partial u}{\partial y} \equiv$ the differential coefficient of u with respect to y only, calculated on the assumption that z, x are constant;

$\dfrac{\partial u}{\partial z} \equiv$ the differential coefficient of u with respect to z only, calculated on the assumption that x, y are constant.

Thus, if
$$u \equiv x^4 y^3 z^2,$$

then
$$\frac{\partial u}{\partial x} = 4x^3 y^3 z^2, \quad \frac{\partial u}{\partial y} = 3x^4 y^2 z^2, \quad \frac{\partial u}{\partial z} = 2x^4 y^3 z.$$

As another illustration, suppose that

$$u \equiv \cos(ax + by^2)$$

is a function of the two variables x, y. Then

$$\frac{\partial u}{\partial x} = -a\sin(ax + by^2), \quad \frac{\partial u}{\partial y} = -2by\sin(ax + by^2).$$

The functions
$$\frac{\partial u}{\partial x}, \quad \frac{\partial u}{\partial y}, \quad \frac{\partial u}{\partial z}$$

are called the *partial differential coefficients* of u with respect to x, y, z respectively.

These partial differential coefficients are, in their turn, also functions of the three variables x, y, z, and have their own partial differential coefficients. We write

$$\frac{\partial}{\partial x}\left(\frac{\partial u}{\partial x}\right) \equiv \frac{\partial^2 u}{\partial x^2},$$

$$\frac{\partial}{\partial y}\left(\frac{\partial u}{\partial y}\right) \equiv \frac{\partial^2 u}{\partial y^2},$$

$$\frac{\partial}{\partial z}\left(\frac{\partial u}{\partial z}\right) \equiv \frac{\partial^2 u}{\partial z^2}.$$

The 'mixed' coefficients are a little more awkward. We write

$$\frac{\partial}{\partial x}\left(\frac{\partial u}{\partial y}\right) \equiv \frac{\partial^2 u}{\partial x\,\partial y}, \quad \frac{\partial}{\partial x}\left(\frac{\partial u}{\partial z}\right) \equiv \frac{\partial^2 u}{\partial x\,\partial z},$$

$$\frac{\partial}{\partial y}\left(\frac{\partial u}{\partial z}\right) \equiv \frac{\partial^2 u}{\partial y\,\partial z}, \quad \frac{\partial}{\partial y}\left(\frac{\partial u}{\partial x}\right) \equiv \frac{\partial^2 u}{\partial y\,\partial x},$$

$$\frac{\partial}{\partial z}\left(\frac{\partial u}{\partial x}\right) \equiv \frac{\partial^2 u}{\partial z\,\partial x}, \quad \frac{\partial}{\partial z}\left(\frac{\partial u}{\partial y}\right) \equiv \frac{\partial^2 u}{\partial z\,\partial y}.$$

In practice, however, it may be proved that for 'ordinary' functions (a term which we do not attempt to make more precise) interchange of the order of partial differentiation leaves the result unaltered. Thus

$$\frac{\partial^2 u}{\partial y\,\partial z} = \frac{\partial^2 u}{\partial z\,\partial y},$$

$$\frac{\partial^2 u}{\partial z\,\partial x} = \frac{\partial^2 u}{\partial x\,\partial z},$$

$$\frac{\partial^2 u}{\partial x\,\partial y} = \frac{\partial^2 u}{\partial y\,\partial x}.$$

For example, returning to our function

$$u \equiv x^4 y^3 z^2,$$

we have the relations

$$\frac{\partial^2 u}{\partial x\,\partial y} = \frac{\partial}{\partial x}(3x^4y^2z^2) = 12x^3y^2z^2,$$

$$\frac{\partial^2 u}{\partial y\,\partial x} = \frac{\partial}{\partial y}(4x^3y^2z^2) = 12x^3y^2z^2;$$

$$\frac{\partial^2 u}{\partial y\,\partial z} = \frac{\partial}{\partial y}(2x^4y^3z) = 6x^4y^2z,$$

$$\frac{\partial^2 u}{\partial z\,\partial y} = \frac{\partial}{\partial z}(3x^4y^2z^2) = 6x^4y^2z;$$

$$\frac{\partial^2 u}{\partial z\,\partial x} = \frac{\partial}{\partial z}(4x^3y^3z^2) = 8x^3y^3z,$$

$$\frac{\partial^2 u}{\partial x\,\partial z} = \frac{\partial}{\partial x}(2x^4y^3z) = 8x^3y^3z.$$

EXAMPLES I

Find $\dfrac{\partial u}{\partial x}, \dfrac{\partial u}{\partial y}, \dfrac{\partial^2 u}{\partial x^2}, \dfrac{\partial^2 u}{\partial x\,\partial y}, \dfrac{\partial^2 u}{\partial y\,\partial x}, \dfrac{\partial^2 u}{\partial y^2}$ for each of the following functions:

1. x^4y^3.
2. xy^2.
3. x^2+y^2.
4. $e^x \cos y$.
5. $\log(x+y)$.
6. $\log(xy)$.
7. $x^3 \sin^2 y$.
8. $e^{xy} \sin x$.
9. $x \tan^{-1} y$.
10. $\sec x + \sec y$.
11. $e^x \sin^2 2y$.
12. xe^y.
13. Prove that, if $f(x, y)$ is any *polynomial* in x, y, then

$$\frac{\partial^2 f}{\partial x\,\partial y} = \frac{\partial^2 f}{\partial y\,\partial x}.$$

14. Prove that, if $f(x,y,z)$ is any *polynomial* in x,y,z, then

$$\frac{\partial^2 f}{\partial y\,\partial z} = \frac{\partial^2 f}{\partial z\,\partial y}.$$

ILLUSTRATION 1. *The 'homogeneous quadratic form'.*

Let $$u \equiv ax^2 + by^2 + cz^2 + 2fyz + 2gzx + 2hxy.$$

Then $$\frac{\partial u}{\partial x} = 2(ax + hy + gz),$$

$$\frac{\partial u}{\partial y} = 2(hx + by + fz),$$

$$\frac{\partial u}{\partial z} = 2(gx + fy + cz).$$

These expressions are probably familiar from analytical geometry.

To show how the partial differential coefficients are linked with the idea of gradient, we use an illustrative example.

Let OX, OY be the axes for a system of rectangular coordinates in a horizontal plane. This is illustrated in the diagram (Fig. 102), where the reader may regard himself as looking 'down' upon axes drawn in the usual position. The straight line OZ is drawn vertically upwards.

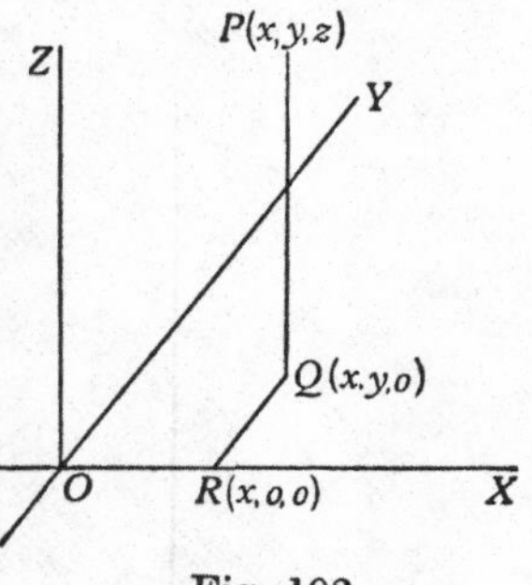

Fig. 102.

Given a point P in space, let the vertical line through it meet the plane XOY in Q; draw QR perpendicular to OX. Denote by x, y, z the lengths $\overrightarrow{OR}, \overrightarrow{RQ}, \overrightarrow{QP}$ respectively; then the triplet x, y, z may be used as coordinates for the point P in space, just as the pair x, y is used for a point in a plane. If P is the point (x, y, z), then Q is the point $(x, y, 0)$ in the horizontal plane; the coordinate z gives the height of P referred to the plane XOY as zero level. (Of course, P may be below the plane, in which case z is negative.)

In particular, if x, y, z are connected by the relation

$$z = f(x, y),$$

where we assume $f(x, y)$ to be a single-valued function defined for each pair of values of x, y, then, as x, y (and consequently z) vary, the point Q moves about the plane XOY, while P describes the surface whose height at any point is equal to the corresponding value of the function. We say that the surface *represents* the function $f(x, y)$.

For instance, it is an easy example on the theorem of Pythagoras to show that the function

$$z = +\sqrt{(1-x^2-y^2)}$$

is represented by the hemisphere of centre O and unit radius lying above the plane XOY.

We now assume, for convenience of language, that $\overrightarrow{OX}$ is due east and $\overrightarrow{OY}$ due north. We regard the surface

$$z = f(x, y)$$

as a hill, and P as the position of a climber on it.

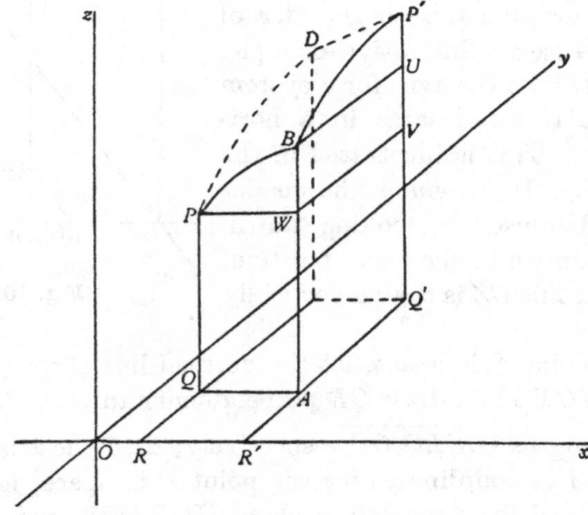

Fig. 103.

Suppose that the climber is at the point P (Fig. 103) defined by the values x, y of the easterly and northerly coordinates, and that he wishes to climb to the point P' defined by $x+h, y+k$. The crux of the difference between functions of one variable and functions of two lies in the fact that, whereas for one variable motion along the CURVE representing the function is defined all the way, for two variables the SURFACE may be traversed by an innumerable choice of paths. Moreover, each way of leaving P will demand a gradient all of its own. The partial differential coefficients are the bases of the mathematical expressions for such gradients corresponding to the various paths.

From the mathematical point of view, the obvious way to pass from P to P' is firstly to move the distance h easterly, to B, and then to move the distance k northerly. The climber thus describes in succession the two arcs PB, BP' shown in the diagram.

Now suppose that P' is very close to P, so that the arcs PB, BP' are very small. The arc PB may be regarded as almost straight, so that the 'rise' between P and B is proportional to the length h, say

$$\delta z(\text{easterly}) = \alpha h.$$

Similarly BP' is almost straight, so that the 'rise' between B and P' is proportional to k, say

$$\delta z(\text{northerly}) = \beta k.$$

If δz is the total 'rise' between P, P', then

$$\begin{aligned}\delta z &= \delta z(\text{easterly}) + \delta z(\text{northerly})\\ &= \alpha h + \beta k.\end{aligned}$$

If the climber had gone first northerly and then easterly, following the course PD, DP' in the diagram, then, for distances so small that the paths may be regarded as straight, $PBP'D$ is approximately a parallelogram, and so, once again,

$$\delta z = \alpha h + \beta k$$

for the SAME values of α, β.

Two simple observations complete the illustration. Geometrically, α, β are the gradients of those curves which are the sections of the hill in the easterly and northerly directions respectively. Analytically, we see by putting $k = 0$ that α is the ratio $\delta z \div h$ calculated on the assumption that y is constant; thus

$$\alpha = \frac{\partial z}{\partial x}$$

evaluated at P. Similarly we see by putting $h = 0$ that β is the ratio $\delta z \div k$ calculated on the assumption that x is constant; thus

$$\beta = \frac{\partial z}{\partial y}$$

calculated at P.

Hence *the partial differential coefficients* $\dfrac{\partial z}{\partial x}, \dfrac{\partial z}{\partial y}$ *are identified as the gradients of the surface* $z = f(x, y)$ *in the x- and y-directions respectively.*

EXAMPLES II

1. Show that the function

$$z = 1 - \sqrt{(2x - x^2 - y^2)},$$

with the z-axis measured vertically upwards, is represented by a hemisphere.

Prove also that the gradients in the x- and y-directions at the point (x, y, z) are in the ratio$(x-1) : y$, and that these gradients are equal only for points on a certain vertical diametral plane.

2. Prove that the function

$$z = x^2 + 4y^2,$$

with the z-axis measured vertically upwards, is represented by a 'bowl-shaped' surface whose horizontal sections are ellipses of eccentricity $\frac{1}{2}\sqrt{3}$.

Prove that the gradient in the x-direction at the point $(1, 2, 17)$ is 2, and that the gradient in the y-direction at the point $(3, 1, 13)$ is 8.

3. Find $\dfrac{\partial u}{\partial x}, \dfrac{\partial^2 u}{\partial y^2}, \dfrac{\partial^2 u}{\partial z\, \partial x}$ for each of the following functions:

(i) $e^{ax} \sin(by + cz)$, (ii) $xyze^{-x^2}$, (iii) $(y^2 + z^2) \log(ax + b)$.

4. Prove that, if

$$u \equiv ax^2 + by^2 + cz^2 + 2fyz + 2gzx + 2hxy,$$

and if
$$\frac{\partial^2 u}{\partial x^2} + \frac{\partial^2 u}{\partial y^2} + \frac{\partial^2 u}{\partial z^2} = 0,$$

then
$$a + b + c = 0.$$

5. Prove that, if $r = \sqrt{(x^2+y^2+z^2)}$,

then
$$\frac{\partial r}{\partial x} = \frac{x}{r}, \quad \frac{\partial^2 r}{\partial x^2} = \frac{1}{r} - \frac{x^2}{r^3}.$$

Deduce that
$$\frac{\partial^2 r}{\partial x^2} + \frac{\partial^2 r}{\partial y^2} + \frac{\partial^2 r}{\partial z^2} = \frac{2}{r}.$$

Prove also that
$$\frac{\partial^2}{\partial x^2}\left(\frac{1}{r}\right) + \frac{\partial^2}{\partial y^2}\left(\frac{1}{r}\right) + \frac{\partial^2}{\partial z^2}\left(\frac{1}{r}\right) = 0.$$

Finally, there is a point of notation which the reader may meet in physical applications. Consider, as an illustration, the transformation
$$x = r\cos\theta, \quad y = r\sin\theta$$
between the Cartesian and the polar coordinates of a point. Four variables are involved, of which two are independent—for example, r and θ. When we form the partial differential coefficient $\frac{\partial x}{\partial r}$, we naturally have in mind that θ is the other independent variable, and the relation
$$x = r\cos\theta$$
thus gives
$$\frac{\partial x}{\partial r} = \cos\theta = \frac{x}{r}.$$
But it is possible to express x in terms of r and y, in the form
$$x = \sqrt{(r^2-y^2)},$$
and then
$$\frac{\partial x}{\partial r} = \frac{r}{\sqrt{(r^2-y^2)}} = \frac{r}{x}.$$
These two values are quite different; they are, indeed, calculated under the quite different hypotheses θ = constant, y = constant respectively.

To make sure what is intended, we often use the notation
$$\left.\frac{\partial x}{\partial r}\right)_\theta$$
to denote the partial differential coefficient of x with respect to r when θ is the other independent variable (being kept constant

during differentiation). With such notation, the two formulæ given above may be expressed in the form

$$\left.\frac{\partial x}{\partial r}\right)_\theta = \frac{x}{r},$$

$$\left.\frac{\partial x}{\partial r}\right)_y = \frac{r}{x}.$$

The following examples should serve to make the notation clear.

EXAMPLES III

Given the relations $x = r\cos\theta$, $y = r\sin\theta$, establish the following formulæ:

1. $\left.\dfrac{\partial x}{\partial r}\right)_\theta = \dfrac{x}{r}.$
2. $\left.\dfrac{\partial y}{\partial r}\right)_\theta = \dfrac{y}{r}.$
3. $\left.\dfrac{\partial x}{\partial \theta}\right)_r = -y.$
4. $\left.\dfrac{\partial y}{\partial \theta}\right)_r = x.$
5. $\left.\dfrac{\partial r}{\partial x}\right)_y = \dfrac{x}{r}.$
6. $\left.\dfrac{\partial r}{\partial y}\right)_x = \dfrac{y}{r}.$
7. $\left.\dfrac{\partial \theta}{\partial x}\right)_y = -\dfrac{y}{r^2}.$
8. $\left.\dfrac{\partial \theta}{\partial y}\right)_x = \dfrac{x}{r^2}.$
9. $\left.\dfrac{\partial x}{\partial r}\right)_y = \dfrac{r}{x}.$
10. $\left.\dfrac{\partial x}{\partial y}\right)_r = -\dfrac{y}{x}.$
11. $\left.\dfrac{\partial \theta}{\partial y}\right)_r = \dfrac{1}{x}.$
12. $\left.\dfrac{\partial \theta}{\partial r}\right)_x = \dfrac{x}{ry}.$

ANSWERS TO EXAMPLES

CHAPTER VII

Examples I:

1. $\log(x+1)$. 2. $\frac{1}{2}\log(2x+1)$. 3. $-\frac{1}{3}\log(2-3x)$.

4. $\frac{1}{2}x^2+\log x$. 5. $\frac{1}{2}\log\frac{x-1}{x+1}$. 6. $\frac{1}{5}x^5+2\log x-\frac{1}{5}x^{-5}$.

7. $\log 2$. 8. $\log 2$. 9. $\frac{1}{3}\log\frac{5}{2}$.

10. $\frac{1}{4}\log 5$. 11. $\frac{1}{4}\log\frac{5}{3}$. 12. $\frac{1}{6}\log\frac{10}{7}$.

13. $\frac{3}{3x+2}$. 14. $2\operatorname{cosec} 2x$. 15. $-\cot x$.

16. $x+2x\log x$. 17. $x^{n-1}+nx^{n-1}\log x$. 18. $2x/(1+x^2)$.

19. $x\log x-x$. 20. $\frac{1}{2}(\log x)^2$. 21. $\log\sin x$.

22. $\frac{1}{2}x^2\log x-\frac{1}{4}x^2$. 23. $\frac{1}{2}\log\left(\frac{1-\cos x}{1+\cos x}\right)$.

24. $\cos x+\frac{1}{2}\log\left(\frac{1-\cos x}{1+\cos x}\right)$. 25. $\log(x^2+5x+12)$.

26. $\log(x^2-3x+7)$.

27. $\frac{7}{4}\tan^{-1}\left(\frac{x-1}{4}\right)+\log(x^2-2x+17)$.

28. $-12\tan^{-1}(x+3)+\log(x^2+6x+10)$.

29. $9\tan^{-1}\left(\frac{x-4}{3}\right)+\frac{5}{2}\log(x^2-8x+25)$.

30. $-\frac{37}{3}\tan^{-1}\left(\frac{x+5}{3}\right)+\frac{7}{2}\log(x^2+10x+34)$.

Examples II:

1. $\frac{1}{y}\frac{dy}{dx}=\frac{2}{1+x}+\frac{3}{1-x}$. 2. $\frac{1}{y}\frac{dy}{dx}=-2\tan x-\frac{2x}{1+x^2}$.

3. $\frac{1}{y}\frac{dy}{dx}=\frac{1}{x}+2\cot x+\frac{6x^2}{1-2x^3}$. 4. $\frac{1}{y}\frac{dy}{dx}=\frac{2}{x}+\frac{2}{1+x}-\frac{8x^3}{1+x^4}$.

5. $\dfrac{1}{y}\dfrac{dy}{dx}=\dfrac{1}{x}+\cot x-\dfrac{3}{1+x}+\dfrac{1}{1-x}$. 6. $\dfrac{1}{y}\dfrac{dy}{dx}=\dfrac{4x}{1+x^2}-\dfrac{1}{x}+2\tan x$.

7. $\dfrac{1}{y}\dfrac{dy}{dx}=\dfrac{4}{x}-\dfrac{3}{1-x}-8\operatorname{cosec}4x$.

8. $\dfrac{1}{y}\dfrac{dy}{dx}=-\dfrac{2\sin x}{1+\cos x}-\dfrac{1+2x}{1+x+x^2}$.

9. $\dfrac{1}{y}\dfrac{dy}{dx}=-\dfrac{1}{1-x}+\dfrac{4}{1+2x}+\dfrac{9}{1-3x}-\dfrac{16}{1+4x}$.

Examples III:

1. $\frac{1}{6}\log\left(\dfrac{x-3}{x+3}\right)$. 2. $x+2\log(x-1)-1/(x-1)$.

3. $\log\dfrac{(x-2)^2}{(x-1)}$. 4. $\frac{1}{3}x^3+\frac{1}{2}x^2+x+\log(x-1)$.

5. $\frac{1}{2}\tan^{-1}x+\frac{1}{4}\log\dfrac{(x+1)^2}{(x^2+1)}$. 6. $\frac{1}{3}\log(x+1)-\frac{16}{3}\log(x+4)+x$.

7. $\frac{1}{4}\log\left(\dfrac{x+1}{x-1}\right)-\dfrac{1}{2(x-1)}$. 8. $-\dfrac{1}{2(x-1)}-\frac{1}{2}\tan^{-1}x$.

9. $\frac{1}{21}\log(x-2)-\frac{1}{42}\log(x^2+4x+9)-\dfrac{4}{21\sqrt{5}}\tan^{-1}\left(\dfrac{x+2}{\sqrt{5}}\right)$.

10. $\log(x-2)-\dfrac{4}{(x-2)}-\dfrac{2}{(x-2)^2}$. 11. $\frac{1}{2}\log(x^2-4)-\dfrac{2}{x^2-4}$.

12. $\frac{11}{250}\tan^{-1}\frac{1}{2}x-\frac{1}{250}\log\dfrac{(x-1)^2}{(x^2+4)}+\dfrac{2}{25(x-1)}-\dfrac{1}{10(x-1)^2}$.

13. $\frac{1}{2}\log\dfrac{(x-1)(x-3)^3}{(x-2)^4}$.

14. $\dfrac{4}{x-2}+\frac{9}{2}\log(x-3)-4\log(x-2)-\frac{1}{2}\log(x-1)$.

15. $\frac{5}{338}\tan^{-1}\left(\dfrac{x-3}{2}\right)+\frac{3}{169}\log\left(\dfrac{x^2}{x^2-6x+13}\right)-\dfrac{1}{13x}$.

16. $\frac{1}{16}\log\left(1+\dfrac{4}{x}\right)-\dfrac{1}{4x}$.

17. $\dfrac{1}{3x^2}-\dfrac{5}{9x}+\frac{5}{27}\log\left(1+\dfrac{3}{x}\right)$. 18. $\frac{1}{2}x+\frac{3}{8}\log(4x+7)$.

19. $x-3\log(x^2+6x+25)-\frac{7}{4}\tan^{-1}\left(\frac{x+3}{4}\right)$.

20. $x+\log(x^2+1)$.

21. $\frac{1}{4}\log x+\frac{3}{8}\log(x^2+4)-\tan^{-1}\frac{1}{2}x$.

22. $\frac{6}{25}\log\left(\frac{x-2}{x+3}\right)-\frac{4}{5(x-2)}$. 23. $\frac{1}{4}\log\frac{(x^2+1)}{(x-1)^2}-\frac{1}{2(x-1)}$.

24. $x-\frac{16}{17(x+2)}-\frac{512}{289}\log(x+2)-\frac{611}{289}\log(x^2+2x+17)$

$$-\tfrac{495}{1156}\tan^{-1}\left(\frac{x+1}{4}\right).$$

Examples IV:

1. $2e^{2x}$.
2. $2xe^{x^2}$.
3. $e^{5x}+5xe^{5x}$.
4. $e^x\cos x-e^x\sin x$.
5. $e^{\sin x}\cos x$.
6. $-e^{-x}(1-x)^2$.
7. $e^{2x}-e^{-2x}$.
8. $e^x\sin x+xe^x\sin x+xe^x\cos x$.
9. $\frac{e^x(1+2x-x^2)}{(1-x^2)^2}$.
10. $e^x\sin x+(1+e^x)\cos x$.
11. $e^{3x}(3\cos 4x-4\sin 4x)$.
12. $e^x(\tan x+\sec^2 x)$.
13. $\frac{1}{2}e^{2x}$.
14. $-\frac{1}{5}e^{-5x}$.
15. e^{x^2}.
16. $-\frac{1}{3}e^{-x^3}$.
17. $(x-1)e^x$.
18. $e^x(x^2-2x+2)$.
19. $e^{\sin x}$.
20. $\frac{1}{2}e^{\sin^2 x}$.
21. $e^{\tan x}$.
22. $\frac{1}{2}e^x(\sin x+\cos x)$.
23. $\frac{1}{25}e^{3x}(3\cos 4x+4\sin 4x)$.
24. $\frac{1}{4}e^{2x}(1+2x)$.

Examples V:

1. $aI_n=x^n e^{ax}-nI_{n-1}$.
2. $(a^2+n^2)I_n=e^{ax}\sin^{n-1}x(a\sin x-n\cos x)+n(n-1)I_{n-2}$.
3. $(a^2+b^2n^2)I_n$
$$=e^{ax}\cos^{n-1}bx(a\cos bx+nb\sin bx)+n(n-1)b^2I_{n-2}.$$
4. $120-44e$.
5. $\frac{1}{85}(41e^{\frac{1}{2}\pi}-24)$.
6. $-\frac{2}{5}(1+e^{\pi})$.

REVISION EXAMPLES III

1. (i) $(1+x)^{n-1}\{1+n\log(1+x)\}$. (ii) $\dfrac{1}{n}+\log(1+x)$.

2. $\dfrac{d^3z}{dx^3}-4\dfrac{dz}{dx}=0$.

3. $\left(\dfrac{1-2t}{1-t}\right)^2$, $\frac{2}{3}$.

4. $2\cot x$, $\dfrac{-2x}{(1-x^2)^{\frac{1}{2}}(1+x^2)^{\frac{3}{2}}}$.

5. $2\sin x-3\sin^3 x$, $2ae^{ax}\cos ax$, $\dfrac{1}{\sqrt{(x^2+1)}}$, $\dfrac{-2}{4+x^2}$.

6. $\dfrac{1}{2\sqrt{x}}$, 1, $\dfrac{-x}{(x^2+1)^{\frac{3}{2}}}$, $\dfrac{-2x}{\sqrt{(1-x^4)}}$, $\dfrac{1}{x\log_e 10}$.

7. $-\dfrac{1}{x^2}$, $2\cos 2x$, $\dfrac{2}{1+x^2}$, $2xe^{x^2}$.

8. $\dfrac{1}{(1-x)^2}$, $3\sin 6x$, $\log_e x$, $\dfrac{1}{1+x^2}$.

9. (i) $\dfrac{1}{(1-x^2)^{\frac{3}{2}}}$, (ii) $\sec x$, (iii) $\dfrac{-4}{5+3\cos x}$.

10. $-\dfrac{2}{x^3}$, $\dfrac{1-x}{(x+1)^3}$, $\dfrac{2x}{1+x^4}$. $4\operatorname{cosec} 4x$.

11. $1-\dfrac{1}{x^2}$, $-3x(a^2-x^2)^{\frac{1}{2}}$, $\dfrac{1}{(1+x)(1+2x)^{\frac{1}{2}}}$.

12. $\dfrac{12}{x^3}\left(1-\dfrac{3}{x^2}\right)$, $\dfrac{1}{(1-x)^{\frac{3}{2}}(1+x)^{\frac{1}{2}}}$, $\dfrac{2}{1+\sin 2x}$.

13. $\dfrac{1}{1+x^2}$, $\dfrac{-1}{1+x^2}$, $\dfrac{4x}{(1-x^2)^2}$, $\frac{1}{2}e^{-\frac{1}{2}x}(4\cos 2x-\sin 2x)$,
$\tan x-\cot x$.

14. $\dfrac{x^2(3+2x)}{(1+x)^2}$, $2\sin x-7\sin^3 x+5\sin^5 x$, $\dfrac{1}{2\sqrt{(x-x^2)}}$.

15. $-\dfrac{4}{x^5}$, $\dfrac{1}{x^2(x^2-1)^{\frac{1}{2}}}$, $\dfrac{7}{(4\cos x+3\sin x)^2}$, $\dfrac{2x^2}{1+x^2}+\log(1+x^2)$.

16. Velocity e^{π}, acceleration $-e^{\pi}$.

18. Tangent: $x\tan t+y-a\sin t=0$.

Normal: $x\cos t-y\sin t-a\cos 2t=0$.

19. $\dfrac{ah\sin\theta}{h-a\cos\theta}$.

20. $\dfrac{3\sqrt{3}}{2}$.

21. $x = 0, y = 1$, minimum;
$x = 1, y = 2$, neither;
$x = -2, y = 29$, maximum.

28. $(\sqrt{a}+\sqrt{b})^2$.

30. Area: $2+\frac{1}{2}\cot\theta+2\tan\theta$; angle: $\tan^{-1}\dfrac{1}{3\sqrt{2}}$.

31. $(-1, -1)$ minimum; $(1, 1)$ maximum; $24x-25y+8 = 0$.

34. (i) $2\log(2x+3)-\log(x-1)$.
(ii) $\frac{1}{3}\cos^3 x-\cos x$.
(iii) $x^2\sin x+2x\cos x-2\sin x$.

35. (i) $\frac{1}{2}x+\frac{1}{8}\sin 4x$.
(ii) $\log(x-1)+\frac{1}{3}\log(3x+1)$.
(iii) $\frac{1}{4}x^4\log x-\frac{1}{16}x^4$.

36. (i) $\frac{1}{3}\sec^3 x-\sec x$.
(ii) $\frac{1}{2}(1+x^2)\tan^{-1}x-\frac{1}{2}x$.
(iii) $\log(4x-1)-\log(x+1)$.

37. (i) $2\log(2x-1)-2\log(x+2)$.
(ii) $\frac{1}{3}\sin^3 x-\frac{1}{5}\sin^5 x$.
(iii) $-\dfrac{1}{(n-1)^2x^{n-1}}\{(n-1)\log x+1\}$.

38. (i) $3\log(x-3)-\frac{2}{3}\log(3x-2)$.
(ii) $\sin x-x\cos x$.
(iii) $\frac{3}{8}a^4\theta+\frac{1}{4}a^4\sin 2\theta+\frac{1}{32}a^4\sin 4\theta$, where $x = a\sin\theta$.

39. $-\sin x-\operatorname{cosec} x$, $\frac{1}{2}e^x(\sin x-\cos x)$, $2{\cdot}24$.

40. $\dfrac{\pi}{3}+\dfrac{\sqrt{3}}{2}$.

41. $\dfrac{1}{\sqrt{(1-x^2)}}$, $\frac{1}{2}x^2-4x+6\log(x+1)$, $2\log 2-\frac{3}{4}$.

42. 1.

43. $x^2\sin x+2x\cos x-2\sin x$, $\frac{2}{3}\log(1+3x)-\log(1-3x)$.

44. $\frac{1}{3}(1+x^2)^{\frac{3}{2}}$, $\quad \frac{1}{2}x\sqrt{(1+x^2)}+\frac{1}{2}\log\{x+\sqrt{(1+x^2)}\}$,
$\frac{1}{2}x\sqrt{(1+x^2)}-\frac{1}{2}\log\{x+\sqrt{(1+x^2)}\}$, $\quad \frac{1}{5}(3-\sqrt{2})$.

45. $\frac{1}{2}x\sqrt{(1-x^2)}+\frac{1}{2}\sin^{-1}x$, $\quad \frac{1}{2}(\sin^{-1}x)^2$, $\quad \dfrac{x}{\sqrt{(1-x^2)}}$, $\quad e^x\cos x$.

46. $2\log(1+\sqrt{x})$, $\quad -\cos x+\frac{2}{3}\cos^3 x-\frac{1}{5}\cos^5 x$,
$x+37\log(x-6)-26\log(x-5)$.

47. $\dfrac{1}{\sqrt{2}}\tan^{-1}\left(\dfrac{1}{\sqrt{2}}\tan x\right)$,
$\frac{1}{3}\log(x+1)-\frac{1}{6}\log(x^2-x+1)+\dfrac{1}{\sqrt{3}}\tan^{-1}\left(\dfrac{2x-1}{\sqrt{3}}\right)$,
$x\sin^{-1}x+\sqrt{(1-x^2)}$.

48. $x-\log(1-x)$, $\quad -\cos x+\frac{2}{3}\cos^3 x-\frac{1}{5}\cos^5 x$, $\quad e^x(x-1)$.

50. $-\dfrac{1}{x}-\tan^{-1}x$, $\quad \frac{1}{3}\tan^3 x-\tan x+x$, $\quad 1$.

51. $x-\log(x+2)$, $\quad x\tan x+\log\cos x$, $\quad \frac{1}{2}$.

52. $\log(2x^2-x-3)$, $\quad \frac{1}{3}\sin^3 x-\frac{1}{5}\sin^5 x$, $\quad \pi-2$.

53. $x-\dfrac{6}{x}-\dfrac{3}{x^3}$, $\quad \frac{2}{15}$, $\quad \frac{1}{2}\log 2$.

54. $\frac{1}{2}\pi$, $\quad 16\log 2-\frac{15}{4}$, $\quad \frac{1}{2}\pi^3-12\pi+24$.

55. (i) $\dfrac{7\sqrt{2}}{8}+\frac{3}{8}\log(1+\sqrt{2})$. $\quad$ (ii) 1, $\quad \frac{2}{3}(4\sqrt{2}-5)$.

56. $(m+n)\displaystyle\int_0^{\frac{1}{2}\pi}\sin^m x\cos^n x\,dx=(n-1)\int_0^{\frac{1}{2}\pi}\sin^m x\cos^{n-2}x\,dx$, $\quad \frac{1}{24}$, $\quad 0$.

57. $\dfrac{e^{ax}}{a^2+c^2}(a\cos cx+c\sin cx)$,
$\dfrac{1}{2a}e^{ax}-\dfrac{e^{ax}}{2a^2+8b^2}(a\cos 2bx+2b\sin 2bx)$.

58. (i) $2(n+1)\displaystyle\int_0^1(1+x^2)^{n+\frac{1}{2}}dx=2^{n+\frac{1}{2}}+(2n+1)\int_0^1(1+x^2)^{n-\frac{1}{2}}dx$,
$\dfrac{67}{24\sqrt{2}}+\frac{5}{16}\log(1+\sqrt{2})$.

CHAPTER VIII

Examples II:

1. $x-\frac{x^3}{3!}+\frac{x^5}{5!}-\ldots+(-1)^n\frac{x^{2n+1}}{(2n+1)!}+\ldots.$

2. $1-\frac{x^2}{2!}+\frac{x^4}{4!}-\ldots+(-1)^n\frac{x^{2n}}{(2n)!}+\ldots.$

3. $1-x+x^2-x^3+\ldots+(-1)^n x^n+\ldots.$

Examples V:

2. $1+3x+\frac{(3x)^2}{2!}+\frac{(3x)^3}{3!}+\ldots+\frac{(3x)^n}{n!}+\ldots.$

3. $2x-\frac{1}{2}(2x)^2+\frac{1}{3}(2x)^3+\ldots+\frac{(-1)^{n+1}}{n}(2x)^n+\ldots.$

4. $2x-\frac{(2x)^3}{3!}+\frac{(2x)^5}{5!}-\ldots+(-1)^n\frac{(2x)^{2n+1}}{(2n+1)!}+\ldots.$

5. $1-x+x^2-x^3+\ldots+(-1)^n x^n+\ldots.$

6. $1+2x+3x^2+\ldots+(n+1)x^n+\ldots.$

7. $1-\frac{(4x)^2}{2!}+\frac{(4x)^4}{4!}-\ldots+(-1)^n\frac{(4x)^{2n}}{(2n)!}+\ldots.$

8. $1-2x+\frac{(2x)^2}{2!}-\frac{(2x)^3}{3!}+\ldots+(-1)^n\frac{(2x)^n}{n!}+\ldots.$

9. $1+x-\frac{1}{2}x^2+\ldots+(-1)^{n+1}\frac{1.3.5\ldots.(2n-3)}{n!}x^n+\ldots.$

10. $-3x-\frac{1}{2}(3x)^2-\ldots-\frac{1}{n}(3x)^n-\ldots.$

Examples VI:

1. 2·005.
2. 2·995.
3. 2·0017.
4. 2·999.
5. 1·9996.
6. 0·3328.

Examples VII:

1. $1680x^4\sin x + 1344x^5\cos x - 336x^6\sin x - 32x^7\cos x + x^8\sin x$.

2. $x^2\sin x - 8x\cos x - 12\sin x$.

3. $e^{2x}(122\cos 3x + 597\sin 3x)$.

4. $3^4 e^{3x}(9x^3 + 54x^2 + 90x + 40)$.

5. $x^3\cos\{\frac{1}{2}n\pi + x\} + 3nx^2\cos\{(n-1)\frac{1}{2}\pi + x\}$
$+ 3n(n-1)x\cos\{(n-2)\frac{1}{2}\pi + x\}$
$+ n(n-1)(n-2)\cos\{(n-3)\frac{1}{2}\pi + x\}$.

6. $2^{n-3}e^{2x}\{8x^3 + 12nx^2 + 6n(n-1)x + n(n-1)(n-2)\}$.

7. $2^5.10.9.8(1-2x)^5(-132x^2 + 55x - 5)$.

8. $\dfrac{2.3^6.12!}{6!}(3x+1)^4(819x^2 + 312x + 28)$.

Examples VIII:

1. $x + \dfrac{1}{2}.\dfrac{x^3}{3} + \dfrac{1}{2}.\dfrac{3}{4}.\dfrac{x^5}{5} + \dfrac{1}{2}.\dfrac{3}{4}.\dfrac{5}{6}.\dfrac{x^7}{7} + \ldots$.

2. $x - \frac{1}{3}x^3 + \frac{1}{5}x^5 - \frac{1}{7}x^7 + \ldots$.

REVISION EXAMPLES IV

1. (i) $ax(1 + 2\log bx)e^{ax^2\log bx}$. (ii) $\dfrac{a}{1+a^2x^2}$.

$ab, \quad a^2b^2$.

2. $\dfrac{-1}{1+x^2}$, $\dfrac{-1}{(1+x)\sqrt{\{2x(1-x)\}}}$, $\sec x$, $\dfrac{2(-x^2-x+2)}{(x^2-8x-2)^2}$,

$x = 1$ maximum, $x = -2$ minimum.

4. $\sec^m x\tan^{n-1}x\{n + (m+n)\tan^2 x\}$.

$840\tan^4 x + 640\tan^2 x + 56$.

5. $\dfrac{x-1}{(x-2)^2(x-3)}$, $\tan^4 x$.

6. $4\sin^3 x\cos x$, $\quad 12\sin^2 x - 16\sin^4 x$,

$24\sin x\cos x - 64\sin^3 x\cos x$, $\quad 256\sin^4 x - 240\sin^2 x + 24$,

$x = 0$ minimum, $\quad x = \frac{1}{2}\pi$ maximum.

8. $n\left(x+\frac{1}{x}\right)^{n-1}\left(1-\frac{1}{x^2}\right)$, $\quad n\sin x\tan^2 x(\cos x+\sec x)^{n-1}$,

$\frac{(-1)^n(n+1)!}{x^{n+2}}$ when $n>2$, $\quad 2+\frac{6}{x^4}$ when $n=2$,

$2x-\frac{2}{x^3}$ when $n=1$.

10. (ii) $x^2\frac{d^2y}{dx^2}-n(n-1)y=0$.

11. $\frac{1}{(x+1)^{\frac{3}{2}}(x-1)^{\frac{1}{2}}}$, $\quad \tan x$, $\quad \frac{2^n n!}{(1-2x)^{n+1}}-\frac{n!}{(1-x)^{n+1}}$.

12. (i) $\frac{(-1)^n n!}{x^{n+1}}$, $\quad (-1)^n\frac{n!}{2}\left\{\frac{1}{(x-1)^{n+1}}-\frac{1}{(x+1)^{n+1}}\right\}$.

(ii) $\sin(x+\frac{1}{2}n\pi)$, $\quad x\sin(x+\frac{1}{2}n\pi)+n\sin\{x+\frac{1}{2}(n-1)\pi\}$.

13. (i) $-2\cos x$.

(ii) $\frac{\cos^3 x-\sin^3 x}{(\cos x+\sin x)^2}$, $\quad -\frac{1-x+2x^2}{\sqrt{(1+x^2)}}$, $\quad \frac{e^x-e^{-x}}{e^x+e^{-x}}$.

14. $-\frac{3}{x^4}$, $\quad -\frac{3}{2x^5}$, $\quad -\tan x$, $\quad \frac{1-2x^2}{\sqrt{(1-x^2)}}$, $\quad \frac{e^x(1-4x+x^2)}{(1+x^2)^3}$.

15. (i) $(-1)^n\frac{(n+1)!}{x^{n+2}}$. $\quad$ (ii) $2^n\sin(2x+\frac{1}{2}n\pi)$.

(iii) $2^n e^{2x}\left\{\sin 2x+n\sin(2x+\frac{1}{2}\pi)+\frac{n(n-1)}{2!}\sin(2x+\pi)+\dots\right.$

$\left.\dots+\sin(2x+n.\frac{1}{2}\pi)\right\}$.

$k=0,1,2$.

16. $\frac{6x-8}{(1+3x)^3}$, $\quad 2\tan x\sec^2 x\, e^{\tan^2 x}$, $\quad \frac{-1}{1+x^2}$.

17. $\sec^2 x\, e^{\tan x}$, $\quad \frac{-3}{2(x-2)^{\frac{1}{2}}(x+1)^{\frac{3}{2}}}$.

18. Velocity, $\frac{1}{2}ak(1-2\sin kt)$; acceleration, $-ak^2\cos kt$.

$$t=\frac{\pi}{6k},\quad x=\frac{a}{12}(\pi+6\sqrt{3}).\quad t=\frac{13\pi}{6k},\quad x=\frac{a}{12}(13\pi+6\sqrt{3}).$$

At rest when $t=\dfrac{5\pi}{6k}$, $x=\dfrac{a}{12}(5\pi-6\sqrt{3})$.

Total distance, $\dfrac{a}{3}(\pi+6\sqrt{3})$.

19. Velocity, $e^t\sqrt{2}$; acceleration, $2e^t$.

20. (i) $v^2=p^2(a^2\sin^2 pt+b^2\cos^2 pt)$.

(ii) $f^2=p^4(a^2\cos^2 pt+b^2\sin^2 pt)$.

21. Velocity, $t^2\cos t-4t\sin t-6\cos t$;

Minimum at $t=(4k+1)\frac{1}{2}\pi$; Maximum at $t=(4k-1)\frac{1}{2}\pi$.

22. Velocity, $-ap(\sin pt+\sin 2pt)$;

Acceleration, $-ap^2(\cos pt+2\cos 2pt)$;

$$x=-\frac{3a}{4},\quad -\frac{a}{2},\quad -\frac{3a}{4}.$$

24. $3x-y=0$ at $(1,3)$; $5x+y=0$ at $(-1,5)$.

25. $x-2y=0$.

27. $\cos^{-1}\left\{\dfrac{a(q-p)}{2pq}\right\}$.

29. Maximum.

32. (i) 0·857. (ii) 30·2.

33. (i) 2·004. (ii) 0·515.

34. 8·03.

35. 1·532.

39. $x-\frac{1}{2}x^2+\frac{1}{6}x^3-\frac{1}{12}x^4+\dots$.

40. $-\frac{1}{2}x^2-\frac{1}{12}x^4$.

43. $xy^{(n+2)}-(2x-1-n)y^{(n+1)}+(x-1-2n)y^{(n)}+ny^{(n-1)}=0$.

44. $c_1=b,\quad c_2=2ab,\quad c_3=3a^2b-b^3$.

45. $y=\alpha,\quad y'=0,\quad y''=\cot\alpha,\quad 45{\cdot}028^\circ$.

46. $1+2x+2x^2+\frac{8}{3}x^3+\dots$, 0·9930.

47. 0·06285.

49. $y' = 1, \quad y'' = 1, \quad y''' = 2, \quad y^{\text{iv}} = 3.$

50. $y''' = 2 + 8y^2 + 6y^4,$

$y^{\text{iv}} = 16y + 40y^3 + 24y^5,$

$y^{\text{v}} = 16 + 136y^2 + 240y^4 + 120y^6,$

$x + \frac{1}{3}x^3 + \frac{2}{15}x^5.$

52. $-\frac{1}{2}e^{-x^2}, \quad \frac{1}{2}(\tan^{-1} x)^2, \quad e^x(x^2 - 2x + 2).$

53. $1 - \frac{1}{4}\pi.$

54. $\frac{1}{2}x^2 \log x - \frac{1}{4}x^2, \quad \log \tan \frac{1}{2}x, \quad \frac{1}{4}\log\left(\frac{1+x}{1-x}\right) - \frac{1}{2}\tan^{-1} x.$

55. $\sin x - \frac{1}{3}\sin^3 x, \quad \log\left(\frac{x}{x+1}\right) + \frac{1}{x} - \frac{1}{2x^2},$

$x\{(\log x)^3 - 3(\log x)^2 + 6\log x - 6\}, \quad \frac{2}{\sqrt{5}}\tan^{-1}(\sqrt{5}\tan\frac{1}{2}x).$

56. $12x + 8\sin 2x + \sin 4x, \quad x^3\{9(\log x)^2 - 6\log x + 2\},$

$\log\left(\frac{1+x}{1-x}\right) + 2\tan^{-1} x - 2x, \quad \frac{a^x}{\log a}.$

57. $4\log x - 2\log(1+x) - \log(1+x^2) - 2\tan^{-1} x,$

$2\log 2 - 1, \quad \frac{1}{2}\tan^{-1}\frac{1}{2}.$

58. $\frac{1}{6}\log\left(\frac{1+x^3}{1-x^3}\right), \quad \sin^{-1}(x-1),$

$x\tan x + \log\cos x, \quad \frac{1}{3}\tan^3 x - \tan x + x.$

59. $\frac{1}{6(1-3x)^2}, \quad \frac{1}{2}\tan 2x - x, \quad (1+x^2)\tan^{-1} x - x, \quad \tan^{-1}(\sin x).$

60. $\frac{8}{15}, \quad \frac{3}{8}\pi, \quad \frac{2}{5}(e^{2\pi} - 1).$

61. (i) $\frac{1}{1+x} + \log\left(\frac{x}{1+x}\right), \quad \log(x^2 + 2x + 2) + \tan^{-1}(x+1).$

62. $\frac{1}{3}\sin 3x - \frac{1}{9}\sin^3 3x, \quad \frac{2}{\sqrt{5}}\tan^{-1}(\sqrt{5}\tan\frac{1}{2}x),$

$-x^3\cos x + 3x^2\sin x + 6x\cos x - 6\sin x,$

$(1-x^2)y - 4x^2 + Ax + B.$

64. $\frac{3}{2}\pi a^2$, $(\frac{5}{6}a, 0)$.

65. $x \sec \xi + y - \sin \xi - \xi \sec \xi = 0$.

67. $\frac{4}{3}\pi a^2 b$.

68. $\frac{a}{3} \cdot \frac{(2ac+3b)}{(ac+2b)}$.

69. $y = 2 + 3x - x^3$; $(1, 4)$, $(-1, 0)$; $\frac{14}{15}$.

71. $\frac{1}{2}x - \frac{1}{4}\sin 2x$, $\frac{1}{3}\cos^3 x - \cos x$, $x^2 \sin x + 2x \cos x - 2 \sin x$.

72. $\frac{4}{3}a^2$, $\frac{8}{3}\pi a^2(2\sqrt{2} - 1)$.

73. $\frac{16}{3}$, $(\frac{21}{5}, -\frac{2}{5})$.

75. $\frac{18}{5}a^2$, $\frac{81}{160}a^2$.

77. $\frac{2555}{558}a^2$.

78. $\left(\frac{128\sqrt{2}}{105}\frac{a}{\pi}, 0\right)$; $\frac{32\sqrt{2}}{105}\pi a^3$.

79. $(\frac{4}{7}, 0)$; $\frac{32\sqrt{2}}{105}\pi$.

80. $\frac{4}{3}a^2$, $(\frac{3}{5}a, \frac{3}{4}a)$; $(\frac{2}{3}a, 0)$, $\frac{8}{3}\pi\rho a^5$.

83. $I_1 = 1$, $I_2 = \frac{1}{4}\pi$, $I_3 = \frac{2}{3}$, $I_4 = \frac{3}{16}\pi$.

84. $\sqrt{\left(\frac{8\pi a^2}{21}\right)}$.

85. $\frac{3}{2}\pi a^2$, $\frac{8}{3}\pi a^3$.

86. -1 and 0, 0 and 1, 2 and 3, $2{\cdot}88$.

87. $2{\cdot}426$.

88. $-1{\cdot}844$.

89. $\frac{2^n}{n!}\sin \frac{1}{3}n\pi$.

90. $x - \frac{3}{2}x^2 + \frac{11}{6}x^3 - \frac{25}{12}x^4 + \ldots$.

CHAPTER IX

Examples I:

1. $3 \cosh 3x$.
2. $4 \cosh 2x \sinh 2x$.
3. $\tanh x + x \operatorname{sech}^2 x$.
4. $4 \cosh (2x+1) \sinh (2x+1)$.
5. $\cosh x \cos x - \sinh x \sin x$.
6. $2 \operatorname{sech} x \sin x \cos x - \operatorname{sech} x \tanh x \sin^2 x$.
7. $3(1+x)^2 \cosh^3 3x + 9(1+x)^3 \cosh^2 3x \sinh 3x$.
8. $2x \tanh^2 4x + 8x^2 \tanh 4x \operatorname{sech}^2 4x$.
9. $\coth x$.
10. 1.
11. $\cosh x e^{\sinh x}$.
12. $e^{-\tanh x}(1 - x \operatorname{sech}^2 x)$.

13. $\frac{1}{4}\cosh 4x$.
14. $\frac{1}{4}\sinh 2x-\frac{1}{2}x$.
15. $\frac{1}{4}\sinh 2x+\frac{1}{2}x$.
16. $x\cosh x-\sinh x$.
17. $\frac{1}{4}e^{2x}+\frac{1}{2}x$.
18. $\frac{1}{8}\cosh 4x+\frac{1}{4}\cosh 2x$.
19. $\frac{1}{4}x\sinh 2x-\frac{1}{8}\cosh 2x-\frac{1}{4}x^2$.
20. $\sinh x+\frac{1}{3}\sinh^3 x$.
21. $x-\tanh x$.
22. $x^2\sinh x-2x\cosh x+2\sinh x$.
23. $\frac{1}{14}e^{7x}+\frac{1}{6}e^{-3x}$.
24. $\frac{1}{2}\tanh {}^2x$.

Examples II:

1. $\cosh^{-1}x\pm\dfrac{x}{\sqrt{(x^2-1)}}$.
2. $\dfrac{2x}{\sqrt{(x^4+2x^2+2)}}$.
3. $\dfrac{1}{1-x^2}$.
4. $\dfrac{\pm 1}{x\sqrt{(1-x^2)}}$
5. $\dfrac{-1}{x\sqrt{(1+x^2)}}$.
6. $\dfrac{\pm 1}{\sqrt{(x^2-1)}\cosh^{-1}x}$.
7. $\cosh^{-1}(x^2+1)\pm\dfrac{2x}{\sqrt{(x^2+2)}}$
8. $\pm\dfrac{2\cosh^{-1}x}{\sqrt{(x^2-1)}}$.
9. $\dfrac{-1}{\sqrt{(x^2+1)}(\sinh^{-1}x)^2}$.
10. $\cosh^{-1}\frac{1}{2}x \quad (x>2)$.
11. $\frac{1}{3}\sinh^{-1}3x$.
12. $\frac{1}{2}\cosh^{-1}\frac{2}{3}x \quad (x>\frac{3}{2})$.
13. $\cosh^{-1}\left(\dfrac{x+1}{2}\right) \quad (x>1)$.
14. $\sinh^{-1}\left(\dfrac{x+1}{2}\right)$.
15. $\frac{1}{2}\cosh^{-1}\left(\dfrac{2x-1}{4}\right) \quad (x>\frac{5}{2})$.

CHAPTER X

Examples II:

1. $x^2-y^2=a^2$.
2. $x^2-y^2=a^2$.
3. $x^2-y^2=a^2$.
4. $x^2+y^2=a^2$.
5. $4(x-2a)^3-27ay^2=0$.
6. $x=\frac{1}{2}c\left(3t+\dfrac{1}{t^3}\right),\quad y=\frac{1}{2}c\left(\dfrac{3}{t}+t^3\right)$.

Examples III:

1. $x = a(2+3t^2), \quad y = -2at^3.$

2. $x = \dfrac{1}{a}(a^2-b^2)\cos^3 t, \quad y = -\dfrac{1}{b}(a^2-b^2)\sin^3 t.$

3. $x = 2a\sec^3 t, \quad y = -2a\tan^3 t.$

REVISION EXAMPLES V

1. Tangent, $x\sin\psi - y\cos\psi - 2a\psi\sin\psi = 0$;

   normal, $x\cos\psi + y\sin\psi - 2a\psi\cos\psi - 2a\sin\psi = 0.$

2. $t = \sinh^{-1}\left(\dfrac{s}{\sqrt{2}}\right)$

3. $\dfrac{2}{\sqrt{\lambda}}(c+\lambda)^{\frac{3}{2}}.$

4. $(1, \frac{1}{4}).$

6. $\dot{x} = -2a\sin 2t - 2a\sin t, \quad \dot{y} = 2a\cos 2t + 2a\cos t;$

   speed, $4a\cos\frac{1}{2}t$ (numerical value).

9. $\dfrac{t(2+t)}{1+t}.$

10. $\frac{3}{8} + \frac{1}{2}\cosh 2\theta + \frac{1}{8}\cosh 4\theta.$

13. $2\pi a\left(x\sinh\dfrac{x}{a} - a\cosh\dfrac{x}{a} + a\right).$

14. $8a.$

15. $\frac{27}{2}.$

16. $1, \quad (0, 2).$

17. $(b^2/2a).$

18. $\dfrac{1}{2a}.$

20. $\dfrac{13^{\frac{3}{2}}a}{6}.$

21. $\dfrac{13^{\frac{3}{2}}a}{4}.$

22. $\sqrt{2}(1+2x+2x^2)^{\frac{3}{2}}, \quad (-1, -1).$

24. $\frac{1}{12}\pi a^2.$

26. $-\frac{2}{9}\sqrt{3}$ at $x = \frac{1}{2}\sqrt{2}.$

REVISION EXAMPLES VI

6. $f_{n+2}(x) = f_n''(x) + 4xf_n'(x) + 2(1+2x^2)f_n(x).$

7. $y = \displaystyle\sum_{n=0}^{\infty} \frac{1}{(2n+1)!}(2n-1)^2(2n-3)^2 \ldots 3^2 x^{2n+1}.$

13. $f(x)$, $f'(x)$, $f''(x)$ continuous everywhere in the range;
$f'''(x)$ has a discontinuity at $x = 1$;
Maximum.

16. $\frac{1}{4}$.

17. 0·55.

18. $$y_{n+2} + a\left\{y_1 y_{n+1} + n y_2 y_n + \frac{n(n-1)}{2!} y_3 y_{n-1} + \ldots + y_{n+1} y_1\right\} = 0.$$

19. $$1 + \sqrt{3}x + x^2 - \tfrac{1}{3}x^4 - \tfrac{2}{15}\sqrt{3}x^5 - \tfrac{4}{45}x^6 + \ldots + \left(\frac{2^r}{r!}\cos\frac{r\pi}{6}\right)x^r + \ldots;$$
$$2^{-n} e^{x\sqrt{3}} \cos\left(x - \frac{n\pi}{6}\right) + \text{arbitrary polynomial of degree } n-1.$$

21. 1·8, 4·5 (radians).

23. $r^k = a^k \sin k\theta$.

26. $-\dfrac{1}{27} \pm \dfrac{1}{\sqrt{10}}$.

27. $$\frac{dy}{dx} = \frac{dy}{dt}\bigg/\frac{dx}{dt}, \quad \frac{d^2y}{dx^2} = \frac{\dfrac{dx}{dt}\dfrac{d^2y}{dt^2} - \dfrac{d^2x}{dt^2}\dfrac{dy}{dt}}{\left(\dfrac{dx}{dt}\right)^3},$$
$$\frac{dx}{dy} = \frac{dx}{dt}\bigg/\frac{dy}{dt}, \quad \frac{d^2x}{dy^2} = \frac{\dfrac{d^2x}{dt^2}\dfrac{dy}{dt} - \dfrac{dx}{dt}\dfrac{d^2y}{dt^2}}{\left(\dfrac{dy}{dt}\right)^3}.$$

28. $\frac{1}{3}\{b + \sqrt{(3a^2 + b^2)}\}$.

29. Equality when $e^{x-1} = y$.

30. (ii) Minimum.

35. $\frac{1}{3}\pi$.

37. $\frac{8}{3}\pi$.

38. $x = -\frac{11}{12}\pi$, $y = -0{\cdot}65$, maximum;
$x = -\frac{13}{20}\pi$, $y = -1{\cdot}19$, minimum;
$x = -\frac{1}{4}\pi$, $y = 0$, inflexion;
$x = \frac{3}{20}\pi$, $y = 1{\cdot}19$, maximum;
$x = \frac{5}{12}\pi$, $y = 0{\cdot}65$, minimum;
$x = \frac{11}{20}\pi$, $y = 0{\cdot}73$, maximum;
$x = \frac{19}{20}\pi$, $y = 0{\cdot}44$, minimum.

40. $(1, 2)$, $\left(\frac{1}{\sqrt{2}}, 3-\frac{1}{\sqrt{2}}\right)$; $\sqrt{2}-1$.

41. (i) Maximum, $y = \frac{1}{2}(\sqrt{2}-1)$; minimum, $y = -\frac{1}{2}(\sqrt{2}+1)$.

(ii) Inflexions, $(1, 0)$, $\{-2+\sqrt{3}, -\frac{1}{4}(3-\sqrt{3})\}$, and $\{-2-\sqrt{3}, -\frac{1}{4}(3+\sqrt{3})\}$.

42. $A = \dfrac{p}{\sqrt{(p^2+k^2)}}$.

47. $$\frac{dy}{dx} = \frac{\tan\alpha + \dfrac{d\eta}{d\xi}}{1-\dfrac{d\eta}{d\xi}\tan\alpha}, \quad \frac{d^2y}{dx^2} = \frac{\dfrac{d^2\eta}{d\xi^2}}{\left(\cos\alpha - \dfrac{d\eta}{d\xi}\sin\alpha\right)^3}.$$

49. $\dfrac{5\sqrt{5}}{6}$.

50. $x = 0$, maximum, $\kappa = -16$;

$x = \frac{4}{5}$, minimum, $\kappa = \frac{144}{25}$.

51. $3a\cos t\sin t$ (numerical value of).

52. $(0, 0)$, $(2, 2)$.

53. $\frac{1}{2}$, $2\sqrt{2}$.

56. Minimum.

58. $\frac{1}{16}\log\left(\dfrac{x^2+2x+2}{x^2-2x+2}\right) + \frac{1}{8}\tan^{-1}(x+1) + \frac{1}{8}\tan^{-1}(x-1)$,

$\dfrac{e^{ax}}{a^2+b^2}(a\cos bx + b\sin bx)$,

$\frac{1}{2}x^2 - \frac{1}{2}x\sqrt{(x^2-1)} + \frac{1}{2}\cosh^{-1}x$.

60. (b) $\dfrac{A}{x}$, A an arbitrary constant.

61. $a^2(\alpha + 3\sin\alpha)$, where $\cos\alpha = -\frac{1}{3}$ $(\frac{1}{2}\pi < \alpha < \pi)$.

62. (i) $\frac{8}{15}\log\sin x - \frac{1}{15}x\cos x(3\operatorname{cosec}^5 x + 4\operatorname{cosec}^3 x + 8\operatorname{cosec} x) - \frac{1}{20}\operatorname{cosec}^4 x - \frac{2}{15}\operatorname{cosec}^2 x$.

(ii) $\dfrac{b^5}{a^6}\log\left(\dfrac{a+bx}{x}\right) - \dfrac{b^4}{a^5x} + \dfrac{b^3}{2a^4x^2} - \dfrac{b^2}{3a^3x^3} + \dfrac{b}{4a^2x^4} - \dfrac{1}{5ax^5}$.

63. $\frac{1}{2}at\sqrt{(1+t^2)} + \frac{1}{2}a\sinh^{-1}t$.

64. (i) $n = 4k, \lambda = 1;\ n = 4k+1, \lambda = n;$
$n = 4k+2, \lambda = -1;\ n = 4k+3, \lambda = -n.$

(ii) $(\frac{1}{3} + \frac{1}{2}\pi)a^2.$

65. $(\tan^{-1} x)^2 - 2x \tan^{-1} x + x \tan^{-1} x \log(1+x^2)$
$+ \log(1+x^2) - \frac{1}{4}\{\log(1+x^2)\}^2.$

68. $\phi(x) = (1+x)(3+x).$
$f(x) = (1+x)^2,\ g(x) = 3(1+x)^2.$

71. $A = \frac{3}{8},\ B = \frac{2}{3},\ C = \frac{5}{8},\ D = \frac{3}{8}.$

72. $A = \frac{1}{16},\ B = \frac{1}{6},\ C = -\frac{1}{16},\ D = \frac{1}{16}.$

76. $a^2 I_n = a(1-x^2)^n \sinh ax + 2nx(1-x^2)^{n-1} \cosh ax$
$- 2n(2n-1) I_{n-1} + 4n(n-1) I_{n-2} \quad (n \geqslant 2).$

$I_3 = \dfrac{888}{e} - 120e.$

77. $\cos^{q-1} x\{p \sin^{p-1} x - (p+q+k^2+pk^2) \sin^{p+1} x$
$+ k^2(p+q+1) \sin^{p+3} x\}.$

$k^2(m-1) I_m - (1+k^2)(m-2) I_{m-2} + (m-3) I_{m-4} = 0.$

79. Volume $5\pi^2 a^3$, area $\frac{64}{3}\pi a^2$.

81. $x = a(\cos t \cos 3t + 3 \sin t \sin 3t),$
$y = a(\cos t \sin 3t - 3 \sin t \cos 3t).$

84. $\dfrac{1}{x^2} + \dfrac{1}{y^2} = \dfrac{1}{a^2};\quad \frac{2}{3}a.$

85. $P(-8at^3, -6at);$
Normal, $4t^2x + y + 6at + 32at^5 = 0;$
Inflexion at origin.

86. Normal, $2x + 3ty - 3at^4 - 2at^2 = 0;$
Centre of curvature, $(-\frac{9}{2}at^4 - at^2, 4at^3 + \frac{4}{3}at);$
Radius $\frac{1}{6}at(4+9t^2)^{\frac{3}{2}}$.

87. Envelope, $x = -f'(t), \quad y = f(t) - tf'(t);$
$\rho = f''(t)(1+t^2)^{\frac{3}{2}}.$

88. Envelope, $x = a \sin t(3 - 2\sin^2 t),$
$y = a \cos t(3 - 2\cos^2 t).$

89. Tangent, $x \sin t - y \cos t + \cos 2t = 0;$
Normal, $x \cos t + y \sin t - 2 \sin 2t = 0.$

90. Tangent, $t(3+t^2)x-2y-t^3=0$;
Normal, $2x+t(3+t^2)y-t^2(2+t^2)=0$.

92. $\pi(ab-cdk)$.

93. $\frac{16}{5}ab$.

94. $\dfrac{1}{8e^2}$.

95. $\frac{6}{5}\sqrt{3}$.

98. $2-\frac{1}{2}\pi$.

102. $2c\sinh\dfrac{a}{c}$; $\pi c\left(2a+c\sinh\dfrac{2a}{c}\right)$.

103. $\frac{64}{3}\pi a^2$.

105. Volume, $2\pi^2 a^2 h$; area $4\pi^2 ah$.

106. $\dfrac{4\sqrt{2}}{\pi}a$, $\dfrac{2}{\pi a}\log(1+\sqrt{2})$.

107. $f+\dfrac{a^2}{3f}$.

108. $\dfrac{1}{a(a^2-f^2)}$ if $f<a$, $\dfrac{1}{f(f^2-a^2)}$ if $f>a$.

109. $\frac{6}{5}a$.

CHAPTER XI

Examples I:

1. 1 and 3; $2\pm\sqrt{3}$; $2\pm i$.
2. -5 and -3; $-4\pm\sqrt{5}$; $-4\pm 2i$.
3. -1 and 3; $1\pm\sqrt{5}$; $1\pm 3i$.

Examples II:

1. $10+3i$. 2. $-8+8i$. 3. $34+22i$. 4. $-16-3i$. 5. $2+7i$.
6. $0+0i$. 7. $0+2i$. 8. $2+11i$. 9. $-2-16i$. 10. $-10+0i$.

Examples III:

1. $-7+22i$.
2. $26+2i$.
3. $7-i$.
4. a^2+b^2.
5. $-3+4i$.
6. $\cos(A+B)+i\sin(A+B)$.
7. 10.
8. $-46+9i$.

Examples IV:

1. i.
2. $\frac{24}{25}+\frac{7}{25}i$.
3. $\frac{5}{7}-\frac{6}{7}i$.
4. $\frac{5}{17}-\frac{14}{17}i$.
5. $-\frac{27}{37}+\frac{23}{37}i$.
6. $\cos\theta+i\sin\theta$.

Examples V:

1. $5-2i$, $-1+8i$, $21-i$, $-\frac{9}{34}+\frac{19}{34}i$.

2. $-7+9i$, $-1-5i$, $-2-34i$, $\frac{13}{29}+\frac{11}{29}i$.

3. $4+2i$, $4-2i$, $8i$, $-2i$.

4. $3+i$, $3+3i$, $2-3i$, $-2+3i$.

5. 4, $6i$, 13, $-\frac{5}{13}+\frac{12}{13}i$.

6. -6, $8i$, 25, $-\frac{7}{25}-\frac{24}{25}i$.

7. $\frac{3}{25}-\frac{4}{25}i$.

8. $-\frac{5}{169}-\frac{12}{169}i$.

9. $-\frac{1}{6}i$.

10. $-\frac{3}{50}+\frac{2}{25}i$.

11. $-2\pm 3i$.

12. $1\pm i$.

13. $-3\pm i$.

14. $\pm\frac{3}{2}i$.

15. $4\pm 3i$.

16. $-2\pm i$.

Examples VII:

3. $2, -30°$; $5, 53° 7'$; $13, 112° 36'$; $3, 0°$; $10, -53° 7'$; $2, -90°$.

5. Straight line, $4x+10y-21=0$.

Examples VIII:

1. (*a*) $(3,2)$; (*b*) $(2,1)$; (*c*) $(4,7)$.

2. (i) (*a*) $(1,0)$; (*b*) $(3,-5)$; (*c*) $(3,-5)$.

   (ii) (*a*) $(2,-1)$; (*b*) $(2,1)$; (*c*) $(5,0)$.

Examples IX:

1. (i) $-0{\cdot}5+0{\cdot}866i$.

   (ii) $\pm(0{\cdot}866+0{\cdot}5i)$.

   (iii) $0{\cdot}940+0{\cdot}342i$, $-0{\cdot}766+0{\cdot}643i$, $-0{\cdot}174-0{\cdot}985i$.

   (iv) $\pm(0{\cdot}966+0{\cdot}259i)$, $\pm(0{\cdot}259-0{\cdot}966i)$.

2. (i) $7+24i$.

(ii) $\pm(2{\cdot}121+0{\cdot}707i)$.

(iii) $1{\cdot}671+0{\cdot}364i,\ -1{\cdot}151+1{\cdot}265i,\ -0{\cdot}520-1{\cdot}629i$.

(iv) $\pm(1{\cdot}476+0{\cdot}239i),\ \pm(0{\cdot}239-1{\cdot}476i)$.

3. (i) $119-120i$.

(ii) $\pm(3{\cdot}535+0{\cdot}708i)$.

(iii) $2{\cdot}331+0{\cdot}308i,\ -1{\cdot}432+1{\cdot}865i,\ -0{\cdot}899-2{\cdot}173i$.

(iv) $\pm(1{\cdot}890+0{\cdot}187i),\ \pm(0{\cdot}187-1{\cdot}890i)$.

Examples X:

1. $4(\cos\pi+i\sin\pi)$.

2. $\dfrac{1}{4\sqrt{2}}\left(\cos\dfrac{5\pi}{4}+i\sin\dfrac{5\pi}{4}\right)$.

3. $64(\cos 2\pi+i\sin 2\pi)$.

4. $2^{\frac{1}{3}}\left\{\cos\dfrac{\pi}{18}(12k-1)+i\sin\dfrac{\pi}{18}(12k-1)\right\}$.

5. $2^{\frac{1}{12}}\left\{\cos\dfrac{\pi}{24}(8k-1)+i\sin\dfrac{\pi}{24}(8k-1)\right\}$.

6. $2^{\frac{1}{4}}\left\{\cos\dfrac{7\pi}{24}(12k+1)+i\sin\dfrac{7\pi}{24}(12k+1)\right\}$.

7. (i) $1,\ \cos\dfrac{2\pi}{5}\pm i\sin\dfrac{2\pi}{5},\ \cos\dfrac{4\pi}{5}\pm i\sin\dfrac{4\pi}{5}$.

(ii) $\pm 1,\ \pm\tfrac{1}{2}(1\pm i\sqrt{3})$.

Examples XI:

1. $e^{-(\frac{1}{4}\pi+2k\pi)}\{\cos(\tfrac{1}{2}\log 2)+i\sin(\tfrac{1}{2}\log 2)\}$.

2. $-2e^{(2k\pi-\frac{1}{4}\pi)}\{\sin(\tfrac{1}{2}\log 2)+i\cos(\tfrac{1}{2}\log 2)\}$.

3. $e^{-(\frac{1}{12}\pi+k\pi)}\{\cos(\tfrac{1}{2}\log 2)+i\sin(\tfrac{1}{2}\log 2)\}$.

4. $-8e^{-(4k\pi-\frac{3}{2}\pi)}\{\cos(\log 4)+i\sin(\log 4)\}$.

5. $2e^{-(k\pi-\frac{1}{12}\pi)}\{\cos(\tfrac{1}{2}\log 2-\tfrac{1}{6}\pi)+i\sin(\tfrac{1}{2}\log 2-\tfrac{1}{6}\pi)\}$.

6. $-4e^{-(k\pi+\frac{1}{8}\pi)}\{\cos(\tfrac{1}{4}\log 2)+i\sin(\tfrac{1}{4}\log 2)\}$.

Examples XII:

1. $\dfrac{1-\cos\theta+\cos n\theta-\cos(n+1)\theta}{2(1-\cos\theta)}$.

2. $\dfrac{\sin\theta+(-1)^{n+1}x^n\{x\sin n\theta+\sin(n+1)\theta\}}{1+2x\cos\theta+x^2}$.

3. $\cos x+x\sin x$.

4. $\frac{1}{2}e^x(\sin x-\cos x)$.

5. $\frac{1}{25}e^{4x}(4\sin 3x-3\cos 3x)$.

6. $\frac{1}{5}xe^{2x}(2\sin x-\cos x)+\frac{1}{25}e^{2x}(4\cos x-3\sin x)$.

7. $\frac{1}{9}\sin 3x-\frac{1}{3}x\cos 3x$.

8. $-\frac{1}{25}xe^{-4x}(4\cos 3x-3\sin 3x)-\frac{1}{625}e^{-4x}(7\cos 3x-24\sin 3x)$.

REVISION EXAMPLES VII

3. $A=\dfrac{\sin a\cos a}{\cos^2 a+\sinh^2 b}$, $\quad B=\dfrac{-\sinh b\cosh b}{\cos^2 a+\sinh^2 b}$.

5. $-\tan\dfrac{\pi}{4n}(4k-1)$, $k=1,\ldots,n$.

7. $\cos^2 x+\sinh^2 y$.

8. $X=\dfrac{(x_1^2+y_1^2)x_2+(x_2^2+y_2^2)x_1-(x_1+x_2)}{(x_1^2+y_1^2)(x_2^2+y_2^2)-2(x_1x_2-y_1y_2)+1}$.

   $Y=-\dfrac{(x_1^2+y_1^2)y_2+(x_2^2+y_2^2)y_1+(y_1+y_2)}{(x_1^2+y_1^2)(x_2^2+y_2^2)-2(x_1x_2-y_1y_2)+1}$.

10. $x-iy, r(\cos\theta-i\sin\theta)$; $\pm 1\pm 3i$.

11. $z_1=2(\cos\frac{1}{4}\pi+i\sin\frac{1}{4}\pi), z_2=8(\cos\frac{3}{4}\pi+i\sin\frac{3}{4}\pi)$;

    $2(\cos\frac{11}{12}\pi+i\sin\frac{11}{12}\pi), 2(\cos\frac{19}{12}\pi+i\sin\frac{19}{12}\pi)$.

16. (i) $(-1,0)$, $(3,0)$;

    (ii) circle, centre $(2,0)$, radius 3;

    (iii) ellipse, foci $(1,0)$, $(2,0)$, eccentricity $\frac{1}{3}$.

17. $1, \omega, \omega^2, \omega^3, \omega^4$ where $\omega=\cos\frac{2}{5}\pi+i\sin\frac{2}{5}\pi$;

    $x=\frac{1}{4}(1+i\cot\frac{1}{5}k\pi)$, $k=1,2,3,4$.

19. $(1,\sqrt{3})$, distance 2.

20. $z_3 = 7+2\sqrt{3}+i(4+3\sqrt{3})$ or $7-2\sqrt{3}+i(4-3\sqrt{3})$.
Vertices $(4+4\sin\frac{1}{3}k\pi+6\cos\frac{1}{3}k\pi,\ 6+6\sin\frac{1}{3}k\pi-4\cos\frac{1}{3}k\pi)$, $k = 0, 1, \ldots, 5$.

21. $\cos\frac{2}{5}k\pi+i\sin\frac{2}{5}k\pi,\quad k = 0, 1, \ldots, 4$.

22. $\dfrac{4-5\cos\frac{2}{5}k\pi-3i\sin\frac{2}{5}k\pi}{5-4\cos\frac{2}{5}k\pi}\quad (k = 0, 1, 2, 3, 4)$.

23. $(\sqrt{3}+i)(a+ib)$ and $a+ib$ subtend an angle $\frac{1}{6}\pi$ at the origin, and the distance of $(\sqrt{3}+i)(a+ib)$ from the origin is twice that of $a+ib$ from the origin. C' is at $\pm 2\sqrt{3}+2i$.

24. $\pm(3-2i)$; $1+i$, $\sqrt{2}\cos\frac{11}{12}\pi+i\sqrt{2}\sin\frac{11}{12}\pi$,
$\sqrt{2}\cos\frac{19}{12}\pi+i\sqrt{2}\sin\frac{19}{12}\pi$.

26. $-2300-2100i$.

27. (i) $(1, 1)$, $(1, -1)$,
(ii) all points with $x \geqslant 0$,
(iii) interior of circle, centre $(-\frac{5}{3}, 0)$, radius $\frac{4}{3}$,
(iv) interior of ellipse, foci $(\pm 1, 0)$, eccentricity $\frac{1}{2}$.

29. (i) $\frac{1}{5}$, $-\frac{2}{5}$; $\dfrac{1}{\sqrt{5}}$.
(ii) $\dfrac{2ab}{a^2+b^2}$, $\dfrac{b^2-a^2}{a^2+b^2}$; 1,
(iii) $\dfrac{\cos\frac{1}{2}\alpha\cos\frac{1}{2}(\alpha-\beta)}{\cos\frac{1}{2}\beta}$, $\dfrac{\cos\frac{1}{2}\alpha\sin\frac{1}{2}(\alpha-\beta)}{\cos\frac{1}{2}\beta}$; $\left|\dfrac{\cos\frac{1}{2}\alpha}{\cos\frac{1}{2}\beta}\right|$.

30. $i\pi$, $\frac{1}{2}\log_e 5-i\tan^{-1}\frac{1}{2}$, $(x+iy)\log_e 10$.

32. $1{\cdot}67-0{\cdot}36i$, $-0{\cdot}52+1{\cdot}63i$, $-1{\cdot}15-1{\cdot}27i$.

33. $\frac{1}{2}(z_1+z_2)$; $(2+2\sqrt{3})i$.

34. $\dfrac{1}{2}\left(z+\dfrac{1}{z}\right)$, $\dfrac{1}{2i}\left(z-\dfrac{1}{z}\right)$, $\dfrac{1}{2}\left(z^n+\dfrac{1}{z^n}\right)$, $\dfrac{1}{2i}\left(z^n-\dfrac{1}{z^n}\right)$;
$\dfrac{-1}{2^6}(\sin 7\theta-7\sin 5\theta+21\sin 3\theta-35\sin\theta)$.

37. Circle, centre $(5, -1)$, radius 3; 8, 2.

38. $\dfrac{\sqrt{(10+6y)}}{|10-6y|}$.

39. Amplitude increases from $-\pi$ to π.

40. $2^{\frac{4}{3}}\cos\frac{1}{9}\pi,\ -2^{\frac{4}{3}}\cos\frac{2}{9}\pi,\ -2^{\frac{4}{3}}\cos\frac{4}{9}\pi.$

41. $\dfrac{\cos x\cosh y+i\sin x\sinh y}{\cos^2 x+\sinh^2 y}.$

42. $(a+ib)(a-ib),\ (c+id)(c-id).$

CHAPTER XIII

Examples I:

1. 1. 2. $\sqrt{2}$. 3. 2. 4. $\frac{1}{4}\pi$.

5. $\frac{3}{4}\pi$. 6. $\frac{1}{2}$. 7. $-1/a\quad(a<0)$.

REVISION EXAMPLES VIII

1. $\frac{1}{2}\log\dfrac{x-3}{x-1};\ x+\frac{1}{2}\log\dfrac{(x-3)^9}{x-1};\ \cos^{-1}(2-x).$

2. $x+\log\dfrac{(x-3)^5}{x-1};\ \sqrt{(x^2+2x+2)}-\sinh^{-1}(x+1).$

3. $\frac{1}{2}(a+b)\pi.$

4. $x+\frac{1}{5}\log\{(x-2)^8(x^2+1)\}-\frac{1}{5}\tan^{-1}x;$
$\frac{3}{7}(x+b)^{\frac{7}{3}}+\frac{3}{4}(a-b)(x+b)^{\frac{4}{3}};\ \frac{2}{3}\cos^3 x-\cos x-\frac{1}{5}\cos^5 x.$

5. $\frac{1}{4}\log\dfrac{1+x^2}{(1-x)^2}-\frac{1}{2}\tan^{-1}x;$
$\sqrt{(x^2+4x+5)}-2\log\{x+2+\sqrt{(x^2+4x+5)}\};$
$\frac{1}{2}(a^2+x^2)-a^2\log(a^2+x^2)-\frac{1}{2}a^4/(a^2+x^2).$

6. $\frac{1}{6}\log\dfrac{(x-1)^{10}}{x^9(x+2)};\ \sin^{-1}\left(\dfrac{x-1}{x\sqrt{2}}\right);\ e^x/(x+1).$

7. $\frac{1}{30}\log\dfrac{(x-3)^8}{x^5(x+2)^3};\ \frac{3}{8}\sinh^{-1}\left(\dfrac{2x+1}{\sqrt{3}}\right)+\frac{1}{4}(2x+1)\sqrt{(x^2+x+1)};$
$x\{(\log x)^2-2\log x+2\}.$

8. $u_n=\dfrac{2n+1}{2n+2}u_{n-1}+\dfrac{1}{n+1}\cdot 2^{n-\frac{1}{2}};\ \frac{67}{48}\sqrt{2}+\frac{5}{16}\log(1+\sqrt{2}).$

REVISION EXAMPLES IX

3. (ii) $\dfrac{1}{\sqrt{3}}\tan^{-1}\left\{\dfrac{t\sqrt{3}}{\sqrt{(t^2+4)}}\right\}-\dfrac{1}{2\sqrt{3}}\log\left\{\dfrac{\sqrt{(t^2+4)}+\sqrt{3}}{\sqrt{(t^2+4)}-\sqrt{3}}\right\}.$

4. $\dfrac{\pi^2}{72}+\dfrac{\pi\sqrt{3}}{6}-1$; $2\log(2+\sqrt{3})-\frac{2}{3}\pi$.

6. $\dfrac{2}{\sqrt{5}}\log\dfrac{\sqrt{5}+1}{\sqrt{5}-1}$; $\frac{3}{8}\pi$.

7. $\frac{1}{8}\pi$; $\dfrac{\pi}{2a(a+b)}$; $\dfrac{1}{\sqrt{2}}\log\dfrac{\sqrt{2}+\sqrt{(1+x)}}{\sqrt{2}-\sqrt{(1+x)}}$.

8. $\frac{1}{4}\pi$; $-x/\sqrt{(x^2-1)}$; $\frac{1}{8}x^4-\frac{1}{8}(2x^3-3x)\sin 2x-\frac{3}{16}(2x^2-1)\cos 2x$.

9. $I_n=2\pi$.

10. $a-\dfrac{a}{\sqrt{2}}\log(1+\sqrt{2})$; $\dfrac{\sqrt{2}}{16}\left\{7\tan^{-1}\left(\dfrac{1}{\sqrt{2}}\right)+\sqrt{2}\right\}$.

11. $1-\frac{1}{4}\pi$; $\frac{3}{4}\pi$; $-ae^a/(1-e^a)$.

12. $\frac{1}{8}(\frac{4}{5}-\log 3)$.

13. $\Sigma\dfrac{\pi}{2a(a^2-b^2)(a^2-c^2)}$.

14. $\frac{1}{4}\pi\sec\frac{1}{2}\alpha$, $\frac{1}{2}\pi\sec\frac{1}{2}\alpha$.

16. $\sqrt{2}\log\dfrac{\sqrt{2}-1}{2-\sqrt{3}}$; $\frac{1}{3}\pi-\frac{1}{3}\log 2-\frac{1}{6}$.

17. $$\begin{cases} 2\tan^{-1}\left\{\left(\dfrac{1+r}{1-r}\right)\tan\dfrac{\delta}{2}\right\}+\delta, & r<1, \\ -2\tan^{-1}\left\{\left(\dfrac{r+1}{r-1}\right)\tan\dfrac{\delta}{2}\right\}+\delta, & r>1. \end{cases}$$

Limits $\pi+\delta, r<1$; $-\pi+\delta, r>1$; $I=\delta$ when $r=1$.

18. $x\cos\alpha+\sin\alpha\log\sin(x-\alpha)$;

$\frac{1}{2}\log\dfrac{t}{t-1}-\dfrac{1}{\sqrt{7}}\tan^{-1}\dfrac{2t+1}{\sqrt{7}}$, where $t=\tan\frac{1}{2}x$.

19. $\log\left\{x+\dfrac{1}{x}+\sqrt{(x^2+1+x^{-2})}\right\}$.

20. $\dfrac{1}{\sqrt{(1-b^2)}}\log\left\{\dfrac{x\sqrt{(1-b^2)}+\sqrt{(x^2+b^2)}}{\sqrt{(x^2+1)}}\right\}$; $\dfrac{3(b-a)^4\pi}{128}$; $\dfrac{\pi}{2a}\log\dfrac{1+a}{1-a}$.

21. $2+\frac{3}{2}\log 3$; $\log(2+\sqrt{3})-\frac{1}{2}\sqrt{3}$; $\sqrt{3}-\frac{1}{3}\pi$.

22. $\frac{1}{7}\log\dfrac{x-2}{x+2}+\dfrac{\sqrt{3}}{7}\tan^{-1}\left(\dfrac{x}{\sqrt{3}}\right)$; $e^x(x\log x-1)$; $\dfrac{\pi}{4\sqrt{2}}$.

23. $\frac{1}{4}\pi$.

24. $\frac{1}{2}\log\frac{1+t^2}{2+t^2}+\frac{1}{\sqrt{2}}\tan^{-1}\left(\frac{t}{\sqrt{2}}\right)-\tan^{-1}c+\frac{1}{2\sqrt{2}}\log\frac{\sqrt{2}+s}{\sqrt{2}-s}$;

$\tan\theta-\sec\theta\equiv-\frac{2}{1+\tan\frac{1}{2}\theta}$, differing by a constant.

25. $\frac{4}{x+1+\sqrt{(x^2-1)}}+\log\{x+\sqrt{(x^2-1)}\}$.

26. $9(n-1)I_n-5(2n-3)I_{n-1}+(n-2)I_{n-2}=-\frac{4\sin x}{(5+4\cos x)^{n-1}}$.

27. $(n-2)u_{n-2}-(n-1)u_n=0,\ n>2$;
$u_1=\frac{1}{2}\pi,\ u_2=1$.

28. $\frac{8}{\alpha(\alpha+2)(\alpha+4)}$.

31. If the given integral is u_{n+1}, then

$2n(ab-h^2)u_{n+1}-(2n-1)au_n=\frac{ax+h}{(ax^2+2hx+b)^n}$.

33. $2(n+1)u_{n+2}-(2n+1)u_{n+1}+nu_n=-\frac{(1+x^2)^{\frac{1}{2}}}{(1+x)^{n+1}}$.

34. (i) $\frac{1}{12}\sqrt{3}\log\frac{2\sqrt{(1+x^2)}+x\sqrt{3}}{2\sqrt{(1+x^2)}-x\sqrt{3}}$.

(ii) $u_n-\frac{2n(2n-1)}{a^2}u_{n-1}+\frac{2n(2n-2)}{a^2}u_{n-2}$

$=\frac{1}{a}e^{ax}(1+x^2)^n-\frac{2nx}{a^2}e^{ax}(1+x^2)^{n-1}$.

35. $\pi/\sqrt{(a^2-1)}$; $u_0=\log(1+\operatorname{cosec}\frac{1}{2}\alpha)$, $u_1=u_0\cos\alpha-1+2\sin\frac{1}{2}\alpha$.

36. $u_n=nu_{n-1}$, $u_n=n!$; $u_n+\frac{n(n-1)}{4}u_{n-2}=\frac{\pi^{n+1}}{2(n+1)}$.

38. $\frac{1}{6}-\frac{\sqrt{2}}{8}\tan^{-1}\left(\frac{\sqrt{2}}{4}\right)$.

40. $(n-1)u_n=(n-2)u_{n-2}+2^{(n-2)/2}$; $u_1=\log(1+\sqrt{2})$, $u_2=1$.

41. $\frac{x(8x^4+8x^2+3)\sqrt{(x^2+1)}}{3(2x^2+1)^3}$.

42. Formula: see p. 208.

$I_3=\frac{1}{2}\log\{x+1+\sqrt{(x^2+2x+2)}\}+\frac{1}{6}(2x^2-5x+7)\sqrt{(x^2+2x+2)}$.

APPENDIX

Examples I:

1. $4x^3y^3$, $3x^4y^2$, $12x^2y^3$, $12x^3y^2$, $12x^3y^2$, $6x^4y$.

2. y^2, $2xy$, 0, $2y$, $2y$, $2x$.

3. $2x$, $2y$, 2, 0, 0, 2.

4. $e^x \cos y$, $-e^x \sin y$, $e^x \cos y$, $-e^x \sin y$, $-e^x \sin y$, $-e^x \cos y$.

5. $\dfrac{1}{x+y}$, $\dfrac{1}{x+y}$, $-\dfrac{1}{(x+y)^2}$, $-\dfrac{1}{(x+y)^2}$, $-\dfrac{1}{(x+y)^2}$, $-\dfrac{1}{(x+y)^2}$.

6. $\dfrac{1}{x}$, $\dfrac{1}{y}$, $-\dfrac{1}{x^2}$, 0, 0, $-\dfrac{1}{y^2}$.

7. $3x^2 \sin^2 y$, $2x^3 \sin y \cos y$, $6x \sin^2 y$, $6x^2 \sin y \cos y$,
   $6x^2 \sin y \cos y$, $2x^3(\cos^2 y - \sin^2 y)$.

8. $ye^{xy} \sin x + e^{xy} \cos x$, $xe^{xy} \sin x$,
   $y^2 e^{xy} \sin x + 2ye^{xy} \cos x - e^{xy} \sin x$,
   $e^{xy} \sin x + xye^{xy} \sin x + xe^{xy} \cos x$,
   $e^{xy} \sin x + xye^{xy} \sin x + xe^{xy} \cos x$, $x^2 e^{xy} \sin x$.

9. $\tan^{-1} y$, $\dfrac{x}{1+y^2}$, 0, $\dfrac{1}{1+y^2}$, $\dfrac{1}{1+y^2}$, $-\dfrac{2xy}{(1+y^2)^2}$.

10. $\sec x \tan x$, $\sec y \tan y$, $\sec x \tan^2 x + \sec^3 x$, 0, 0,
    $\sec y \tan^2 y + \sec^3 y$.

11. $e^x \sin^2 2y$, $4e^x \sin 2y \cos 2y$, $e^x \sin^2 2y$, $4e^x \sin 2y \cos 2y$,
    $4e^x \sin 2y \cos 2y$, $8e^x \cos^2 2y - 8e^x \sin^2 2y$.

12. e^y, xe^y, 0, e^y, e^y, xe^y.

Examples II:

3. (i) $ae^{ax} \sin(by+cz)$, $-b^2 e^{ax} \sin(by+cz)$, $ac\, e^{ax} \cos(by+cz)$;

   (ii) $yze^{-x^2} - 2x^2yze^{-x^2}$, 0, $ye^{-x^2} - 2x^2ye^{-x^2}$;

   (iii) $\dfrac{a(y^2+z^2)}{ax+b}$, $2\log(ax+b)$, $\dfrac{2az}{ax+b}$.

INDEX

T0292071

CAMBRIDGE LIBRARY COLLECTION

Books of enduring scholarly value

Physical Sciences

From ancient times, humans have tried to understand the workings of the world around them. The roots of modern physical science go back to the very earliest mechanical devices such as levers and rollers, the mixing of paints and dyes, and the importance of the heavenly bodies in early religious observance and navigation. The physical sciences as we know them today began to emerge as independent academic subjects during the early modern period, in the work of Newton and other 'natural philosophers', and numerous sub-disciplines developed during the centuries that followed. This part of the Cambridge Library Collection is devoted to landmark publications in this area which will be of interest to historians of science concerned with individual scientists, particular discoveries, and advances in scientific method, or with the establishment and development of scientific institutions around the world.

History and Root of the Principle of the Conservation of Energy

The Austrian scientist Ernst Mach (1838–1916) carried out work of importance in several fields of enquiry, including physics, physiology and psychology. In this short work, first published in German in 1872 and translated here into English in 1911 by Philip E.B. Jourdain (1879–1919) from the 1909 second edition, Mach discusses the formulation of one of science's most fundamental theories. He provides his interpretation of the principle of the conservation of energy, claiming its foundations are not in mechanical physics. Mach's 1868 work on the definition of mass – one of his most significant contributions to mechanics – has been incorporated here. His perspective on the topic as a whole remains relevant to those interested in the history of science and the theory of knowledge. Also reissued in this series in English translation are Mach's *The Science of Mechanics* (1893) and *Popular Scientific Lectures* (1895).

Cambridge University Press has long been a pioneer in the reissuing of out-of-print titles from its own backlist, producing digital reprints of books that are still sought after by scholars and students but could not be reprinted economically using traditional technology. The Cambridge Library Collection extends this activity to a wider range of books which are still of importance to researchers and professionals, either for the source material they contain, or as landmarks in the history of their academic discipline.

Drawing from the world-renowned collections in the Cambridge University Library and other partner libraries, and guided by the advice of experts in each subject area, Cambridge University Press is using state-of-the-art scanning machines in its own Printing House to capture the content of each book selected for inclusion. The files are processed to give a consistently clear, crisp image, and the books finished to the high quality standard for which the Press is recognised around the world. The latest print-on-demand technology ensures that the books will remain available indefinitely, and that orders for single or multiple copies can quickly be supplied.

The Cambridge Library Collection brings back to life books of enduring scholarly value (including out-of-copyright works originally issued by other publishers) across a wide range of disciplines in the humanities and social sciences and in science and technology.

History and Root of the Principle of the Conservation of Energy

Ernst Mach
Translated by
Philip E.B. Jourdain

CAMBRIDGE
UNIVERSITY PRESS

University Printing House, Cambridge, CB2 8BS, United Kingdom

Published in the United States of America by Cambridge University Press, New York

Cambridge University Press is part of the University of Cambridge.
It furthers the University's mission by disseminating knowledge in the pursuit of education, learning and research at the highest international levels of excellence.

www.cambridge.org
Information on this title: www.cambridge.org/9781108066662

This edition first published 1911
This digitally printed version 2014

ISBN 978-1-108-06666-2 Paperback

HISTORY AND ROOT

OF THE PRINCIPLE OF THE

CONSERVATION OF ENERGY

BY

ERNST MACH

TRANSLATED FROM THE GERMAN AND ANNOTATED BY

PHILIP E. B. JOURDAIN, M.A. (Cantab.)

CHICAGO
THE OPEN COURT PUBLISHING CO
LONDON
KEGAN PAUL, TRENCH, TRÜBNER & CO., LTD.
1911

CONTENTS

TRANSLATOR'S PREFACE

The pamphlet of fifty-eight pages entitled *Die Geschichte und die Wurzel des Satzes von der Erhaltung der Arbeit.*[1] *Vortrag gehalten in der k. böhm. Gesellschaft der Wissenschaften am 15. Nov. 1871 von E. Mach, Professor der Physik an der Universität Prag* was published at Prague in 1872, and a second—unaltered—edition at Leipzig (Barth) in 1909. To this second edition (pp. iv, 60) were added a short preface and a few notes by Mach himself. This preface is translated below.

Quite apart from the interest which must attach to the first sketch of a way of regarding science which has become of such great importance to students both of science and of the theory of knowledge, this pamphlet is quite essential to the thorough understanding of Mach's work. In the first place, it contains a reprint of Mach's article (1868) on the definition of mass, which is, perhaps, his most important contribution to mechanics; and, in the second place, the discussion of the logical root of the principle of the conservation of energy is fuller than that in any of his later publications.[2]

[1] In the title of this translation, *Arbeit* is translated by *Energy*, as this word conveys a better idea, at the present time, than the older and more literal equivalent of *Work*. In the text, on the other hand, the word *Work* will always be used, as it corresponds more closely to the terminology of science at the time of the first publication of this essay.

[2] Thus, the questions connected with the uniqueness of determination of events are discussed and illustrated very fully in this essay,

It is proper here to give some references to discussions of Mach's point of view in science.

A fairly good general account of Mach's various works was given in Harald Höffding's lectures on modern philosophers held at the University of Copenhagen in 1902;[3] and another account, with a hostile criticism, was given by T. Case in his article "Metaphysics" in the new volumes which make up the tenth edition of the *Encyclopaedia Britannica*.[4] Often valuable criticisms of Mach's position are to be found in the reviews of the first and second editions of the *Analyse der Empfindungen* written by C. Stumpf,[5] Elsas,[6] Lucien Arréat,[7] and W. R. Boyce Gibson.[8]

The last-named writer speaks[9] of the "generous

and it was this essay that formed the starting-point of Petzoldt's development of the view involved.

The essay "On the Principle of the Conservation of Energy" in Mach's *Popular Scientific Lectures* (3d ed., Open Court Publishing Co., 1898, pp. 137–185), though in many respects like the pamphlet of 1872, is not nearly so complete as it is—a remark made by Hans Kleinpeter (*Die Erkenntnistheorie der Naturforschung der Gegenwart*, Leipzig, 1905, p. 150), who therefore pointed out the need for a reprint of this rare pamphlet.

3 In the German translation, by F. Bendixen, of these lectures under the title: *Moderne Philosophen* (Leipzig, 1905), the part relating to Mach is on pp. 104–110. The section devoted to Maxwell, Mach, Hertz, Ostwald, and Avenarius is on pp. 97–127.

4 Vol. XXX, pp. 665–667. Cf. also the references to Mach's work in Ludwig Boltzmann's article "Models" (*ibid.*, pp. 788–790.)

5 *Deutsche Litteraturzeitung*, Nr. 27, 3. Juli, 1886.

6 *Philosophische Monatshefte*, Vol. XXIII, p. 207.

7 *Revue Philosophique*, 1887, p. 80.

8 *Mind*, N.S., Vol. X, pp. 246–264 (No. 38, April, 1901).

9 *Ibid.*, p. 253.

recognition he [Mach] is always ready to give to anyone who succeeds in improving upon his own attempts," and "his still more eager readiness to put fact before theory. With this eagerness to find out the truth is associated a corresponding ardour in developing and applying it when found."

But philosophers seem hardly to have done justice to Mach's work. Mach himself, indeed, has repeatedly disclaimed for himself the name of philosopher; yet, in a sense, any man who forms a general position from which to regard, say, science, is a philosopher.[10] It must be acknowledged that the least satisfactory parts of Mach's writings are those in which he discusses mathematical conceptions, such as numbers and the continuum; and in which he implies that logic is to be founded on a psychological basis; but such things are unconnected with the greater part of his valuable work.

There are three sets of notes to this translation. The first set, referred to by numerals in the body of the text, consists of the notes added by the author to the

[10] Through a reference in the *Jahrbuch über die Fortschritte der Mathematik* for 1904 (Bd. XXXV, p. 78) I learn that D. Wiktorov has published, in Russian, an exposition of Mach's philosophical views, in the periodical whose name, translated, is *Questions of Philosophy and Psychology*, No. 73 (1904, No. 3), pp. 228–313.

J. Baumann (*Archiv für systematische Philos.*, IV, 1897–1898, Heft 1, October, 1897) gave an account of "Mach's philosophy." Cf. also Hönigswald, *Zur Kritik der Mach'schen Philosophie*, Berlin, 1903; and Mach, *Erkenntnis und Irrtum*, 1906, pp. vii–ix. Adolfo Levi ("Il fenomenismo empiristico," *Riv. di Fil.*, T. I., 1909) analyzed the theories of knowledge of Mill, Avenarius, Mach, and Ostwald.

original edition; the second set consists of those added by the author to the reprint of 1909;[11] and the third set, which contains some account of later work by the author and others on subjects connected with the history and root of the principle of the conservation of work, has been added by the translator. Any other notes by the translator, added for the purpose of giving fuller references, are enclosed in square brackets.

Professor Mach has been most kind in carefully reading my manuscript; and so I trust that not all of the freshness, the force of conviction, and the humour of the original are lost in the present translation.[12]

PHILIP E. B. JOURDAIN

THE MANOR HOUSE
BROADWINDSOR
BEAMINSTER, DORSET
November, 1909

[11] These notes are translated, with the exception of one correcting a misprint in the original edition.

AUTHOR'S PREFACE TO THE SECOND EDITION

In this pamphlet, which appeared in 1872, I made the first attempt to give an adequate exposition of my epistemological standpoint—which is based on a study of the physiology of the senses—with respect to science as a whole, and to express it more clearly in so far as it concerns physics. In it both every *metaphysical* and every one-sided *mechanical* view of physics were kept away, and an arrangement, according to the principle of economy of thought, of facts—of what is ascertained by the senses—was recommended. The investigation of the dependence of phenomena on one another was pointed out as the aim of natural science. The digressions, connected with this, on causality, space, and time, may then have appeared far from the point and hasty; but they were developed in my later writings, and do not, perhaps, lie so far from the science of to-day. Here, too, are to be found the fundamental ideas of the *Mechanik* of 1883,[12] of the *Analyse der Empfindungen* of 1886,[13] which was addressed prin-

[12] [*Die Mechanik in ihrer Entwickelung historisch-kritisch dargestellt*, Leipzig, five editions from 1883 to 1904; English translation by T. J. McCormack under the title *The Science of Mechanics*, Open Court Publishing Co., Chicago, three editions from 1893 to 1907 (the third edition of this is quoted hereafter as *Mechanics*).]

[13] [*Beiträge zur Analyse der Empfindungen*, Jena, 1886; Eng. trans. by C. M. Williams under the title *Contributions to the Analysis of the Sensations*, Open Court Publishing Co., Chicago, 1897. A

cipally to biologists, in the *Wärmelehre* of 1896,[14] and in the *Erkenntnis und Irrtum*—a book which treats at length questions of the epistemology of physics—of 1905.[15]

Certainly it is right that, in response to repeated demands, this work, which was out of print twelve years ago, should appear in an *unaltered* form. I could not have entertained sanguine expectations as to the immediate result of my little work; indeed, many years before, Poggendorff had refused for his *Annalen* my short essay on the definition of mass, which definition is now generally accepted. When Max Planck wrote, fifteen years after I did, on the conservation of energy,[16] he had a remark directed against one of my developments, without which remark one would have supposed that he had not seen my pamphlet at all. But it was a ray of hope for me when Kirchhoff[17] pronounced, in 1874, the problem of mechanics to be the complete and simplest description of motions, and this nearly corre-

second, much enlarged, German edition was published at Jena in 1900 under the title: *Die Analyse der Empfindungen und das Verhältnis des Physischen zum Psychischen;* and a fifth edition appeared in 1906.]

14 [*Die Principien der Wärmelehre historisch-kritisch entwickelt*, Leipzig, 1896; 2d ed., 1900. The 2d edition is hereafter referred to as *Wärmelehre.*]

15 [*Erkenntnis und Irrtum. Skizzen zur Psychologie der Forschung*, Leipzig, 1905; 2d ed., 1906.]

16 [*Das Prinzip der Erhaltung der Energie*, Leipzig, 1887; 2d ed., 1909. The reference to Mach's work of 1872 is on p. 156 of the second edition.]

17 [*Vorlesungen über mathematische Physik, Bd. I, Mechanik*, Leipzig, 1874; 4th ed., 1897.]

sponded to the economical representation of facts. Helm esteemed the principle of the economy of thought and the tendency of my little treatise towards a general science of energetics. And, finally, though H. Hertz did not give an open expression of his sympathy, yet the utterances in his *Mechanik* of 1894[18] coincide as exactly as is possible with my own,[19] considering that Hertz was a supporter of the mechanical and atomic physics and a follower of Kant. So those whose positions are near to mine are not the worst of men. But since, even at the present time, when I have almost reached the limit of human age, I can count on my fingers those whose standpoint is more or less near to my own—men like Stallo,[20] W. K. Clifford, J Popper, W. Ostwald, K. Pearson,[21] F. Wald, and P. Duhem, not to speak of the younger generation—it is evident that in this connexion we have to do with a very small minority. I cannot, then, share the apprehension that appears to lie behind utterances like that of M. Planck,[22] that

18 [*Die Prinzipe der Mechanik*, Vol. III of Hertz's *Ges. Werke*, Leipzig, 1894; Eng. trans. by D. E. Jones and J. T. Walley, under the title *The Principles of Mechanics*, London, 1899.]

19 [On Hertz's mechanics, see Mach, *Mechanics*, pp. 548–555.]

20 [*The Concepts and Theories of Modern Physics*, 4th ed., London, 1900.]

21 [*The Grammar of Science*, London, 1892; 2d ed., 1900. The account of the laws of motion in W. K. Clifford's book: *The Common Sense of the Exact Sciences* (London, 1885, 5th ed., 1907), which was completed by Pearson, agrees with Mach's views; but this statement was due, not to Clifford, but to Pearson, whose (see pp viii–ix of the work just mentioned) views were developed independently.]

22 [*Die Einheit des physicalischen Weltbildes*, Leipzig, 1909, pp. 31–38.]

orthodox physics has need of such a powerful speech in its defence. Rather do I fear that, with or without such speeches, the simple, natural, and indeed inevitable reflections which I have tried to stir up will only come into their rights very late.

"Not every physicist is an epistemologist, and not everyone must or can be one. Special investigation claims a whole man, so also does the theory of knowledge."[23] This must be my answer to the excessively naïve demand of a physicist who was justly celebrated and is now dead, that I should wait with my analysis of the sensations until we knew the paths of the atoms in the brain, from which paths all would easily result. The physicist who thinks under the guidance of a working hypothesis usually corrects his concepts sufficiently by accurate comparison of the theory with observation, and has little occasion to trouble himself with the psychology of knowledge. But whoever wishes to criticize a theory of knowledge or instruct others about it, must know it and have thought it out. I cannot admit that my physicist critics have done this, as I will show without difficulty at the proper place.

E. MACH

VIENNA
May, 1909

[23] *Analyse der Empfindungen*, 5th ed., p. 255.

THE HISTORY AND THE ROOT OF THE PRINCIPLE OF THE CONSERVATION OF ENERGY

I
INTRODUCTION

HE who calls to mind the time when he obtained his first view of the world from his mother's teaching will surely remember how upside-down and strange things then appeared to him. For instance, I recollect the fact that I found great difficulties in two phenomena especially. In the first place, I did not understand how people could like letting themselves be ruled by a king even for a minute. The second difficulty was that which Lessing so deliciously put into an epigram, which may be roughly rendered:

> "One thing I've often thought is queer,"
> Said Jack to Ted, "the which is
> "That wealthy folk upon our sphere,
> "Alone possess the riches."*

The many fruitless attempts of my mother to help me over these two problems must have led her to form a very poor opinion of my intelligence.

Everybody will remember similar experiences in his own youth. There are two ways of reconciling oneself with actuality: either one grows accustomed to the puzzles and they trouble one no more, or one learns to

> *"Es ist doch sonderbar bestellt,"
> Sprach Hänschen Schlau zu Vetter Fritzen,
> "Dass nur die Reichen in der Welt
> "Das meiste Geld besitzen."

understand them by the help of history and to consider them calmly from that point of view.

Quite analogous difficulties lie in wait for us when we go to school and take up more advanced studies, when propositions which have often cost several thousand years' labour of thought are represented to us as self-evident. Here too there is only one way to enlightenment: historical studies.

The following considerations, which, if I except my reading of Kant and Herbart, have arisen quite independently of the influence of others, are based upon some historical studies. The reason why, in discussion of these thoughts with able colleagues of mine, I could not, as a rule, come to agreement, and why my colleagues always tended to seek the ground of such "strange" views in some confusion of mine, was, without doubt, that historical studies are not so generally cultivated as they should be.[1]

However this may be, these thoughts, which, as the notes and quotations from my earlier writings show, are not of very recent date, but which I have held since the year 1862, were not suited for discussion with my colleagues—I, at least, soon tired of such discussions. With the exception of some short notices written on the occasion of other works and in journals little read by physicists, but which may suffice to prove my independence, I have published nothing about these thoughts.

[1] In fact, I have known only one man, Josef Popper, with whom I could discuss the views exposed here without rousing a horrified opposition. Popper and I, indeed, arrived at similar views on many points of physics independently of one another, which fact I take pleasure in mentioning here.

But now, since some renowned investigators have begun to set foot in this province, perhaps I, too, may bring my small contribution to the classification of the questions with which we are concerned. I must protest at once against this investigation being considered a metaphysical one. We are accustomed to call concepts metaphysical, if we have forgotten how we reached them. One can never lose one's footing, or come into collision with facts, if one always keeps in view the path by which one has come. This pamphlet merely contains straightforward reflections on some facts belonging both to natural science and to history.

Perhaps the following lines will also show the value of the historical method in teaching. Indeed, if from history one learned nothing else than the variability of views, it would be invaluable. Of science, more than anything else, Heraclitus's words are true: "One cannot go up the same stream twice." Attempts to fix the fair moment by means of textbooks have always failed. Let us, then, early get used to the fact that science is unfinished, variable.

Whoever knows only one view or one form of a view does not believe that another has ever stood in its place, or that another will ever succeed it; he neither doubts nor tests. If we extol, as we often do, the value of what is called a classical education, we can hardly maintain seriously that this results from an eight-years' discipline of declining and conjugating. We believe, rather, that it can do us no harm to know the point of view of another eminent nation, so that we can, on occasion, put ourselves in a different position from that in which we have been brought up. The essence of classical education is historical education.

But if this is correct, we have a much too narrow idea of classical education. Not the Greeks alone concern us, but all the cultured people of the past. Indeed there is, for the investigator of nature, a special classical education which consists in the knowledge of the historical development of his science.

Let us not let go the guiding hand of history. History has made all; history can alter all. Let us expect from history all, but first and foremost, and I hope this of my historical investigation, that it may not be too tedious.

II

ON THE HISTORY OF THE THEOREM OF THE CONSERVATION OF WORK

THE place given to the law of the conservation of energy in modern science is such a prominent one that the question as to its validity, which I will try to answer, obtrudes itself, as it were, of itself. I have allowed myself, in the headline, to call the law that of the conservation of work, because it appeared to me to be a name which is understood by all and prevents wrong ideas. Let us call to mind the considerations, laden with misunderstandings, of the great Faraday on the "law of the conservation of force," and a well-known controversy which was not much poorer in obscurities. One should say "law of the conservation of force" only when one, with J. R. Mayer, calls "force" what Euler called "*effort*" and Poncelet "*travail.*" Of course, one cannot find fault with Mayer, who did not get his concepts from the schools, for using his own peculiar names.

Usually the theorem of the conservation of work is expressed in two forms:

1. $\frac{1}{2}\Sigma mv^2 - \frac{1}{2}\Sigma mv_0{}^2 = \int \Sigma(Xdx + Ydy + Zdz)$; or

2. It is impossible to create work out of nothing, or to construct a ***perpetuum mobile.***

This theorem is usually considered as the flower of the mechanical view of the world, as the highest and most general theorem of natural science, to which the thought of many centuries has led.

I will now try to show:

Firstly, that this theorem, in the second form, is by no means so new as one tends to believe; that, indeed, almost all eminent investigators had a more or less confused idea of it, and since the time of Stevinus and Galileo, it has served as the foundation of the most important extensions of the physical sciences.

Secondly, that this theorem by no means stands and falls with the mechanical view of the world, but that its logical root is incomparably deeper in our mind than that view.

In the first place, as for the first part of my assertion, the proof must be drawn from original sources. Although, now, Lagrange, in his celebrated historical introductions to the sections of the *Mécanique analytique*,[1] repeatedly refers to the development of our theorem, one soon finds, if one takes the trouble to consult the originals themselves, that in his exposition this theorem does not play the part which it played in fact.

Although, now, the following facts, with the exception of a few, coincide with those mentioned by Lagrange, we derive from the important passages, given *in extenso*, another view than that which is found in Lagrange's work.

[1] [The first edition of this work was published at Paris in 1788 (1 vol.) under the title *Méchanique analitique*, the second at Paris, 1811–1813 (2 vols.), the third (ed. J. Bertrand), 1853, and the fourth (*Œuvres*, XI, XII, ed. G. Darboux), 1892.]

Let me emphasize only some points:

Simon Stevinus, in his work *Hypomnemata mathematica*, Tom. IV, *De statica*, of 1605,[2] treats of the equilibrium of bodies on inclined planes.

Over a triangular prism $A\,B\,C$, one side of which, $A\,C$, is horizontal, an endless cord or chain is slung, to which at equal distances apart fourteen balls of equal weight are attached, as represented in cross-section in Fig. 1. Since we can imagine the lower symmetrical part of the cord $A\,B\,C$ taken away, Stevinus concludes that the four balls on $A\,B$ hold in equilibrium the two balls on $B\,C$. For if the equilibrium were for a moment disturbed, it could never subsist: the cord would keep moving round forever in the same direction—we should have a perpetual motion. He says:

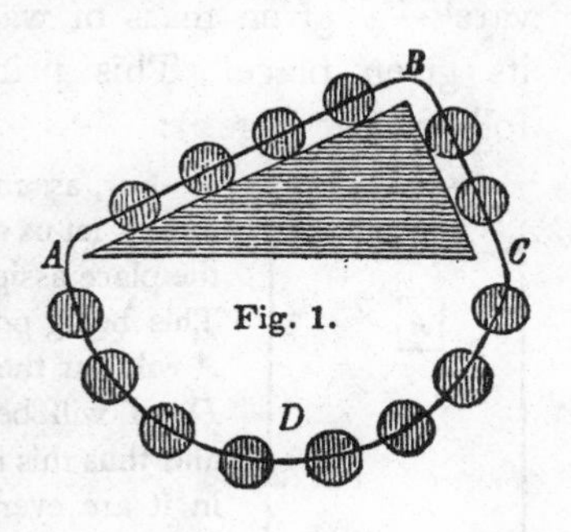

Fig. 1.

But if this took place, our row or ring of balls would come once more into their original position, and from the same cause the eight globes to the left would again be heavier than the six to the right, and therefore those eight would sink a second time and these six rise, and all the globes would keep up, of themselves, *a continuous and unending motion, which is false.*[3]

[2] Leiden, 1605, p. 34. [According to Moritz Cantor (*Vorlesungen über Geschichte der Mathematik*, II, 2. Aufl., Leipzig, 1900, p. 572), this work was first published in 1586, and a Latin translation, by Snellius, appeared in 1608. Cf. also Cantor, *ibid.*, pp. 576–577.]

[3] "Atqui hoc si sit, globorum series sive corona eundem situm cum priore habebit, eademque de causa octo globi sinistri pondero-

Stevinus, now, easily derives from this principle the laws of equilibrium on the inclined plane and numerous other fruitful consequences.

In the chapter "Hydrostatics" of the same work, p. 114, Stevinus sets up the following principle: "Aquam datam, datum sibi intra aquam locum servare"—a given mass of water preserves within water its given place. This principle is demonstrated as follows (see Fig. 2):

Fig. 2.

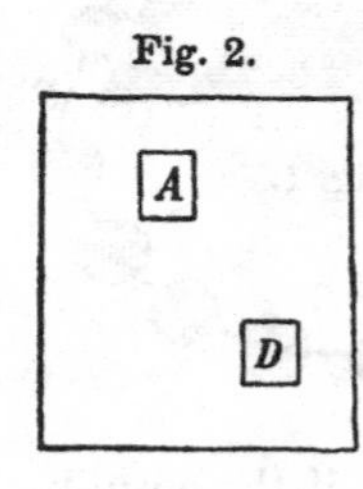

For, assuming it to be possible by natural means, let us suppose that *A* does not preserve the place assigned to it, but sinks down to D. This being posited, the water which succeeds *A* will, for the same reason, also flow down to *D*; *A* will be forced out of its place in *D*; and thus this body of water, for the conditions in it are everywhere the same, ***will set up a perpetual motion, which is absurd.***[4]

From this all the principles of hydrostatics are deduced. On this occasion Stevinus also first develops the thought so fruitful for modern analytical mechanics that the equilibrium of a system is not destroyed by the addition of rigid connexions. As we know, the principle of the conservation of the centre of gravity is now sometimes deduced from d'Alembert's principle

siores erunt sex dextris, ideoque rursus octo illi descendent, sex illi ascendent, istique globi ex sese ***continuum et aeternum motum efficient, quod est falsum.***"

[4] "A igitur (si ullo modo per naturam fieri possit), locum sibi tributum non servato, ac delabatur in *D*; quibus positis aqua quae ipsi *A* succedit eandem ob causam deffluet in *D*, eademque ab alia istinc expelletur, atque adeo aqua haec (cum ubique eadem ratio sit) ***motum instituet perpetuum, quod absurdum fuerit.***"

with the help of that remark. If we were to reproduce Stevinus's demonstration to-day, we should have to change it slightly. We find no difficulty in imagining the cord on the prism possessed of unending uniform motion, if all hindrances are thought away, but we should protest against the assumption of an accelerated motion or even against that of a uniform motion, if the resistances were not removed. Moreover, for greater precision of proof, the string of balls might be replaced by a heavy homogeneous cord of infinite flexibility. But all this does not affect in the least the historical value of Stevinus's thoughts. It is a fact that Stevinus deduces apparently much simpler truths from the principle of an impossible perpetual motion.

In the process of thought which led Galileo to his discoveries at the end of the sixteenth century, the following principle plays an important part: that a body in virtue of the velocity acquired in its descent can rise exactly as high as it fell. This principle, which appears frequently and with much clearness in Galileo's thought, is simply another form of the principle of excluded perpetual motion, as we shall see it is also with Huygens.

Galileo, as we know, arrived at the law of uniformly accelerated motion by *a priori* considerations, as that law which was the "simplest and most natural," after having first assumed a different law which he was compelled to reject. To verify his law he performed experiments with falling bodies on inclined planes, measuring the times of descent by the weights of the water which flowed out of a small orifice in a large vessel. In this experiment he assumes, as a fundamental principle, that the velocity acquired in descent down an inclined plane always corresponds to the vertical height descended through, a conclusion which for him is the immediate outcome of the fact that a body which has fallen down one inclined plane can, with the velocity it has acquired, rise on another plane of any inclination only to the same vertical height. This principle of the height of ascent also led him, as it seems, to the law of inertia. Let us hear his own masterly words in the *Dialogo terzo* (*Opère*, Padova, 1744, Tom. III). On page 96 we read:

I take it for granted that the velocities acquired by a body in descent down planes of different inclinations are equal if the heights of those planes are equal.[5]

Then he makes Salviati say in the dialogue:[6]

What you say seems very probable, but I wish to go farther and by an experiment so to increase the probability of it that it

[5] "Accipio, gradus velocitatis ejusdem mobilis super diversas planorum inclinationes acquisitos tunc esse aequales, cum eorundem planorum elevationes aequales sint."

[6] "Voi molto probabilmente discorrete, ma oltre al veri simile voglio con una esperienza crescer tanto la probabilità, che poco gli manchi all'agguagliarsi ad una ben necessaria dimostrazione. Figuratevi questo foglio essere una parete eretta al orizzonte, e da un chiodo fitto in essa pendere una palla di piombo d'un'oncia, o due, sospesa dal sottil filo *A B* lungo due, o tre braccia perpendicolare all'orizzonte, e nella parete segnate una linea orizzontale *D C* segante a squadra il perpendicolo *A B*, il quale sia lontano dalla parete due dita in circa, trasferendo poi il filo *A B* colla palla in *A C*, làsciata essa palla in libertà, la quale primieramente vedrete scendere descrivendo l'arco *C B D*, e di tanto trapassare il termine *B*, che scorrendo per l'arco *B D* sormonterà fino quasi alla segnata parallela *C D*, restando di per vernirvi per piccolissimo intervallo, toltogli il precisamente

shall amount almost to absolute demonstration. Suppose this sheet of paper to be a vertical wall, and from a nail driven in it a ball of lead weighing two or three ounces to hang by a very fine thread *A B* four or five feet long. (Fig. 3.) On the wall mark a horizontal line *D C* perpendicular to the vertical *A B*,

arrivarvi dall'impedimento dell'aria, e del filo. Dal che possiamo veracemente concludere, che l'impeto acquistato nel punto *B* dalla palla nello scendere per l'arco *C B*, fu tanto, che bastò a risospingersi per un simile arco *B D* alla medesima altezza; fatta, e più volte reiterata cotale esperienza, voglio, che fiechiamo nella parete rasente al perpendicolo *A B* un chiodo come in *E*, ovvero in *F*, che sporga in fuori cinque, o sei dita, e questo acciocchè il filo *A C* tornando come prima a riportar la palla *C* per l'arco *C B*, giunta che ella sia in *B*, inoppando il filo nel chiodo *E*, sia costretta a camminare per la circonferenza *B G* descritta intorno al centro *E*, dal che vedremo quello, che potrà far quel medesimo impeto, che dianzi concepizo nel medesimo termine *B*, sospinse l'istesso mobile per l'arco *E D* all'altezza dell'orizzontale *C D*. Ora, Signori, voi vedrete con gusto condursi la palla all'orizzontale nel punto *G*, e l'istesso accadere, l'intoppo si metesse più basso, come in *F*, dove la palla descriverebbe l'arco *B J*, terminando sempre la sua salita precisamente nella linea *C D*, e quando l'intoppe del chiodo fusse tanto basso, che l'avanzo del filo sotto di lui non arivasse all'altezza di *C D* (il che accaderebbe, quando fusse più vicino all punto *B*, che al segamento dell' *A B* coll'orizzontale *C D*), allora il filo cavalcherebbe il chiodo, e segli avolgerebbe intorno. Questa esperienza non lascia luogo di dubitare della verità del supposto: imperocchè essendo li due archi *C B*, *D B* equali e similmento posti, l'acquisto di momento fatto per la scesa nell'arco *C B*, è il medesimo, che il fatto per la scesa dell'arco *D B;* ma il momento acquistato in *B* per l'arco *C B* è potente a risospingere in su il medesimo mobile per l'arco *B D;* adunque anco il momento acquistato nella scesa *D B* è eguale a quello, che sospigne l'istesso mobile pel medesimo arco da *B* in *D*, sicche universalmente ogni momento acquistato per la scesa dun arco è eguale a quello, che può far risalire l'istesso mobile pel medesimo arco: ma i momenti tutti che fanno risalire per tutti gli archi *B D*, *B G*, *B J* sono eguali, poichè son fatti dal istesso medesimo momento acquistato per la scesa *C B*, come mostra l'esperienza: adunque tutti i momenti, che si acquistano per le scese negli archi *D B*, *G B*, *J B* sono eguali."

which latter ought to hang about two inches from the wall. If now the thread *A B* with the ball attached take the position *A C* and the ball be let go, you will see the ball first descend through the arc *C B* and passing beyond *B* rise through the arc *B D* almost to the level of the line *C D*, being prevented from reaching it exactly by the resistance of the air and the thread. From this we may truly conclude that its impetus at the point *B*, acquired by its descent through the arc *C B*, is sufficient to urge it through a similar arc *B D* to the same height. Having performed this experiment and repeated it several times, let us drive in the wall, in the projection of the vertical *A B*, as at *E* or at *F*, a nail five or six inches long, so that the thread *A C*, carrying as before the ball through the arc *C B*, at the moment it reaches the position *A B*, shall strike the nail *E*, and the ball be thus compelled to move up the arc *B G* described about *E* as centre. Then we shall see what the same impetus will here accomplish, acquired now as before at the same point *B*, which then drove the same moving body through the arc *B D* to the height of the horizontal *C D*. Now, gentlemen, you will be pleased to see the ball rise to the horizontal line at the point *G*, and the same thing also happen if the nail be placed lower as at *F*, in which case the ball would describe the arc *B J*, always terminating its ascent precisely at the line *C D*. If the nail be placed so low that the length of thread below it does not reach to the height of *C D* (which would happen if *F* were nearer *B* than to the intersection of *A B* with the horizontal *C D*), then the thread will wind itself about the nail. This experiment leaves no room for doubt as to the truth of the supposition. For as the two arcs *C B*, *D B* are equal and similarly situated, the

Fig. 3.

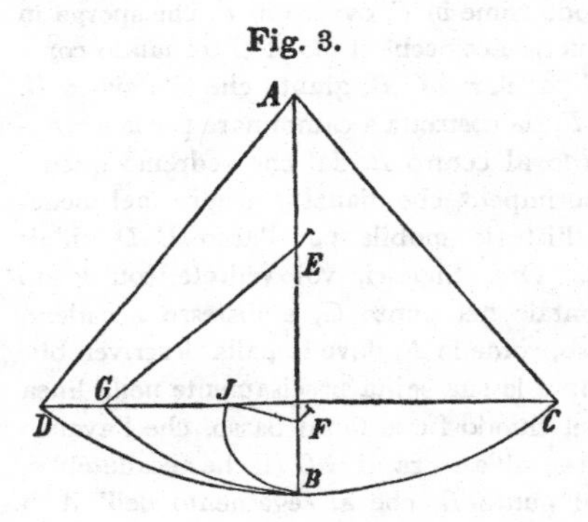

momentum acquired in the descent of the arc CB is the same as that acquired in the descent of the arc DB; but the momentum acquired at B by the descent through the arc CB is capable of driving up the same moving body through the arc BD; hence also the momentum acquired in the descent DB is equal to that which drives the same moving body through the same arc from B to D, so that in general every momentum acquired in the descent of an arc is equal to that which causes the same moving body to ascend through the same arc; but all the momenta which cause the ascent of all the arcs BD, BG, BJ are equal since they are made by the same momentum acquired in the descent CB, as the experiment shows: therefore all the momenta acquired in the descent of the arcs DB, GB, JB are equal.

The remark relative to the pendulum may be applied to the inclined plane and leads to the law of inertia. We read on p. 124:[7]

It is plain now that a movable body, starting from rest at A and descending down the inclined plane AB, acquires a velocity proportional to the increment of its time: the velocity possessed at B is the greatest of the velocities acquired, and by its nature immutably impressed,

Fig. 4.

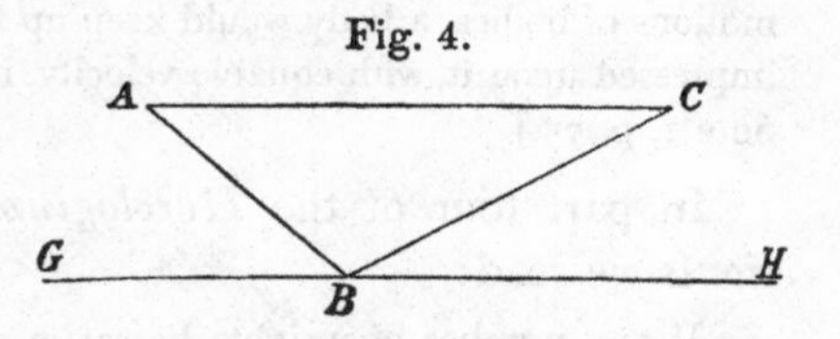

provided all causes of new acceleration or retardation are taken away: I say acceleration, having in view its possible further progress along the plane extended; retardation, in view of the

[7] "Constat jam, quod mobile ex quiete in A descendens per AB, gradus acquirit velocitatis juxta temporis ipsius incrementum: gradum vero in B esse maximum acquisitorum, et suapte natura immutabiliter impressum, sublatis scilicet causis accelerationis novae, aut retardationis: accelerationis inquam, si adhuc super extenso plano ulterius progrederetur; retardationis vero, dum super planum acclive BC fit reflexio: in horizontali autem GH aequabilis motus juxta gradum velocitatis ex A in B acquisitae in infinitum extenderetur.

possibility of its being reversed and made to mount the ascending plane BC. But in the horizontal plane GH its equable motion, according to its velocity as acquired in the descent from A to B, will be continued *ad infinitum.* (Fig. 4.)

Huygens, upon whose shoulders the mantle of Galileo fell, formed a sharper conception of the law of inertia and generalized the principle respecting the heights of ascent which was so fruitful in Galileo's hands. He employed this principle in the solution of the problem of the centre of oscillation and was perfectly clear in the statement that the principle respecting the heights of ascent is identical with the principle of excluded perpetual motion.

The following important passages then occur (Hugenii, *Horologium oscillatorium, pars secunda*). *Hypotheses:*

If gravity did not exist, nor the atmosphere obstruct the motions of bodies, a body would keep up forever the motion once impressed upon it, with equable velocity, in a straight line.[8] [See note 1, p. 75.]

In part four of the *Horologium; De centro oscillationis* we read:

If any number of weights be set in motion by the force of gravity, the common centre of gravity of the weights as a whole cannot possibly rise higher than the place which it occupied when the motion began.

That this hypothesis of ours may arouse no scruples, we will state that it simply means, what no one has ever denied, that heavy bodies do not move *upwards.*—And truly if the devisers of the new machines who make such futile attempts to

[8] "Si gravitas non esset, neque aër motui corporum officeret, unumquodque eorum, acceptum semel motum continuaturúm velocitate aequabili, secundum lineam rectam."

construct a perpetual motion would acquaint themselves with this principle, they would easily be brought to see their errors and to understand that the thing is utterly impossible by mechanical means.[9]

There is possibly a Jesuitical mental reservation contained in the words "mechanical means." One might be led to believe from them that Huygens held a non-mechanical perpetual motion as possible.

The generalization of Galileo's principle is still more clearly put in Prop. IV of the same chapter:

If a pendulum, composed of several weights, set in motion from rest, complete any part of its full oscillation, and from that point onwards, the individual weights, with their common connexions dissolved, change their acquired velocities upwards and ascend as far as they can, the common centre of gravity of all will be carried up to the same altitude with that which it occupied before the beginning of the oscillation.[10]

On this last principle, now, which is a generalization, applied to a system of masses (see note 2, p. 80),

[9] "Si pondera quotlibet, vi gravitatis suae, moveri incipiant; non posse centrum gravitatis ex ipsis compositae altius, quam ubi incipiente motu reperiebatur, ascendere.

"Ipsa vero hypothesis nostra quominus scrupulum moveat, nihil aliud sibi velle ostendemus, quam quod nemo unquam negavit, gravia nempe sursum non ferri.—Et sane, si hac eadem uti scirent novorum operum machinatores, qui motum perpetuum irrito conatu moliuntur, facile suos ipsi errores deprehenderent, intelligerentque rem eam mechanica ratione haud quaquam possibilem esse."

[10] "Si pendulum e pluribus ponderibus compositum, atque e quiete dimissum, partem quamcunque oscillationis integrae confecerit, atque inde porro intelligantur pondera ejus singula, relicto communi vinculo, celeritates acquisitas sursum convertere, ac quousque possunt ascendere; hoc facto centrum gravitatis ex omnibus compositae, ad eandem altitudinem reversum erit, quam ante inceptam oscillationem obtinebat."

of one of Galileo's ideas respecting a single mass, and which from Huygens's explanation we recognize as the principle of excluded perpetual motion, Huygens grounds his theory of the centre of oscillation. Lagrange characterizes this principle as precarious and is rejoiced at James Bernoulli's successful attempt, in 1681, to reduce the theory of the centre of oscillation to the laws of the lever, which appeared to him clearer. All the great inquirers of the seventeenth and eighteenth centuries broke a lance on this problem; and it led ultimately, in conjunction with the principle of virtual velocities, to the principle enunciated by d'Alembert in 1743 in his *Traité de dynamique*, though previously employed in a somewhat different form by Euler and Hermann.

Furthermore, the Huygenian principle respecting the heights of ascent became the foundation of the "law of the conservation of living force," as that was enunciated by John and Daniel Bernoulli and employed with such signal success by the latter in his *Hydrodynamics*. The theorems of the Bernoullis differ in form only from Lagrange's expression in the *Analytical Mechanics*.

The manner in which Torricelli reached his famous law of efflux for liquids leads again to our principle. Torricelli assumed that the liquid which flows out of the basal orifice of a vessel cannot by its velocity of efflux ascend to a greater height than its level in the vessel.

Let us next consider a point which belongs to pure mechanics, the history of the principle of *virtual mo-*

tions or *virtual velocities*. This principle was not first enunciated, as is usually stated, and as Lagrange also asserts, by Galileo, but earlier, by Stevinus. In his *Trochleostatica* of the above-cited work, p. 172, he says:

> Observe that this axiom of statics holds good here:
> As the space of the body acting is to the space of the body acted upon, so is the power of the body acted upon to the power of the body acting.[11]

Galileo, as we know, recognized the truth of the principle in the consideration of the simple machines, and also deduced the laws of the equilibrium of liquids from it.

Torricelli carried the principle back to the properties of the centre of gravity. The condition controlling equilibrium in a simple machine, in which power and load are represented by weights, is that the common centre of gravity of the weights shall not sink. Conversely, if the centre of gravity cannot sink, equilibrium obtains, because heavy bodies of themselves do not move upwards. In this form the principle of virtual velocities is identical with Huygens's principle of the impossibility of a perpetual motion.

John Bernoulli, in 1717, first perceived the universal import of the principle of virtual movements for all systems; a discovery stated in a letter to Varignon. Finally, Lagrange gave a general demonstration of the principle and founded upon it his whole *Analytical Mechanics*. But this general demonstration is based

[11] "Notato autem hic illud staticum axioma etiam locum habere:
"Ut spatium agentis ad spatium patientis
Sic potentia patientis ad potentiam agentis."

after all upon Huygens's and Torricelli's remarks. Lagrange, as is known, conceived simple pulleys arranged in the directions of the forces of the system, passed a cord through these pulleys, and appended to its free extremity a weight which is a common measure of all the forces of the system. With no difficulty, now, the number of elements of each pulley may be so chosen that the forces in question shall be replaced by them. It is then clear that if the weight at the extremity cannot sink, equilibrium subsists, because heavy bodies cannot of themselves move upwards. If we do not go so far, but wish to abide by Torricelli's idea, we may conceive every individual force of the system replaced by a special weight suspended from a cord passing over a pulley in the direction of the force and attached at its point of application. Equilibrium subsists then when the common centre of gravity of all the weights together cannot sink. The fundamental supposition of this demonstration is plainly the impossibility of a perpetual motion.

Lagrange tried in every way to supply a proof free from extraneous elements and fully satisfactory, but without complete success. Nor were his successors more fortunate.

The whole of mechanics is thus based upon an idea, which, though unequivocal, is yet unwonted and not coequal with the other principles and axioms of mechanics. Every student of mechanics, at some stage of his progress, feels the uncomfortableness of this state of affairs; everyone wishes it removed; but seldom is the difficulty stated in words. Accordingly, the zealous pupil of the science is greatly rejoiced when he reads in a master like Poinsot (*Théorie générale de l'équilibre et du mouvement des systèmes*) the following passage, in which that author is giving his opinion of the *Analytical Mechanics:*

In the meantime, because our attention in that work was first wholly engrossed with the consideration of its beautiful development of mechanics, which seemed to spring complete from a single formula, we naturally believed that the science was completed, and that it only remained to seek the demonstration of the principle of virtual velocities. But that quest brought back all the difficulties that we had overcome by the principle itself. That law so general, wherein are mingled the vague and unfamiliar ideas of infinitely small movements and of perturbations of equilibrium, only grew obscure upon examination; and the work of Lagrange supplying nothing clearer than the march of analysis, we saw plainly that the clouds had appeared lifted from the course of mechanics only because they had, so to speak, been gathered at the very origin of that science.

At bottom, a general demonstration of the principle of virtual velocities would be equivalent to the establishment of the whole of mechanics upon a different basis: for the demonstration of a law which embraces a whole science is neither more nor less than the reduction of that science to another law just as general, but evident, or at least more simple than the first, and which, consequently, would render that useless.[12]

[12] "Cependant, comme dans cet ouvrage on ne fut d'abord attentif qu'à considérer ce beau développement de la mécanique qui semblait sortir tout entière d'une seule et même formule, on crut naturellement que la science était faite, et qu'il ne restait plus qu'à chercher la démonstration du principe des vitesses virtuelles. Mais cette recherche ramena toutes les difficultés qu'on avait franchies par le principe même. Cette loi si générale, où se mêlent des idées vagues et étrangères de mouvements infinement petits et de perturbation d'équilibre, ne fit en quelque sorte que s'obsurcir à l'examen; et le livre de Lagrange n'offrant plus alors rien de clair que la marche des calculs,

According to Poinsot, therefore, a proof of the principle of virtual movements is tantamount to a total rehabilitation of mechanics.

Another circumstance of discomfort to the mathematician is that, in the historical form in which mechanics at present exists, dynamics is founded on statics, whereas it is desirable that in a science which pretends to deductive completeness the more special statical theorems should be deducible from the more general dynamical principles.

In fact, a great master, Gauss, gave expression to this desire in his presentment of the principle of least constraint (Crelle's *Journal für reine und angewandte Mathematik*, Vol. IV, p. 233) in the following words: "Proper as it is that in the gradual development of a science, and in the instruction of individuals, the easy should precede the difficult, the simple the complex, the special the general, yet the mind, when once it has reached a higher point of view, demands the contrary course, in which all statics shall appear simply as a special case of mechanics." Gauss's own principle, now, possesses all the requisites of universality, but its difficulty is that it is not immediately intelligible,

on vit bien que les nuages n'avaient paru levé sur le cours de la mécanique que parcequ'ils étaient, pour ainsi dire, rassemblés à l'origine même de cette science.

"Une démonstration générale du principe des vitesses virtuelles devait au fond revenir a établir le mécanique entière sur une autre base: car la démonstration d'une loi qui embrasse toute une science ne peut être autre chose que la réduction de cette science à une autre loi aussi générale, mais évidente, ou du moins plus simple que la première, et qui partant la rende inutile" (Poinsot, *Éléments de statique*, 10. éd., Paris, 1861, pp. 263–264).

and that Gauss deduced it with the help of d'Alembert's principle, a procedure which left matters where they were before.

Whence, now, is derived this strange part which the principle of virtual motion plays in mechanics? For the present I shall only make this reply. It would be difficult for me to tell the difference of impression which Lagrange's proof of the principle made on me when I first took it up as a student and when I subsequently resumed it after having made historical researches. It first appeared to me insipid, chiefly on account of the pulleys and the cords which did not fit in with the mathematical view, and whose action I would much rather have discovered from the principle itself than have taken for granted. But now that I have studied the history of the science I cannot imagine a more beautiful demonstration.

In fact, through all mechanics it is this selfsame principle of excluded perpetual motion which accomplishes almost everything that displeased Lagrange, but which he still had to employ, at least tacitly, in his own demonstration. If we give this principle its proper place and setting, the paradox is explained.

Let us consider another department of physics, the theory of heat.

S. Carnot, in his *Réflexions sur la puissance motrice du feu,*[13] established the following theorem: Whenever work is performed by means of heat, a certain quantity of heat passes from a warmer to a colder body (supposing that a permanent alteration in the state of the

[13] Paris, 1824. [Cf. a note to p. 38 below.]

acting body does not take place). To the performance of work corresponds a transference of heat. Inversely, with the same amount of the work obtained, one can again transfer the heat from the cooler body to the warmer one. Carnot, now, found that the quantity of heat flowing from the temperature t to the temperature t_1, for a definite performance of work, cannot depend upon the chemical nature of the bodies in question, but only upon these temperatures. If not, a combination of bodies, which would continually generate work out of nothing, could be imagined. Here, then, an important discovery is founded on the principle of excluded perpetual motion. This is without doubt the first extra-mechanical application of the theorem.

Carnot considered the quantity of heat as invariable. Clausius, now, found that with the performance of work, heat not merely flows over from t to t_1, but also a part of it, which is always proportional to the work performed, is lost. By a continued application of the principle of excluded perpetual motion, he found that

$$-\frac{Q}{T}+Q_1\left(\frac{1}{T_1}-\frac{1}{T}\right)=0,$$

where Q denotes the quantity of heat transformed into work and Q_1, that which flows from the absolute temperature T to the absolute temperature T_1.

Special weight has been laid on this vanishing of heat with the performance of work and the formation of heat with the expenditure of mechanical work—which processes were confirmed by the considerations of J. R. Mayer, Helmholtz, and W. Thomson, and by the experiments of Rumford, Joule, Favre, Silbermann, and many others. From this it was concluded that, if heat can be transformed into mechanical work, heat consists in mechanical processes—in motion. This conclusion, which has spread over the whole cultivated world like wild-fire, had, as an effect, a huge mass of literature on this subject, and now people are everywhere eagerly bent on explaining heat by means of motions; they determine the velocities, the average distances, and the paths of the molecules, and there is hardly a single problem which could not, people say, be completely solved in this way by means of sufficiently long calculations and of different hypotheses. No wonder that in all this clamour the voice of one of the most eminent, that of the great founder of the mechanical theory of heat, J. R. Mayer, is unheard:

> Just as little as, from the connexion between the tendency to fall (*Fallkraft*) and motion, we can conclude that the essence of this tendency is motion, just so little does this conclusion hold for heat. Rather might we conclude the opposite, that, in order to become heat, motion—whether simple or vibrating, like light or radiant heat—must cease to be motion.[14]

We will see later what is the cause of the vanishing of heat with the performance of work.

The second extra-mechanical application of the theorem of excluded perpetual motion was made by Neumann for the analytical foundation of the laws of electrical induction. This is, perhaps, the most talented work of this kind.

[14] *Mechanik der Wärme*, Stuttgart, 1867, p. 9.

Finally, Helmholtz[15] attempted to carry the law of the conservation of work through the whole of physics, and, from this point onwards, the applications of this law to the extension of science are innumerable.

Helmholtz carried the principle through in two ways. We can, said he, set out from the fundamental theorem that work cannot be created out of nothing, and thereby bring physical phenomena into connexion, or we can consider physical processes as molecular processes which are produced by central forces alone—thus by forces which have a potential. For the latter processes, the

[15] [A convenient edition of H. Helmholtz's paper *Ueber die Erhaltung der Kraft* of 1847, together with the notes that Helmholtz himself added to its reprint in his *Wissenschaftliche Abhandlungen* (Vol. I, pp. 12–75), is that in Nr. 1 of Ostwald's *Klassiker der exakten Wissenschaften.* This same series of *Klassiker* also includes, in German translations, and with notes that are often valuable, the following works, which are referred to by Mach in the present work: Galileo's *Discorsi* (notes by Arthur von Oettingen), Nr. 11, 24, and 25; Carnot's work of 1824 (notes by W. Ostwald), Nr. 37; F. E. Neumann's papers on induced electric currents (notes by C. Neumann), Nr. 10 and 36; Clausius's paper of 1850 on thermodynamics (notes by M. Planck), Nr. 99; and Coulomb's papers on the torsion balance (notes by Walter König), Nr. 13. In the same series are some papers of Helmholtz and Kirchhoff on thermodynamics (notes by M. Planck) in Nr. 124 and 101 respectively; and Huygens's *Traité de la lumière* of 1678, in which certain views as to mechanical physics (cf. Mach, *Pop. Sci. Lect.*, 3d ed., Open Court Publishing Co., Chicago, 1898, pp. 155–156) are given, is annotated by E. Lommel and A. von Oettingen in Nr. 20.

We may also add here that Clausius's papers on thermodynamics have been translated into English by W. R. Browne (*The Mechanical Theory of Heat by R. Clausius*, London, 1879; reviewed in *Nature*, February 19, 1880. The German edition was published in Braunschweig, 3 vols., Vol. I, 3d ed., 1887, Vol. II, 2d ed., 1879, Vol. III, 2d ed., 1889–91).]

mechanical law of the conservation of work, in Lagrange's form, of course holds.

As regards the first thought, we must regard it as an important one as containing the generalization of the attempts of Carnot, Mayer, and Neumann to apply the principle outside mechanics. Only we must combat the view, to which Helmholtz inclined, that the principle first came to be accepted through the development of mechanics. In fact, it is older than the whole of mechanics.

This view, now, seems to have been the leading motive in occasioning the second manner of treatment, against which, as I hope to show, very much can be urged.

However this may be, the view that physical phenomena can be reduced to processes of motion and equilibrium of molecules is so universally spread that, at the present time, one can only let people know that one's convictions are opposed to it, with caution, guardedly, and at the risk of rousing the opinion that one is not up to date and has not grasped the trend of modern culture.

To illustrate this point, I will quote a passage from a tract of 1866 on the physical axioms by Wundt,[16] for Wundt is a representative of the modern natural scientific tendency, and his way of thinking is probably that of a great majority of the investigators of natural science. Wundt lays down the following axioms:

1. All causes in nature are motional causes.

[16] *Die physikalischen Axiome und ihre Beziehungen zum Kausalprincip*, Erlangen, 1866.

2. Every motional cause lies outside the object moved.

3. All motional causes act in the direction of the straight line of junction, and so on.

4. The effect of every cause persists.

5. To every effect corresponds an equal counter-effect.

6. Every effect is equivalent to its cause.

Thus, there is no doubt that here all phenomena are thought of as a sum of mechanical events. And, so far as I know, no objection has been raised to Wundt's view. Now, however valuable Wundt's work may be in so far as it relates to mechanics, especially for what concerns the derivation of the axioms, and however much it agrees in that with the thoughts which I have held for many years, I can regard his theorems as mechanical only and not as physical. I will return to this question later.

Thus we have seen, in this historical sketch of many centuries, that our principle of the conservation of work has played a great part as an instrument of research. The second theorem of excluded perpetual motion was always leading to the discovery of mechanical—and later other physical—truths, and can also be considered as the historical foundation of the first theorem. On the other hand, the attempt to regard the whole of physics as mechanics and to make the first theorem the foundation of the second, or to extend the first to the second, is not capable of being misunderstood. Now, this circle is objectionable and rouses one's suspicions. It calls urgently for an investigation.

In the first place it is clear that the principle of excluded perpetual motion cannot be founded on mechanics, since its validity was felt long before the edifice of mechanics was raised. The principle must have another foundation. This view will now be supported if, on closer consideration of the mechanical conception of physics, we find that the latter suffers from being a doubtful anticipation and from one-sidedness, neither of which accusations can be laid against our principle. We will, then, first of all, examine the mechanical view of nature, in order to prove that the said principle is independent of it.

III
MECHANICAL PHYSICS

THE attempt to extend the mechanical theorem of the conservation of work to the theorem of excluded perpetual motion is connected with the rise of the mechanical conceptions of nature, which again was especially stimulated by the progress of the mechanical theory of heat. Let us, then, glance at the theory of heat.

The modern mechanical theory of heat and its view that heat is motion principally rest on the fact that the quantity of heat present decreases in the measure that work is performed and increases in the measure that work is used, provided that this work does not appear in another form. I say the modern theory of heat, for it is well known that the explanation of heat by means of motion had already more than once been given and lost sight of.

If, now, people say, heat vanishes in the measure that it performs work, it cannot be material, and consequently must be motion.

S. Carnot found that whenever heat performs work, a certain quantity of heat goes from a higher temperature-level to a lower one. He supposed in this that the quantity of heat remains constant. A simple analogy is this: If water (say, by means of a water-mill) is to perform work, a certain quantity of it must flow from a higher to a lower level; the quantity of water remains constant during the process.

When wood swells with dampness, it can perform work, burst open rocks, for example; and some people, as the ancient Egyptians, have used it for that purpose. Now, it would have been easy for an Egyptian wiseacre to have set up a mechanical theory of humidity. If wetness is to do work, it must go from a wetter body to one less wet. Evidently the wiseacre could have added that the quantity of wetness remains constant.

Electricity can perform work when it flows from a body of higher potential to one of lower potential; the quantity of the electricity remains constant.

A body in motion can perform work if it transfers some of its *vis viva* to a body moving more slowly. *Vis viva* can perform work by passing from a higher velocity-level to a lower one; the *vis viva* then decreases.

It would not be difficult to produce such an analogy from every branch of physics. I have intentionally chosen the last, because complete analogy breaks down.

When Clausius brought Carnot's theorem into connexion with the reflexions and experiments of Mayer, Joule, and others, he found that the addition "the quantity of heat remains constant" must be given up. One must, on the other hand, say that a quantity of heat proportional to the work performed vanishes.

"The quantity of water remains constant while work is performed, because it is a substance. The quantity of heat varies because it is not a substance."

These two statements will appear satisfactory to most scientific investigators; and yet both are quite worthless and signify nothing.

We will make this clear by the following question which bright students have sometimes put to me. Is there a mechanical equivalent of electricity as there is a mechanical equivalent of heat? Yes, and no. There is no mechanical equivalent of *quantity* of electricity as there is an equivalent of *quantity* of heat, because the same quantity of electricity has a very different capacity for work, according to the circumstances in which it is placed; but there *is* a mechanical equivalent of electrical energy.

Let us ask another question. Is there a mechanical equivalent of water? No, there is no mechanical equivalent of quantity of water, but there is a mechanical equivalent of weight of water multiplied by its distance of descent.

When a Leyden jar is discharged and work thereby performed, we do not picture to ourselves that the quantity of electricity disappears as work is done, but we simply assume that the electricities come into different positions, equal quantities of positive and negative electricity being united with one another.

What, now, is the reason of this difference of view in our treatment of heat and of electricity? The reason is purely historical, wholly conventional, and, what is still more important, is wholly indifferent. I may be allowed to establish this assertion.

In 1785 Coulomb constructed his torsion balance, by which he was enabled to measure the repulsion of electrified bodies. Suppose we have two small balls, A, B, which over their whole extent are similarly electrified. These two balls will exert on one another, at a certain distance r of their centres from one another, a certain repulsion p. We bring into contact with B, now, a ball C, suffer both to be equally electrified, and then measure the repulsion of B from A and of C from A at the same distance r. The sum of these repulsions is again p. Accordingly something has remained constant. If we ascribe this effect to a substance, then we infer naturally its constancy. But the essential point of the exposition is the divisibility of the electric force p and not the simile of substance.

In 1838 Riess constructed his electrical air-thermometer (the thermoelectrometer). This gives a measure of the quantity of heat produced by the discharge of jars. This quantity of heat is not proportional to the quantity of electricity contained in the jar by Coulomb's measure, but if q be this quantity and s be the capacity, is proportional to q^2/s, or, more simply still, to the energy of the charged jar. If, now, we discharge the jar completely through the thermometer, we obtain a certain quantity of heat, W. But if we make the discharge through the thermometer into a second jar, we obtain a quantity less than W. But we may obtain the remainder by completely discharging both jars through the air-thermometer, when it will again be proportional to the energy of the two jars. On the first, incomplete discharge, accordingly, a part of the electricity's capacity for work was lost.

When the charge of a jar produces heat, its energy

is changed and its value by Riess's thermometer is decreased. But by Coulomb's measure the quantity remains unaltered.

Now let us imagine that Riess's thermometer had been invented before Coulomb's torsion balance, which is not a difficult feat of imagination, since both inventions are independent of each other; what would be more natural than that the "quantity" of electricity contained in a jar should be measured by the heat produced in the thermometer? But then, this so-called quantity of electricity would decrease on the production of heat or on the performance of work, whereas it now remains unchanged; in the first case, therefore, electricity would not be a *substance* but a *motion*, whereas now it is still a substance. The reason, therefore, why we have other notions of electricity than we have of heat, is purely historical, accidental, and conventional.

This is also the case with other physical things. Water does not disappear when work is done. Why? Because we measure quantity of water with scales, just as we do electricity. But suppose the capacity of water for work were called quantity, and had to be measured, therefore, by a mill instead of by scales; then this quantity also would disappear as it performed the work. It may, now, be easily conceived that many substances are not so easily got at as water. In that case we should be unable to carry out the one kind of measurement with the scales while many other modes of measurement would still be left us.

In the case of heat, now, the historically established

measure of "quantity" is accidentally the work-value of the heat. Accordingly, its quantity disappears when work is done. But that heat is not a substance follows from this as little as does the opposite conclusion that it is a substance. In Black's case the quantity of heat remains constant because the heat passes into no *other* form of energy.

If anyone to-day should still wish to think of heat as a substance, we might allow that person this liberty with little ado. He would only have to assume that that which we call quantity of heat was the energy of a substance whose quantity remained unaltered, but whose energy changed. In point of fact we might much better say, in analogy with the other terms of physics, energy of heat, instead of quantity of heat.

By means of this reflection, the peculiar character of the second principal theorem of the mechanical theory of heat quite vanishes, and I have shown in another place that we can at once apply it to electrical and other phenomena if we put "potential" instead of "quantity of heat" and "potential function" instead of "absolute temperature." (See note 3, p. 85.)

If, then, we are astonished at the discovery that heat is motion, we are astonished at something which has never been discovered. It is quite irrelevant for scientific purposes whether we think of heat as a substance or not.

If a physicist wished to deceive himself by means of the notation that he himself has chosen—a state of things which cannot be supposed to be—he would behave similarly to many musicians who, after they have long forgotten how musical notation and softened pitch arose, are actually of the opinion that a piece marked in the key of six flats (G*b*) must sound differently from one marked in the key of six sharps (F♯).

If it were not too much for the patience of scientific people, one could easily make good the following statement. Heat is a substance just as much as oxygen is, and it is not a substance just as little as oxygen. Substance is possible phenomenon, a convenient word for a gap in our thoughts.

To us investigators, the concept "soul" is irrelevant and a matter for laughter. But matter is an abstraction of exactly the same kind, just as good and just as bad as it is. We know as much about the soul as we do of matter.

If we explode a mixture of oxygen and hydrogen in an eudiometer-tube, the phenomena of oxygen and hydrogen vanish and are replaced by those of water. We, say, now, that water *consists* of oxygen and hydrogen; but this oxygen and this hydrogen are merely two thoughts or names which, at the sight of water, we keep ready, to describe phenomena which are not present, but which will appear again whenever, as we say, we decompose water.

It is just the same case with oxygen as with latent heat. Both can appear when, at the moment, they cannot yet be remarked. If latent heat is not a substance, oxygen need not be one.

The indestructibility and conservation of matter cannot be urged against me. Let us rather say conservation of *weight;* then we have a pure fact, and we see at once that it has nothing to do with any theory. This cannot here be carried out any farther.

One thing we maintain, and that is, that in the investigation of nature, we have to deal only with knowledge of the connexion of appearances with one another. What we represent to ourselves behind the appearances exists *only* in our understanding, and has for us only the value of a *memoria technica* or formula, whose form, because it is arbitrary and irrelevant, varies very easily with the standpoint of our culture.

If, now, we merely keep our hold on the new laws as to the connexion between heat and work, it does not matter how we think of heat itself; and similarly in all physics. This way of presentation does not alter the facts in the least. But if this way of presentation is so limited and inflexible that it no longer allows us to follow the many-sidedness of phenomena, it should not be used any more as a formula and will begin to be a hindrance to us in the knowledge of phenomena.

This happens, I think, in the mechanical conception of physics. Let us glance at this conception that all physical phenomena reduce to the equilibrium and movement of molecules and atoms.

According to Wundt, all changes of nature are mere changes of place. All causes are motional causes.[17] Any discussion of the philosophical grounds on which Wundt supports his theory would lead us deep into the speculations of the Eleatics and the Herbartians. Change of place, Wundt holds, is the *only* change of a thing in which a thing remains identical with

[17] *Op. cit.*, p. 26.

itself. If a thing changed *qualitatively*, we should be obliged to imagine that something was annihilated and something else created in its place, which is not to be reconciled with our idea of the identity of the object observed and of the indestructibility of matter. But we have only to remember that the Eleatics encountered difficulties of exactly the same sort in motion. Can we not also imagine that a thing is destroyed in *one* place and in *another* an exactly similar thing created?

It is a bad sign for the mechanical view of the world that it wishes to support itself on such preposterous things, which are thousands of years old. If the ideas of matter, which were made at a lower stage of culture, are not suitable for dealing with the phenomena accessible to those on a higher plane of knowledge, it follows for the true investigator of nature that these ideas must be given up; not that only those phenomena exist, for which ideas that are out of order and have been outlived are suited.

But let us suppose for a moment that all physical events can be reduced to spatial motions of material particles (molecules). What can we do with that supposition? Thereby we suppose that things which can never be seen or touched and only exist in our imagination and understanding, can have the properties and relations only of things which can be touched. We impose on the creations of thought the limitations of the visible and tangible.

Now, there are also other forms of perception of other senses, and these forms are perfectly analogous to space—for example, the tone-series for hearing, which corresponds to a space of one dimension—and we do not allow ourselves a like liberty with them. We do not think of all things as sounding and do not figure to ourselves molecular events musically, in relations of heights of tones, although we are as justified in doing this as in thinking of them spatially.

This, therefore, teaches us what an unnecessary restriction we here impose upon ourselves. There is no more necessity to think of what is merely a product of thought spatially, that is to say, with the relations of the visible and tangible, than there is to think of these things in a definite position in the scale of tones.

And I will immediately show the sort of drawback that this limitation has. A system of n points is in form and magnitude determined in a space of r dimensions, if e distances between pairs of points are given, where e is given by the following table:

r	e_1	e_2
1	$n-1$	$2n-3$
2	$2n-3$	$3n-6$
3	$3n-6$	$4n-10$
4	$4n-10$	$5n-15$
5	$5n-15$	$6n-21$
r	$rn-\frac{r(r+1)}{2}$	$(r+1)n-\frac{(r+1)(r+2)}{2}$

In this table, the column marked by e_1 is to be used for e if we have made conditions about the sense of the given distances, for example, that in the straight line all points are reckoned according to one direction; in the plane all towards one side of the straight line through the first two points; in space all towards one side of the plane through the first three points; and so on. The column marked by e_2 is to be used if merely the absolute magnitude of the distance is given.

Between n points, combining them in pairs, $\frac{n(n-1)}{1.2}$ distances are thinkable, and therefore in general more than a space of a given number of dimensions can satisfy. If, for example, we suppose the e_1-column to be the one to be used, we find in a space of r dimensions the difference between the number of thinkable distances and those possible in this space to be

$$\frac{n(n-1)}{1.2} - rn + \frac{r(r+1)}{2} = k,$$

or

$$n(n-1) - 2rn + r(r+1) = 2k,$$

which can be brought to the form

$$(r-n)^2 + (r-n) = 2k.$$

This difference is, now, zero, if

$$(r-n)^2 + (r-n) = 0, \text{ or } (r-n) + 1 = 0, \text{ or } n = r + 1.$$

For a space of three dimensions, the number of distances thinkable is greater than the number of distances possible in this space when the number of points is greater than four. Let us imagine, for example, a molecule consisting of five atoms, *A*, *B*, *C*, *D*, and *E*, then between them, ten distances are thinkable, but, in a space of three dimensions, only nine are possible, that is to say, if we choose nine such distances, the tenth thinkable one is determined by means of the nature of this space, and it is no longer arbitrary. If *A B*, *B C*, *C A*, *A D*, *D B*, *D C*, are given me, I get a tetrahedron of fixed form. If, now,

I add E with the distances EA, EB, and EC determined, then DE is determined by them. Thus it would be impossible gradually to alter the distance DE without the other distances being thereby altered. Thus, there might be serious difficulties in the way of imagining many pent-atomic isomeric molecules which merely differ from one another by the relation of D and E. This difficulty vanishes in our example, when we think the pent-atomic molecule in a space of four dimensions; then ten independent distances are thinkable and also ten distances can be set up.

Fig. 5.

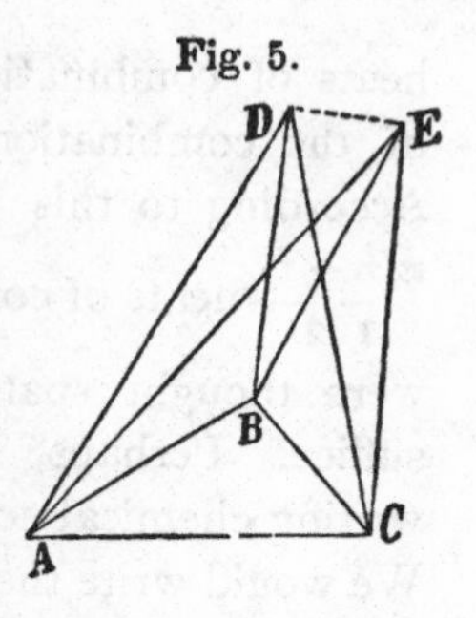

Now, the greater the number of atoms in a molecule, the higher the number of the dimensions of space do we need to make actual all the thinkable possibilities of such combinations. This is only an example to show under what limitations we proceed when we imagine the chemical elements lying side by side in a space of three dimensions, and how a crowd of the relations of the elements can escape us thereby if we wish to represent them in a formula which cannot comprise them. (See note 4, p. 86.)

It is clear how we can study the nature of chemical combinations without giving ourselves up to the conception mentioned, and how, indeed, people have now begun to study them. The heat of combustion generated by a combination gives us a clearer idea of the stability and manner of combination than any pictorial representation. If, then, it were possible, in any molecule composed of n parts, to determine the $\frac{n(n-1)}{1.2}$ heats of combination of every two parts, the nature of the combination would be characterized thereby. According to this view, we would have to determine $\frac{n(n-1)}{1.2}$ heats of combination, whereas, if the molecules were thought spatially, $3n-6$ heats of combination suffice. Perhaps, too, a more rational manner of writing chemical combinations can be founded on this. We would write the components in a circle, draw a line from each to each, and write on the latter the respective heat of combination.

Perhaps the reason why, hitherto, people have not succeeded in establishing a satisfactory theory of electricity is because they wished to explain electrical phenomena by means of molecular events in a space of three dimensions.

Herewith I believe that I have shown that one can hold, treasure, and also turn to good account the results of modern natural science without being a supporter of the mechanical conception of nature, that this conception is not necessary for the knowledge of the phenomena and can be replaced just as well by another theory, and that the mechanical conceptions can even be a hindrance to the knowledge of phenomena.

Let me add a view on scientific theories in general: If all the individual facts—all the individual phenomena, knowledge of which we desire—were immediately accessible to us, a science would never have arisen.

Because the mental power, the memory, of the individual is limited, the material must be arranged. If, for example, to every time of falling, we knew the corresponding space fallen through, we could be satisfied with that. Only, what a gigantic memory would be needed to contain the table of the correspondences of s and t. Instead of this we remember the formula $s=\frac{gt^2}{2}$, that is to say, the rule of derivation by means of which we find, from a given t, the corresponding s, and this replaces the table just mentioned in a very complete, convenient, and compendious manner.

This rule of derivation, this formula, this "law," has, now, not in the least more real value than the aggregate of the individual facts. Its value for us lies merely in the convenience of its use: it has an economical value. (See note 5, p. 88.)

Besides this collection of as many facts as possible in a synoptical form, natural science has yet another problem which is also economical in nature. It has to resolve the more complicated facts into as few and as simple ones as possible. This we call explaining. These simplest facts, to which we reduce the more complicated ones, are always unintelligible in themselves, that is to say, they are not further resolvable. An example of this is the fact that one mass imparts an acceleration to another.

Now, it is only, on the one hand, an economical question, and, on the other, a question of taste, at what unintelligibilities we stop. People usually deceive themselves in thinking that they have reduced the unintelligible to the intelligible. Understanding consists in analysis alone; and people usually reduce uncommon unintelligibilities to common ones. They always get, finally, to propositions of the form: if A is, B is, therefore to propositions which must follow from intuition, and, therefore, are not further intelligible.

What facts one will allow to rank as fundamental facts, at which one rests, depends on custom and on history. For the lowest stage of knowledge there is no more sufficient explanation than pressure and impact.

The Newtonian theory of gravitation, on its appearance, disturbed almost all investigators of nature because it was founded on an uncommon unintelligibility. People tried to reduce gravitation to pressure and impact. At the present day gravitation no longer disturbs anybody: it has become a *common* unintelligibility.

It is well known that action at a distance has caused difficulties to very eminent thinkers. "A body can only act where it is"; therefore there is only pressure and impact, and no action at a distance. But where is a body? Is it only where we touch it? Let us invert the matter: a body is where it acts. A little space is taken for touching, a greater for hearing, and a still greater for seeing. How did it come about that the sense of touch alone dictates to us where a body is? Moreover, contact-action can be regarded as a special case of action at a distance.

It is the result of a misconception, to believe, as people do at the present time, that mechanical facts are more intelligible than others, and that they can provide the foundation for other physical facts. This belief arises from the fact that the history of mechanics is older and richer than that of physics, so that we have been on terms of intimacy with mechanical facts for a longer time. Who can say that, at some future time, electrical and thermal phenomena will not appear to us like that, when we have come to know and to be familiar with their simplest rules?

In the investigation of nature, we always and alone have to do with the finding of the best and simplest rules for the derivation of phenomena from one another. One fundamental fact is not at all more intelligible than another: the choice of fundamental facts is a matter of convenience, history, and custom.

The ultimate unintelligibilities on which science is founded must be facts, or, if they are hypotheses, must be capable of becoming facts. If the hypotheses are so chosen that their subject (*Gegenstand*) can never appeal to the senses and therefore also can never be tested, as is the case with the mechanical molecular theory, the investigator has done more than science, whose aim is facts, requires of him—and this work of supererogation is an evil.

Perhaps one might think that rules for phenomena, which cannot be perceived in the phenomena themselves, can be discovered by means of the molecular theory. Only that is not so. In a complete theory, to all details of the phenomenon details of the hypothesis must correspond, and all rules for these hypothetical things must also be directly transferable to the phenomenon. But then molecules are merely a valueless image.

Accordingly, we must say with J. R. Mayer: "If a fact is known on all its sides, it is, by that knowledge, explained, and the problem of science is ended."[18]

[18] *Mechanik der Wärme*, Stuttgart, 1867, p. 239.

IV

THE LOGICAL ROOT OF THE THEOREM OF EXCLUDED PERPETUAL MOTION

IF the principle of excluded perpetual motion is not based upon the mechanical view—a proposition which must be granted, since the principle was recognized before the development of this view—if the mechanical view is so fluctuating and precarious that it can give no sure foundation for this theorem, and, indeed, if it is likely that our principle is not founded on positive insight, because on it is founded the most important positive knowledge; on what does the principle rest, and whence comes its power of conviction, with which it has always ruled the greatest investigators?

I will now try to answer this question. For this purpose I must go back somewhat, to the foundations of the logic of natural science.

If we attentively observe natural phenomena, we notice that, with the variation of some of them, also variations of others occur, and in this way we have grown used to considering natural phenomena as dependent upon one another. This dependence of phenomena is called the law of causality. Now, people are accustomed to give different forms to the law of causality. Thus, for example, it is sometimes expressed: "Every effect has a cause"; which means that a variation can only occur with, or, as people prefer to say, in consequence of, another. But this expression is too indefinite to be further discussed here. Besides, it can lead to great inaccuracies.

Very clearly, Fechner[19] formulated the law of causality: "Everywhere and at all times, if the same circumstances occur again, the same consequence occurs again; if the same circumstances do not occur again, the same consequence does not." By this means, as Fechner remarked farther on, "a relation is set up between the things which happen in all parts of space and at all times."

I think I must add, and have already added in another publication, that the express drawing of space and time into consideration in the law of causality, is at least superfluous. Since we only recognize what we call time and space by certain phenomena, spatial and temporal determinations are only determinations by means of other phenomena. If, for example, we express the positions of earthly bodies as functions of the time, that is to say, as functions of the earth's angle of rotation, we have simply determined the dependence of the positions of the earthly bodies on *one another*.

The earth's angle of rotation is very ready to our hand, and thus we easily substitute it for other phenomena which are connected with it but less accessible to us; it is a kind of money which we spend to avoid the inconvenient trading with phenomena, so that the proverb "Time is money" has also here a meaning. We can eliminate time from every law of nature by putting

[19] *Berichte der sächs. Ges. zu Leipzig*, Vol. II, 1850.

in its place a phenomenon dependent on the earth's angle of rotation.

The same holds of space. We know positions in space by the affection of our retina, of our optical or other measuring apparatus. And our x, y, z in the equations of physics are, indeed, nothing else than convenient names for these affections. Spatial determinations are, therefore, again determinations of phenomena by means of other phenomena.

The present tendency of physics is to represent every phenomenon as a function of other phenomena and of certain spatial and temporal positions. If, now, we imagine the spatial and temporal positions replaced in the above manner, in the equations in question, we obtain simply *every phenomenon as function of other phenomena*. (See note 6, p. 88.)

Thus the law of causality is sufficiently characterized by saying that it is the presupposition of the mutual dependence of phenomena. Certain idle questions, for example, whether the cause precedes or is simultaneous with the effect, then vanish by themselves.

The law of causality is identical with the supposition that between the natural phenomena $\alpha, \beta, \gamma, \delta, \ldots, \omega$ certain equations subsist. The law of causality says nothing about the number or form of these equations; it is the problem of positive natural investigation to determine this; but it is clear that if the number of the equations were greater than or equal to the number of the $\alpha, \beta, \gamma, \delta, \ldots, \omega$, all the $\alpha, \beta, \gamma, \delta, \ldots, \omega$ would be thereby overdetermined or at least completely determined. The fact of the varying of nature therefore proves that the number of the equations is less than that of the $\alpha, \beta, \gamma, \delta, \ldots, \omega$.

But with this a certain indefiniteness in nature remains behind, and I will at once call attention to it here, because I believe that even investigators of nature have sometimes overlooked it, and have thereby been led to very strange theorems. For instance, such a theorem is that defended by W. Thomson[20] and Clausius,[21] according to which after an infinitely long time the universe, by the fundamental theorems of thermodynamics, must die the death of heat, that is to say, according to which all mechanical motion vanishes and finally passes over into heat. Now such a theorem enunciated about the whole universe seems to me to be illusory throughout.

As soon as a certain number of phenomena is given, the others are co-determined, but the law of causality does not say at what the universe, the totality of phenomena, is aiming, if we may so express it, and this cannot be determined by any investigation; it is no scientific question. This lies in the nature of things.

The universe is like a machine in which the motion of certain parts is determined by that of others, only nothing is determined about the motion of the whole machine.

If we say of a thing in the universe that, after the lapse of a certain time, it undergoes the variation A, we posit it as dependent on another part of the universe,

[20] *Phil. Mag.*, October, 1852; *Math. cnd Phys. Papers*, I, p. 511.

[21] *Pogg. Ann.*, Bd. 93, Dezember, 1854; *Der zweite Hauptsatz d mech. Wärmetheorie*, Braunschweig, 1867.

which we consider as a clock. But if we assert such a theorem for the universe itself, we have deceived ourselves in that we have nothing over to which we could refer the universe as to a clock. For the universe there is no time. Scientific statements like the one mentioned seem to me worse than the worst philosophical ones.

People usually think that if the state of the whole universe is given at one moment, it is completely determined at the next one; but an illusion has crept in there. This next moment is given by the advance of the earth. The position of the earth belongs to the circumstances. But we easily commit the error of counting the same circumstance twice. If the earth advances, this and that occur. Only the question as to *when* it will have advanced has no meaning at all. The answer can be given only in the form: It has advanced farther then, if it has advanced farther.

It may not be unimportant for the investigator of nature to consider and recognize the indetermination which the law of causality leaves over. To be sure, the only value of this for him is to keep him from transgressing its limits. On the other hand, an idle philosopher could perhaps connect his ideas on freedom of the will with this, with better luck than he has had hitherto in the case of other gaps in knowledge. (See note 7, p. 90.) For the investigator of nature there is nothing else to find out but the dependence of phenomena on one another.

Let us call the totality of the phenomena on which a phenomenon *a* can be considered as dependent, *the cause.* If this totality is given, *a* is determined, and determined uniquely. Thus the law of causality may also be expressed in the form: "The effect is determined by the cause."

This last form of the law of causality may well have been that which was already in existence at a very low stage of human culture, and yet existed in full clearness. In general, a lower stage of knowledge may perhaps be distinguished from a higher one not so much by the difference of the conception of causality as by the manner of application of this conception.

He who has no experience will, because of the complication of the phenomena surrounding him, easily suppose a connexion between things which have no perceptible influence on one another. Thus, for example, an alchemist or wizard may easily think that, if he cooks quicksilver with a Jew's beard and a Turk's nose at midnight at a place where roads cross, while nobody coughs within the radius of a mile, he will get gold from it. The man of science of to-day knows from experience that such circumstances do not alter the chemical nature of things, and accordingly he has a smoother path to traverse. Science has grown almost more by what it has learned to ignore than by what it has had to take into account.

If we call to remembrance our early youth, we find that the conception of causality was there very clearly, but not the correct and fortunate application of it. In my own case, for example—I remember this exactly—there was a turning-point in my fifth year. Up to that time I represented to myself everything which I did not understand—a pianoforte, for instance—as simply a motley assemblage of the most wonderful things, to which I ascribed the sound of the notes. That the pressed key struck the cord with the hammer did not occur to me. Then one day I saw a wind-mill. I saw how the cogs on the axle engaged with the cogs which drive the mill-stones, how one tooth pushed on the other; and, from that time on, it became quite clear to me that all is not connected with all, but that, under circumstances, there is a choice. At the present time, every child has abundant opportunities for making this step. But there was a time, as the epidemic of belief in witches, which belief lasted many centuries, proves, in which this step was only permitted to the greatest minds.

By this I only wanted to show that, without positive experiences, the law of causality is empty and barren. This appears still better with another theorem, which we recognize at once as an inverse of the law of causality —with the law of sufficient reason. Let us explain this law by some examples.

Let us take a straight horizontal bar, which we support in its middle and at both ends of which we hang equal weights. Then we perceive at once that equilibrium must subsist, because there is no reason why the bar should turn in one direction rather than in the other. So Archimedes concluded.

If we let four equal forces act at the centre of gravity of a regular tetrahedron in the directions of its vertices, equilibrium reigns. Again there is no reason why motion should result in one direction rather than in another.

Only this is not expressed quite properly: we ought rather to say that there is a reason that, in these cases, *nothing* happens. For the effect is determined by the cause, and the one and only effect which is here determined by the cause is *no* effect at all. In fact, if any effect were to occur, no rule of derivation of it from the circumstances could be given. If, for example, we imagine any resultant in the above tetrahedron of forces and set up a rule for its derivation, there are eleven other resultants which can be found by the same rule. Consequently, nothing is determined. The one and only effect which is determined in this case is the effect which is equal to zero. The law of sufficient reason is not essentially different from the law of causality or from the theorem: "The effect is determined by the cause."

But how is a person who has made no experiments to apply this theorem? Give him a lever with arms of equal length and with its ends loaded with equal weights, but with the weights and arms of different colours and forms. Without experimental knowledge, he will never discover those circumstances which alone are relevant. As an example of how important experience is in such derivations, I will give Galileo's demonstration of the law of the lever. Galileo borrowed it from Stevinus and slightly modified it, and Stevinus somewhat varied Archimedes's demonstration.

A horizontal prism *A B* is hung at the ends by two threads *u* and *v* on to a horizontal bar *a b*, which can be rotated about its middle *c*, or is hung up there by a thread. Such a system is, as we see at once, in equilibrium. If, now, we divide the prism into two parts of lengths $2m$ and $2n$ by a section at E, after we have attached two new threads p and q at both sides of the section, equilibrium still subsists. It will also still subsist if we hang the piece $A\,E$ in its middle by the thread r, and $E\,B$ by s to $a\,b$, and take away p, q, u, v. But then at a distance n from c hangs a prism of weight

Fig. 6.

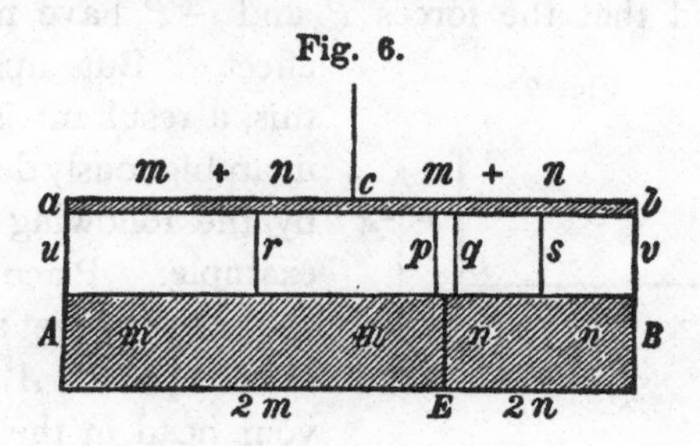

$2m$, and at a distance m from c hangs a prism of weight $2n$. Now the practical physicist knows that the tension of the threads, which alone mediates between the prism and the bar, depends only on the magnitude, and not on the form, of the weight. Therefore we can, again without disturbance of the equilibrium, replace the pieces of the prism by any other weights $2m$ and $2n$; and this gives the known law of the lever.

Now, he who had not had a great deal of experience in mechanical things certainly could not have carried out such a demonstration.

Yet another example. At A and B are the equal and parallel forces P and $-P$ to act. As is well known, they have no resultant. Let us suppose, for example, that $-R$ is a resultant, then we must also suppose that

R is one, for it is determined by the same rule as $-R$, if we turn round the figure through two right angles. Consequently, the one and only resultant completely determined by the circumstances is, in this case, *no* resultant. However, this holds only if we know already that we have to seek the resultant in the plane of symmetry of the system, that is to say, in the plane of P and $-P$, and that the forces P and $-P$ have no lateral effect. But apart from this, a resultant is at once unambiguously determined by the following rule, for example. Place yourself so that your feet are at one of the points A, B, with your head in the direction of the force acting there, and look towards the other point, drawing the resultant towards the right, perpendicularly to the plane $(P, -P)$. In fact, the part of line thus determined has a signification for our case. It is, however, not the resultant, but the axis of the Poinsot's couple represented in Fig. 7.

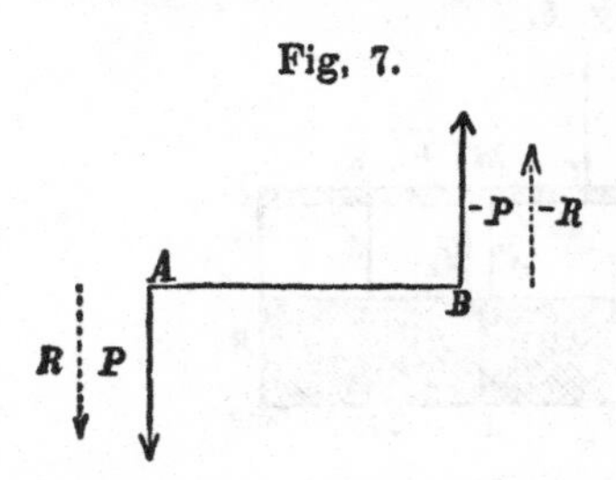

Fig. 7.

If P and $-P$ were not simple forces, but the axes of a Poinsot's couple—if we had, consequently, things affected with a certain lateralness—the direction just determined would represent the direction of the resultant motion, if we choose the axes so that, for an observer with his head at the arrow-head and his feet at B, the rotation takes place in the plane through B perpen dicular to $-P$ in the direction of the hands of a clock.

Now, whether the things we have to consider have

such a lateralness often cannot be determined at the first glance, but can only be so determined by means of much experience. The lateralness of light remained hidden for a long time, and caused great surprise to its discoverer Malus. If an electric current flows in the vertical plane drawn through a magnetic needle, from the south pole towards the north pole, one thinks that all is symmetrical with respect to this plane and that the needle could, at most, move in this plane. One is greatly surprised when one hears for the first time that the north pole deviates to the left of a swimmer in the current, who is looking at the needle.

The law of sufficient reason is an excellent instrument in the hands of an experienced investigator, but is an empty formula in the hands of even the most talented people in whom special knowledge is lacking.

After these considerations, now, it will not be hard for us to discover the source from which the principle of excluded perpetual motion arises. It is again only another form of the law of causality.

"It is not possible to create work out of nothing." If a group of phenomena is to become the source of continual work, this means that it shall become a source of continual variation of another group of phenomena. For, by means of the general connexion of nature, all phenomena are also connected with mechanical phenomena, and therefore with the performance of work. Every source of continual variation of phenomena is a source of work, and inversely.

If, now, the phenomena α, β, γ, . . . depend on the phenomena x, y, z, . . . , certain equations

$$\alpha = f_1(x, y, z, \ldots),$$
$$\beta = f_2(x, y, z, \ldots),$$
$$\gamma = f_3(x, y, z, \ldots),$$

subsist, from which $\alpha, \beta, \gamma, \ldots$ are uniquely determined when $x, y, z, \ldots$ are given. Now, it is clear that:

1. As long as $x, y, z, \ldots$ are constant, $\alpha, \beta, \gamma, \ldots$ are;

2. If $x, y, z, \ldots$ make merely one step, so do $\alpha, \beta, \gamma, \ldots$;

3. If $x, y, z, \ldots$ vary periodically, so do $\alpha, \beta, \gamma, \ldots$;

4. If, finally, $\alpha, \beta, \gamma, \ldots$ are to undergo continual variations, $x, y, z, \ldots$ must necessarily do so.

If a group of phenomena $x, y, z, \ldots$ is to become a source of work, a source of the continual variation of another group $\alpha, \beta, \gamma, \ldots$, the group $x, y, z, \ldots$ itself must be engaged in continual variation. This is a clear form of the theorem of excluded perpetual motion, and one which cannot be misinterpreted. In this abstract form the theorem has nothing to do with mechanics particularly, but can be applied to all phenomena. The theorem of excluded perpetual motion is merely a special case of the theorem here enunciated.

The remark which has been made cannot be inverted. In general, certain systems of continual variations of $x, y, z, \ldots$, which make no difference to $\alpha, \beta, \gamma, \ldots$ can be imagined, that is to say, groups of appearances can be given, which are engaged in continual variation without being sources of continual variation of other groups of phenomena. These are groups shut up in themselves. How such groups can be divided, that is to say, which phenomena depend on one another and in what manner, and which do not, can be taught only by experience, and the law of causality says nothing about it.

The theorem of excluded perpetual motion, without positive experience, is just as empty as the law of sufficient reason and all formal laws of that kind. On this account—and history teaches this—it has found more and more applications in physics as positive knowledge progressed. First it was applied in mechanics alone, then in the theory of heat, and lastly in the theory of electricity. Abstract theorems alone lead to nothing; and Poinsot[22] remarked very correctly: "Rien ne vous dispense d'étudier les choses en elles-mêmes, et de nous bien rendre compte des idées qui font l'objet de nos speculations."

Let us illustrate the theorem of excluded perpetual motion by some examples.

The vibrations of a tuning-fork are periodical variations; they can become a lasting source of work only if they themselves undergo lasting variations—for example, by the diminution of their amplitude. We hear a tuning-fork only because its vibrations thus decrease.

A rotating top can perform work if its angular velocity decreases.

The mere lying side by side of a copper and a zinc plate will generate no electric *current*. From where,

[22] *Théorie nouvelle de la rotation des corps*, Paris, 1851, p. 80.

indeed, would the continual variation come if the plates themselves underwent no such variation? But if a continual chemical change of the plates occurs, we have no further objection to make against the supposition of an electric current.

An example of the unpermissibility of the process of inversion mentioned above is as follows: A top which is protected from resistance can rotate uniformly without becoming a source of work. Its angular velocity remains constant, but its angle of rotation varies continually. This does not contradict the principle. But experience adds—what the principle does not know—that in this case only variations of velocity, and not variations of position, can become a source of other variations But if one were to think that the top's continual variation of position is connected with no other continual variation, it would again be a mistake. It is connected with the increasing angle of rotation of the earth. This view leads, to be sure, to a peculiar conception of the law of inertia, into the further discussion of which we shall not enter.

Though the principle of excluded perpetual motion is very fruitful in the hands of an experienced investigator, it is useless in a department of experience which has not been accurately explored.

People have put a special value on the fact that the sum of the store of the work at our disposal and the *vis viva*, or the energy, is constant. Only, although we must admit that such a commercial or housekeeping expression is very convenient, easily seized, and suitable to human nature, which is planned throughout on economical grounds, we find, on looking into the matter quietly and accurately, that there is nothing essentially more in such a law than in any other law of nature.

The law of causality supposes a dependence between the natural phenomena $\alpha, \beta, \gamma, \ldots$. It is the problem of the investigator of nature to find out the manner of this dependence. Now, it does not matter very much how the equations representing this dependence are written. All will agree that it makes no great difference in which of the three forms an equation is written,

$$f(\alpha, \beta, \gamma, \ldots)=0, \quad \alpha=\psi(\alpha, \beta, \gamma, \ldots), \quad F(\alpha, \beta, \gamma, \ldots)=\text{const.},$$

and that in the last of these forms there lies no specially higher wisdom than in the others.

But it is merely by this form that the law of the conservation of work differs from other laws of nature. We can easily give a similar form to any other law of nature; thus, we can write Mariotte's law, where p is the force of expansion and v is the volume of the unit of mass, in the form $\log p+\log v=\text{const.}$ However beautiful, simple, and perspicuous much in the form of the theorem of the conservation of work looks, I cannot feel any enthusiasm for the mysticism which some people love to push forwards by means of this theorem.

By this I believe that I have shown that the theorem of excluded perpetual motion is merely a special form of the law of causality, which law results immediately from the supposition of the dependence of phenomena on one another—a supposition which precedes every scientific investigation; and which is quite unconnected with the mechanical view of nature, but is consistent with any view, if only it firmly retains a strict rule by laws.

We have, on this occasion, seen that the riches which investigators of nature have, in the course of time, heaped up by their work are of very different kinds. They are in part actual pieces of knowledge, in part also superseded theories, great and small, points of view that were now and then useful at an earlier stage, but are now irrelevant, philosophemes—among them some of the worst kind, by which some people wrongly condemn investigators of nature—and so on. It can only be useful sometimes to hold a review of these treasures; and this gives us the opportunity of putting aside what is worthless, and one does not run the risk of confusing deeds of assignment with property.

The object of natural science is the connexion of phenomena; but the theories are like dry leaves which fall away when they have long ceased to be the lungs of the tree of science.

NOTES

1. (See p. 28.) The law of inertia was afterwards formulated by Newton in the following way:

"Corpus omne perseverare in statu suo quiescendi vel movendi uniformiter in directum nisi quatenus a viribus impressis cogitur statum illum mutare."

Philosophiae Naturalis Principia Mathematica, Amstaelodami, 1714, Tom. I, p. 12 (Lex. I of the "Axiomata sive leges motus"); cf. pp. 2, 358. [The first edition of the *Principia* was published in London in 1687, the second edition at Cambridge in 1713, the third in London in 1726, and an English translation, in two volumes, by Andrew Motte, in London, 1729 (American editions, New York, 1848 and 1850, one vol.). Full bibliographical information as to the various editions and translations of Newton's works is given in George J. Gray's book, *A Bibliography of the Works of Sir Isaac Newton*, 2d ed., Cambridge, 1907.]

Since Newton, this law, which was with Galileo a mere remark, has attained the dignity and intangibleness of a papal dictum. Perhaps the best way to enunciate it is: Every body keeps its direction and velocity as long as they are not altered by outer forces.

Now, I remarked many years ago that there is in this law a great indefiniteness; for which body it is, with respect to which the direction and velocity of the body in motion is determined, is not stated. I first drew attention to this indefiniteness, to a series of paradoxes which can be deduced from it, and to the solution of the difficulty, in my course of lectures "Ueber einige Hauptfragen der Physik" in the summer of 1868, before an audience of about forty persons. I referred regularly to the same subject in the years following, but my investigation was not printed for reasons stated in the next note.

Now, a short while ago, C. Neumann[1] discussed this point, and found exactly the same indefiniteness, difficulties, and paradoxes in the law. Although I was sorry to have lost the priority in this important matter, yet the exact coincidence of my views with those of so distinguished a mathematician gave me great pleasure and richly compensated me for the disdain and surprise which almost all the physicists with whom I discussed this subject showed. Also, I think that I may, without fear, assert my independence in a matter of which I spoke before so large an audience and so long before.

Now, I must add that, although the difficulties which I found in the law of inertia exactly coincide with those of Neumann, yet my solution of them is different. Neumann thought that he had removed the difficulties by considering all motion as absolute and determined by means of a hypothetical body ***a***. Only then everything remains as it was of old. The law of inertia apparently receives a more distinct enunciation, but it did not turn out differently in practice. This appears from the following considerations.

Obviously it does not matter whether we think of the earth as turning round on its axis, or at rest while the celestial bodies revolve round it. Geometrically these

[1] [*Ueber die Principien der Galilei-Newton'schen Theorie* Leipzig, 1870.]

are exactly the same case of a relative rotation of the earth and of the celestial bodies with respect to one another. Only, the first representation is astronomically more convenient and simpler.

But if we think of the earth at rest and the other celestial bodies revolving round it, there is no flattening of the earth, no Foucault's experiment, and so on—at least according to our usual conception of the law of inertia. Now, one can solve the difficulty in two ways: Either all motion is absolute, or our law of inertia is wrongly expressed. Neumann preferred the first supposition, I, the second. The law of inertia must be so conceived that exactly the same thing results from the second supposition as from the first. By this it will be evident that, in its expression, regard must be paid to the masses of the universe.

In ordinary terrestrial cases, it will answer our purposes quite well to reckon the direction and velocity with respect to the top of a tower or a corner of a room; in ordinary astronomical cases, one or other of the stars will suffice. But because we can also choose other corners of rooms, another pinnacle, or other stars, the view may easily arise that we do not need such a point at all from which to reckon. But this is a mistake; such a system of co-ordinates has a value only if it can be determined by means of bodies. We here fall into the same error as we did with the representation of time. Because a piece of paper money need not necessarily be funded by a definite piece of money, we must not think that it need not be funded at all.

In fact, any one of the above points of origin of coordinates answers our purposes as long as a sufficient number of bodies keep fixed positions with respect to one another. But if we wish to apply the law of inertia in an earthquake, the terrestrial points of reference would leave us in the lurch, and, convinced of their uselessness, we would grope after celestial ones. But, with these better ones, the same thing would happen as soon as the stars showed movements which were very noticeable. When the variations of the positions of the fixed stars with respect to one another cannot be disregarded, the laying down of a system of co-ordinates has reached an end. It ceases to be immaterial whether we take this or that star as point of reference; and we can no longer reduce these systems to one another. We ask for the first time which star we are to choose, and in this case easily see that the stars cannot be treated indifferently, but that because we can give preference to none, the influence of all must be taken into consideration.

We can, in the application of the law of inertia, disregard any particular body, provided that we have enough other bodies which are fixed with respect to one another. If a tower falls, this does not matter to us; we have others. If Sirius alone, like a shooting-star, shot through the heavens, it would not disturb us very much; other stars would be there. But what would become of the law of inertia if the whole of the heavens began to move and the stars swarmed in confusion? How would we apply it then? How would it have to be expressed then? We do not inquire after one body as long as we have others enough; nor after one piece of money as long as we have others enough. Only in the case of a shattering of the universe, or a bankruptcy, as the case may be, we learn that *all* bodies, each with its share, are of importance in the law of inertia, and all money, when paper money is funded, is of importance, each piece having its share.

Yet another example: A free body, when acted upon by an instantaneous couple, moves so that its central ellipsoid with fixed centre rolls without slipping on a tangent-plane parallel to the plane of the couple. This is a motion in consequence of inertia. Here the body makes very strange motions with respect to the celestial bodies. Now, do we think that these bodies, without which one cannot describe the motion imagined, are without influence on this motion? Does not that to which one must appeal explicitly or implicitly when one wishes to describe a phenomenon belong to the most essential conditions, to the causal nexus of the phenomenon? The distant heavenly bodies have, in our example, no influence on the acceleration, but they have on the velocity.

Now, what share has every mass in the determination of direction and velocity in the law of inertia? No definite answer can be given to this by our experiences. We only know that the share of the nearest masses vanishes in comparison with that of the farthest. We would, then, be able completely to make out the facts known to us if, for example, we were to make the simple supposition that all bodies act in the way of determination proportionately to their masses and independently of the distance, or proportionately to the distance, and so on. Another expression would be: In so far as bodies are so distant from one another that they contribute no noticeable acceleration to one another, all distances vary proportionately to one another.

I will return to the subject on another occasion.

2. (See p. 29.) Perhaps I may mention here that I tried to get my bearings with respect to the concept of mass by the help of the principle of excluded perpetual motion. My note on this subject was returned as unusable by Poggendorff, the then editor of the *Annalen der Physik und der Chemie*, after he had had it about a year, and it appeared later in the fourth volume of Carl's *Repertorium.*[2] This rejection was also the reason why I did not publish my investigations on the law of inertia. If I ran up against the physics of the schools in so simple and clear a matter, what could I expect in a more difficult question? The *Annalen* often contain pages of fallacies about Torricelli's theorem and the blush of dawn—written, to be sure, in "physical language"; but the inclusion of a short note which is not wholly written in that jargon would obviously greatly lower the value of the *Annalen* in the eyes of the public.

The following is a complete reproduction of the note in question:

ON THE DEFINITION OF MASS

The circumstance that the fundamental propositions of mechanics are neither wholly *a priori* nor can wholly be discovered by means of experience—for sufficiently numerous and

[2] *Ueber die Definition der Masse, Repertorium für physikalische Technik* . . . . , Bd. IV, 1868, pp. 355 sqq.

accurate experiments cannot be made—results in a peculiarly inaccurate and unscientific treatment of these fundamental propositions and conceptions. Rarely is distinguished and stated clearly enough what is *a priori*, what empirical, and what is hypothesis.

Now, I can only imagine a scientific exposition of the fundamental propositions of mechanics to be such that one regards these theorems as hypotheses to which experience forces us, and that one afterwards shows how the denial of these hypotheses would lead to contradictions with the best-established facts.

As evident *a priori* we can only, in scientific investigations, consider the law of causality or the law of sufficient reason, which is only another form of the law of causality. No investigator of nature doubts that under the same circumstances the same always results, or that the effect is completely determined by the cause. It may remain undecided whether the law of causality rests on a powerful induction or has its foundation in the psychical organization (because in the psychical life, too, equal circumstances have equal consequences).

The importance of the law of sufficient reason in the hands of an investigator was proved by Clausius's works on thermodynamics and Kirchhoff's researches on the connexion of absorption and emission. The well-trained investigator accustoms himself in his thought, by the aid of this theorem, to the same definiteness as nature has in its actions, and then experiences which are not in themselves very apparent suffice, by exclusion of all that is contradictory, to discover very important laws connected with the said experiences.

Usually, now, people are not very chary of asserting that a proposition is immediately evident. For example, the law of inertia is often stated to be such a proposition, as if it did not need the proof of experience. The fact is that it can only have grown out of experience. If masses imparted to one another, not acceleration, but, say, velocities which depended on the distance, there would be no law of inertia; but whether we have the one state of things or the other, only experience teaches. If we had merely sensations of heat, there would be merely equalizing velocities (*Ausgleichungsgeschwindigkeiten*), which vanish with the differences of temperature.

One can say of the motion of masses: "The effect of every cause persists," just as correctly as the opposite: "Cessante causa cessat effectus"; it is merely a matter of words. If we call the resulting velocity the "effect," the first proposition is true, if we call the acceleration the "effect," the second is true.

Also people try to deduce *a priori* the theorem of the parallelogram of forces; but they must always bring in tacitly the supposition that the forces are independent of one another. But by this the whole derivation becomes superfluous.

I will now illustrate what I have said by *one* example, and show how I think the conception of mass can be quite scientifically developed. The difficulty of this conception, which is pretty generally felt, lies, it seems to me, in two circumstances: (1) in the unsuitable arrangement of the first conceptions and theorems of mechanics; (2) in the silent passing over important presuppositions lying at the basis of the deduction.

Usually people define $m=\frac{p}{g}$ and again $p=mg$. This is either a very repugnant circle, or it is necessary for one to conceive force as "pressure." The latter cannot be avoided if, as is customary, statics precedes dynamics. The difficulty, in this case, of defining magnitude and direction of a force is well known.

In that principle of Newton, which is usually placed at the head of mechanics, and which runs: "Actioni contrariam semper et aequalem esse reactionem: sive corporum duorum actiones in se mutuo semper esse aequales et in partes contrarias dirigi," the "actio" is again a pressure, or the principle is quite unintelligible unless we possess already the conception of force and mass But pressure looks very strange at the head of the quite phoronomical mechanics of today. However, this can be avoided.

If there were only one kind of matter, the law of sufficient reason would be sufficient to enable us to perceive that two completely similar bodies can impart to each other only *equal* and *opposite* accelerations. This is the one and only effect which is completely determined by the cause.

Now, if we suppose the mutual independence of forces, the following easily results. A body A, consisting of m bodies a, is in the presence of another body B, consisting of m' bodies a. Let the acceleration of A be ϕ and that of B be ϕ'. Then we have $\phi:\phi'=m':m$.

If we say that a body A has the mass m if it contains the body a m times, this means that the accelerations vary as the masses.

To find by experiment the mass-ratio of two bodies, let us allow them to act on one another, and we get, when we pay attention to the sign of the acceleration, $\frac{m}{m'}=-\left(\frac{\phi'}{\phi}\right)$.

If the one body is taken as unit of mass, the calculation gives the mass of the other body. Now, nothing prevents us from applying this definition in cases in which two bodies of different matter act on one another. Only, we cannot know *a priori* whether we do not obtain other values for a mass when we consult other bodies used for purposes of comparison and other forces. When it was found that A and B combine chemically in the ratio $a:b$ of their weights and that A and C do so in the ratio $a:c$ of their weights, it could not be known beforehand that B and C combine in the ratio $b:c$. Only experience can teach us that two bodies which behave to a third as equal masses will also behave to one another as equal masses.

If a piece of gold is opposed to a piece of lead, the law of sufficient reason leaves us completely. We are not even justified in expecting contrary motions: both bodies might accelerate in the same direction. The calculation would then lead to negative masses.

But that two bodies, which behave as equal masses to a third, behave as such to one another with respect to any forces, is very likely, because the contrary would not be reconcilable with the law of the conservation of work (*Kraft*), which has hitherto been found to be valid.

Imagine three bodies A, B, and C movable on an absolutely smooth and absolutely fixed ring. The bodies are to act on one another with any forces. Further, both A and B, on the one hand, and A and C, on the other, are to behave to one another as equal masses. Then the same must hold between B and C. If, for example, C behaved to B as a greater mass to a lesser one, and we gave B a velocity in the direction of the arrow, it would give this velocity wholly to A by impact, and A would give it wholly to C. Then C would communicate to B a greater velocity and yet keep some itself. With every revolution in the direction of the arrow, then, the *vis viva* in the ring would increase; and the contrary would take place if the original motion were in a direction opposite to that of the arrow. But this would be in glaring contradiction with the facts hitherto known.

Fig. 8.

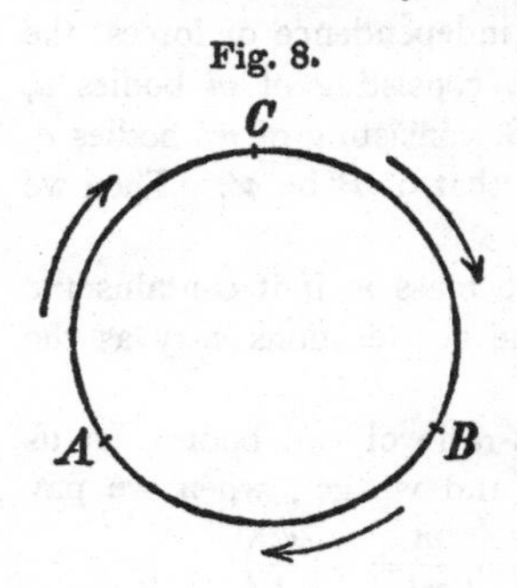

If we have thus defined mass, nothing prevents us from keeping the old definition of force as product of mass and acceleration. The law of Newton mentioned above then becomes a mere identity.

Since all bodies receive from the earth an equal acceleration, we have in this force (their weight) a convenient measure of their masses, again, however, only under the two suppositions that bodies which behave as equal masses to the earth do so to one another, and with respect to every force. Consequently, the following arrangement of the theorems of mechanics would appear to me to be the most scientific.

Theorem of experience.—Bodies placed opposite to one another communicate to each other accelerations in opposite senses in the direction of their line of junction. The law of inertia is included in this.

Definition.—Bodies which communicate to each other equal and opposite accelerations are said to be of equal mass. We get the mass-value of a body by dividing the acceleration which it gives the body with which we compare others, and choose as the unit, by the acceleration which it gets itself

Theorem of experience.—The mass-values remain unaltered when they are determined with reference to other forces and to another body of comparison which behaves to the first one as an equal mass.

Theorem of experience.—The accelerations which many masses communicate to one another are mutually independent. The theorem of the parallelogram of forces is included in this.

Definition.—Force is the product of the mass-value of a body into the acceleration communicated to that body.

PRAGUE
November 15, 1867

3. (See p. 47.) The note in question appeared in the number for February, 1871, of the Prague journal, *Lotos*, but was, however, drawn up a year earlier. This is a complete reproduction of it:

The second law of thermodynamics can, as is well known, be expressed for a simple case by the equation

$$-\frac{Q}{T}+Q'\left(\frac{1}{T'}-\frac{1}{T}\right)=0,$$

where Q denotes the quantity of heat transformed into work, at the absolute temperature T, and Q' the quantity of heat which simultaneously sunk from the higher temperature T to the temperature T'.

Now, we have not far to seek for the observation that this theorem is not limited to the phenomena of heat, but can be transferred to other natural phenomena, if, instead of the quantity of heat, we put the potential of whatever is active in the phenomenon, and, instead of the absolute temperature, the potential function. Then the theorem may be expressed thus:

If a certain potential-value P of an agent at the potential-level V passes over into another form—for example, if the potential of an electrical discharge is transferred into heat—then another potential-value, P', of the same agent sinks simultaneously from the higher potential-level V to the lower one V'. And the said values are connected with one another by the equation

$$-\frac{P}{V}+P'\left(\frac{1}{V'}-\frac{1}{V}\right)=0.$$

In the application of the theorem, the only questions are, what is to be conceived as potential (as equivalent of mechanical work), and what is the potential-function. In many cases this is self-evident and long established, in others it can easily be found. If, for example, we wish to apply the theorem to the impact of inert masses, obviously the *vis viva* of these masses is to be conceived as the potential, and their velocity as the potential-function. Masses of equal velocity can communicate no *vis viva* to one another—they are at the same potential-level.

I must reserve for another occasion the development of these theorems.

PRAGUE
February 16, 1870

4. (See p. 53.) The manner in which I was led to the view that we need not necessarily represent to ourselves molecular-processes spatially, at least not in a space of three dimensions, was as follows:

In the year 1862, I drew up a compendium of physics for medical men, in which, because I strove after a certain philosophical satisfaction, I carried out rigorously the mechanical atomic theory. This work first made me conscious of the insufficiency of this theory, and this was clearly expressed in the preface and at the end of the book, where I spoke of a total reformation of our views on the foundations of physics.

I was busied, at the same time, with psychophysics and with Herbart's works, and so I became convinced that the intuition of space is bound up with the organization of the senses, and, consequently, that we are not justified in ascribing spatial properties to things which are not perceived by the senses. In my lectures on psychophysics,[3] I already stated clearly that we are not justified in thinking of atoms spatially. Also, in my theory of the organ of hearing,[4] I brought before my readers the series of tones as an analogue of space of one dimension. At the same time the quite arbitrary and, on this account, faulty limitation of the number of dimensions in Herbart's derivation of "intelligible" space struck me. By that, now, it became clear to me that, for the understanding, relations like those of space, and of any number of dimensions, are thinkable.

My attempts to explain mechanically the spectra of the chemical elements and the divergence of the theory with experience strengthened my view that we must not represent to ourselves the chemical elements in a space of three dimensions. I did not venture, however, to speak of this candidly before orthodox physicists. My notices in Schlömilch's *Zeitschrift* of 1863 and 1864 contained only an indication of it.

All the views on space and time developed in this pamphlet were first communicated in my course of lectures on mechanics in the summer of 1864 and in my course on psychophysics delivered in the winter of 1864–1865, which latter course was attended by large audiences, and also by many professors of the University of Graz. The most important and most general results of these considerations were published by me in the form of short notes in Fichte's *Zeitschrift für Philosophie* of

3 *Oesterr. Zeitschr. für praktische Heilkunde*, 1863.

4 *Sitzber. der Wiener Akademie*, 1863.

1865 and 1866. In this, external stimuli were entirely lacking, for Riemann's paper, which first appeared in 1867,[5] was quite unknown to me.

5. (See p. 55.) The view that in science we are chiefly concerned with the convenience and saving of thought, I have maintained since the beginning of my work as a teacher. Physics, with its formulae and potential-function, is especially suited to put this clearly before me. The moment of inertia, the central ellipsoid, and so on, are simply examples of substitutes by means of which we conveniently save ourselves the consideration of the single mass-points. I also found this view developed with especial clearness in the case of my friend the political economist E. Herrmann. From him I have taken what seems to me a very suitable expression: "Science has a problem of economy or thrift."

6. (See p. 61) From my essay on the development of presentations of space in Fichte's *Zeitschrift* for 1866,[6] I permit myself to extract the following passage:

> Now, I think that we can go still farther in the scale of presentations of space and thus attain to presentations whose totality I will call *physical space.*
>
> It cannot be my intention here to criticize our conceptions of matter, whose insufficiency is, indeed, generally felt. I will merely make my thoughts clear. Let us imagine, then, a something behind (*unter*) matter in which different states can occur;

[5] [Riemann's work *Ueber die Hypothesen, welche der Geometrie zu Grunde liegen* was written and read to a small circle in 1854, first published posthumously in 1867, and reprinted in his *Ges. Werke*, pp. 255–268.]

[6] "Ueber die Entwicklung der Raumvorstellungen," *Zeitschr. für Philosophie und philosophische Kritik*, 1866.

say, for simplicity, a pressure in it, which can become greater or smaller.

Physics has long been busied in expressing the mutual action, the mutual attraction (opposite accelerations, opposite pressures) of two material particles as a function of their distance from each other—therefore of a spatial relation. Forces are functions of the distance. But now, the spatial relations of material particles can, indeed, only be recognized by the forces which they exert on one another.

Physics, then, does not strive, in the first place, after the discovery of the fundamental relations of the various pieces of matter, but after the derivation of relations from other, already given, ones. Now, it seems to me that the fundamental law of force in nature need not contain the spatial relations of the pieces of matter, but must only state a dependence between the states of the pieces of matter.

If the positions in space of the material parts of the whole universe and their forces as functions of these positions were once known, mechanics could give their motions completely,[7] that is to say, it could make all the positions discoverable at any time, or put down all positions as functions of time.

But, what does time mean when we consider the universe? This or that "is a function of time" means that it depends on the position of the vibrating pendulum, on the position of the rotating earth, and so on. Thus, "All positions are functions of time" means, for the universe, that all positions depend upon one another.

But since the positions in space of the material parts can be recognized only by their states, we can also say that all the states of the material parts *depend upon one another*.

The physical space which I have in mind—and which, at the same time, contains time in itself—is thus nothing other than *dependence of phenomena on one another*. A complete physics, which would know this fundamental dependence, would have

7 [For this purpose, it would be necessary also to know the *velocities* of the various parts at that instant.—Tr.]

no more need of special considerations of space and time, for these latter considerations would already be included in the former knowledge.

My researches on the time-sense of the ear[8] contain the following passage:

Physics sets out to represent every phenomenon as a function of time. The motion of a pendulum serves as the measure of time. Thus, physics really expresses every phenomenon as a function of the length of the pendulum. We may remark that this also happens when forces, say, are represented as functions of the distance; for the conception of force (acceleration) already contains that of time. If one were to succeed in expressing every phenomenon—physical and psychical—as a function of the phenomenon of pendulum motion, this would only prove that all phenomena are so connected that any one of them can be represented as a function of any other. Physically, then, time is the representability of any phenomenon as a function of any other one.

This view of time, now, also plays a part in my discussion of the law of inertia. To this view, too, Neumann, in his discussion of the law of inertia, seems to incline.

7. (See p. 63.) Fechner believed that he could reconcile the law of causality with the freedom of the will, in the following manner:

It is at once evident that our law, in spite of the fact that it would be binding for all space and all time, for all matter and all spirit, yet, in its essence, leaves behind an indetermination—indeed, the greatest that can be imagined. For it says, to be sure, that, if the same circumstances occur again, the same consequence must occur again, and if not, not; but there is nothing in its expression to determine in any way the manner

[8] "Ueber den Zeitsinn des Ohres," *Sitzb. der Wien. Akad.*, 1865

of the first consequence at any place and with any circumstances, nor the manner of the occurrence of the first circumstances themselves.

Farther on, Fechner remarked that the same circumstances never occur again, nor, therefore, ever exactly the same consequences.

As regards the first point, the indefiniteness is put back to the moment of creation, but the second seems to me to be merely an indeterministic subterfuge.

The indefiniteness to which I have drawn attention is essentially different; it is always present and results immediately from the law of causality by the elimination of space and time.

GENERAL REMARKS

We learn very soon to distinguish our presentations from our sensations (perceptions). Now, the problem of science can be split into three parts:

1. The determination of the connexion of presentations. This is psychology.

2. The discovery of the laws of the connexion of sensations (perceptions). This is physics.

3. The clear establishment of the laws of the connexion of sensations and presentations. This is psychophysics.

If we think of the laws of connexion as mathematical, the establishment of those laws presupposes the measurability of all that they embrace. In that there still remains, to be sure, much to be desired. Fechner, in his *Psychophysik*, succeeded in measuring even the single sensations, but it is possible to be in doubt as to the meaning of this measure. A sensation of greater intensity is always also of another quality, and then Fechner's measure is more physical than psychical. However, these difficulties turn out to be not insurmountable.

AUTHOR'S NOTES TO THE SECOND EDITION (1909)

To p. 19.—The confusion caused by the use of the expression "force" in a different signification is also shown in a communication of Faraday's of 1857 (*Phil. Mag.*, Ser. 4, Vol. XIII, p. 225 [in a paper "On the Conservation of Force"; also *Proc. Roy. Inst.*, February 27, 1857]). The same fault was committed by many of the most eminent investigators of that time. [Cf. also *Wärmelehre*, p. 206; and, on the history of the use of such terms as "work" and "energy," cf. A. Voss, *Encykl. der math. Wiss.*, IV, 1, 1901, pp. 102–104; and Mach, *Mechanics*, p. 499, note.]

To pp. 28 and 75.—The question of the law of inertia was treated at length in my *Mechanics* [pp. 140–141, 142–143, 523–525, 542–547, 560–574], where all the literature of the subject is noticed. The last important work that is known to me is J. Petzoldt's article, "Die Gebiete der absoluten und relativen Bewegung" (Ostwald's *Annalen der Naturphilosophie*, VII, p. 29).

To pp. 29 and 80—Further developments in my *Mechanics* [pp. 194–197, 198–222, 243, 536–537, 539–540, 555–560].

To pp. 35–37, 47, 85–86.—The publications which contain analogous considerations—partly coincident, partly allied—are my *Mechanics;* Josef Popper, *Die physikalischen Grundsätze der elektrischen Kraftüber-*

tragung, Wien, Pest, Leipzig, 1884; Helm, *Die Lehre von der Energie*, Leipzig, 1887; Wronsky, *Das Intensitätsgesetz*, Frankfurt a: O., 1888; Mach, "Geschichte und Kritik des Carnot'schen Wärmegesetzes" (*Sitzb. der Wien. Akad.*, 1892); and *Wärmelehre*. As regards pp. 85–86 in particular, such considerations were made mention of, first after Carnot, by Zeuner, *Grundzüge der mechanischen Wärmetheorie*, Leipzig, 2. Aufl., 1866. In the text of p. 86 the double resolution $MV^2/2$, $MV \cdot V/2$ is held to be possible, and, on this account, I have retained the general expression velocity instead of the square of the velocity as, still later, Ostwald did (*Berichte der kgl. sächs. Gesellschaft zu Leipzig*, Bd. XLIV, 1892, pp. 217–218). But I soon recognized that the potential-level is a scalar $V^2/2$ and cannot be a vector V or $V/2$. I did not speak of this further, since Popper had given a sufficient exposition of the correspondence between masses and quantities. This was also done by Friedrich Wolfgang Adler, "Bemerkungen über die Metaphysik in der Ostwald'schen Energetik" (*Vierteljahrsschr. für wiss. Philosophie und Soziologie*, Jahrg. 29, 1905, pp. 287–333).

To pp. 51–53.—Spaces of many dimensions seem to me not so essential for physics. I would only uphold them if things of thought like atoms are maintained to be indispensable, and if, then, also the freedom of working hypotheses is upheld.

To pp. 55 and 88.—The principle of the economy of thought is developed in detail in my later writings.

To p. 57.—I have repeatedly expressed the thought that the foundation of physics may be thermal or electric, in my *Mechanics* and *Analysis of the Sensations.* This thought seems to be becoming an actuality.

To pp. 60–64 and 88–90.—Space and time are not here conceived as independent entities, but as forms of the dependence of the phenomena on one another. I subscribe, then, to the principle of relativity, which is also firmly upheld in my *Mechanics* and *Wärmelehre.* Cf. "Zeit und Raum physikalisch betrachtet," in *Erkenntnis und Irrtum*, Leipzig, 1905 [(2d ed., 1906), pp. 434–448]; H. Minkowski, *Raum und Zeit*, Leipzig, 1909.

To p. 91.—The general remarks indicate the sensationalistic standpoint which I attained by studies in the physiology of the senses. Further developments in my *Bewegungsempfindungen* of 1875, *Analysis of the Sensations*, and *Erkenntnis und Irrtum.* I have also clearly shown there that the nervous, subjectivistic apprehensions which many physicists have for the physics of the inhabitants of Mars are quite groundless.

TRANSLATOR'S NOTES

To p. 15.—On the influence which Kant's *Prolegomena* exerted on Mach when a boy of fifteen, see note on p. 23 of *Analysis of the Sensations*, 1897.

To p. 17.—The investigators referred to on this page are not Kirchhoff and Helmholtz, whose works appeared at a later date (cf. *Mechanics*, p. x). Yet Kirchhoff is still regarded by many as the pioneer of descriptive physics. Cf. Mach's lecture "On the Principle of Comparison in Physics" in *Popular Scientific Lectures* (1898), pp. 236–258.

To p. 21.—On Stevinus's work, see, further, *Mechanics*, pp. 24–35, 49–51, 88–90, 500–501, 515–517; on Galileo's discussions of the laws of falling bodies, *ibid.*, pp. 128–155, 162–163, 247–250, 520–527, 563–567, and Ostwald's *Klassiker der exakten Wissenschaften*, Nr. 24, pp. 18–20, 57–59; on Huygens's researches on the centre of oscillation, *Mechanics*, pp. 173–186; on d'Alembert's principle, *ibid.*, pp. 331–343; on the principle of *vis viva, ibid.*, pp. 343–350; on Torricelli's theorem, *ibid.*, pp. 402–403; and, on the principle of virtual velocities, *ibid.*, pp. 49–77; A. Voss in his article, "Die Prinzipien der rationellen Mechanik," *Encykl. der math. Wiss.*, IV, 1 (1901), pp. 66–76; and, for a historical and critical review of the various proofs of the principle, R. Lindt, "Das Prinzip der virtuellen Geschwindigkeiten," *Abhdl. zur Gesch. der Math.*, Bd. XVIII, 1904, pp. 147–196.[9]

To p. 34.—See the reprint of Gauss's paper in Ostwald's *Klassiker*, Nr. 167; especially p. 28.

Gauss's principle is discussed in Mach's *Mechanics*, pp. 350–364; Voss's above article in the *Encykl. der math. Wiss.*, pp. 84–87; and in the notes (by myself) to Nr. 167 of Ostwald's *Klassiker*, pp. 46–48, 59–68.

To p. 35.—From the *Wärmelehre:* On Carnot's principle and its developments, pp. 211–237; on the principle of Mayer and Joule, pp. 238–268; and on the uniting of the principles, by W. Thomson and Clausius, in particular, pp. 269–301.

An account of the development, meaning, and so on, of the principle of energy, which is, in essentials, the same as that in the *Popular Scientific Lectures* (3d ed., Chicago, 1898, pp. 137–185), is given in *Wärmelehre*, pp. 315–346. Cf. also the end of the note to pp. 51, 94, below.

To pp. 51, 94.—On many dimensional spaces as mathematical helps, cf. *Mechanics*, pp. 493–494.

In H. Weber's edition of Riemann's *Partielle Differential-Gleichungen,*[10] use was made of the idea of a particle in a space of n dimensions to represent what Hertz called "the position of a system" in ordinary

[9] Also separately as an Inaugural Dissertation. Cf. E. Lampe, *Jahrb über die Fortschr. der Math.*, 1904, pp. 691–692.

[10] *Die partiellen Differential-Gleichungen der mathematischen Physik. Nach Riemann's Vorlesungen in vierter Auflage neu bearbeitet von Heinrich Weber.* Two vols., Braunschweig, 1900–1901. The passage referred to occurs in the second part of the first volume.

space; the "position of a system" being the totality of the positions of the points of the system.

To p. 56.—On impact and other theories of gravitation, see J. B. Stallo, *The Concepts and Theories of Modern Physics*, 4th ed., London, 1900, pp. 52–65, v–vi, vii, xxi–xxiv (the three last references are to the "Preface to the Second Edition," which is not contained in the German translation by Hans Kleinpeter, published at Leipzig in 1901 under the title: *Die Begriffe und Theorieen der modernen Physik*, although this translation was made from the third English edition. This is the more regrettable as the preface referred to contains some indications of great value of Stallo's view—which closely resembled that of Mach—of the various forms of the law of causality; cf. below).

To pp. 60, 69, 73.—Clerk Maxwell's (*Matter and Motion*, London, edition of 1908, pp. 20–21) "General Maxim of Physical Science" is similar to Fechner's law of causality. It runs: "The difference between one event and another does not depend on the mere difference of the times or the places at which they occur, but only on differences in the nature, configuration, or motion of the bodies concerned."

The question as to the meaning of "causality" in dynamics is discussed in Bertrand Russell's work on *The Principles of Mathematics*, Vol. I, Cambridge, 1903, pp. 474–481.[11] On p. 478 is the sentence: "Causality, generally, is the principle in virtue of which, from a sufficient number of events at a sufficient number of

[11] Newton's laws of motion are discussed on pp. 482–488.

moments, one or more events at one or more new moments can be inferred."

The various forms of the law of causality were briefly described by J. B. Stallo, *op. cit.*, pp. xxxvi–xli[12]—a discussion not, unfortunately, translated in the German edition.

The present writer ("On Some Points in the Foundation of Mathematical Physics," *Monist*, Vol. XVIII, pp. 217–226, April, 1908) has attempted to formulate Mach's principle of causality and some other principles of physics in the exact mathematical manner to which we have become accustomed by the modern theory of aggregates, and to suggest some new problems in this order of inquiries.[13] It is my belief that this investigation is the only way in which we can become sure that the image of reality at which we aim, by successive approximations, is logically permissible; and also that only in this way can we succeed in formulating exactly the epistemological questions at the basis of physical science, and in answering them.[14] I will here give two illustrations of this.

The postulate as to the "intelligibility of nature," or the existence of a "process of reason in nature" may, it seems to me,[15] be further explained as follows. In our

[12] Cf. Stallo, *op. cit.*, pp. 25–26.

[13] Some of the conceptions and results applied here are contained in my article "On the General Theory of Functions," *Journ. für Math.*, Bd. CXXVIII, 1905, pp. 169–210.

[14] Cf. my article on "The Relevance of Mathematics," *Nature*, May 27, 1909, Vol. LXXX, pp. 282–384.

[15] However, Mr. Russell, who is probably right, tells me that, in his opinion, philosophers mean by this postulate "something much more general and vague."

scientific descriptions, we express elements (in Mach's sense; see the next note, to p. 61) as functions of other elements, determine by observation the character of these functions—whether they are, or may conveniently be considered, continuous, analytic, or so forth—and then deduce purely logically the image of the course of events, that is provided by this mathematical thought-model of nature. Thus, if a function of time, $f(t)$, is analytic, and we know its values for any small period $t_0 \ldots t_1$, we can deduce, in a purely logical fashion, by means of Taylor's theorem, its value for any other value of t whatever. We could not do this if that aspect of nature with which we deal here were not susceptible of this *mimicry by logic*, so to speak; and this is what we mean when we speak of the existence of science implying a conformity of nature to our reason.

In the second place, I will attempt an explanation of the attribute "uniformity" of nature. The difficulty lies in discovering the value of the maxim that like events result from the recurrence of like conditions, if like conditions never *do* recur. The solution seems to me to be as follows: Like conditions probably never do recur in the world around us, but we have learned by experience that we can imitate very closely the course of nature (in certain particulars) by means of a purely mathematical construction or *model*. In *this* model we can, of course, reproduce exactly similar circumstances as often as we wish. The above law applies literally to our model; and that the so-conditioned events in the model approximately coincide with the observed events of nature is, I take it, what we mean when we say that nature is uniform. The point at issue here is quite similar to that discussed, *à propos* of Newton's rotating bucket, by Mach and Ward on the one side and Russell on the other (see my article, quoted above, in the *Monist*, p. 221).

Further references as to the meaning of causality in the light of modern theory of knowledge, and to the views of Mach, Stallo, and others, are as follows:

On the history of Mach's views on mass and on the substitution of the concept of function for that of causation, see *Mechanics*, pp. 555–556. The result of Mach's views which is of the greatest philosophical importance seems to be his disclosure of the character of the mechanical theory of nature (cf. the above translation, and *Mechanics*, pp. 495–501). This theory has been discussed at length and refuted—in many points after Mach's ideas—by James Ward, in the first volume of his *Naturalism and Agnosticism* (2d ed., London, 1903, 2 vols.).

Stallo (*op. cit.*, pp. 68–83) gave a sketch of the evolution of the doctrine of the conservation of energy and expressed views related to those of Mach. Thus, he said (*ibid.*, pp. 68–69): "In a general sense, this doctrine is coeval with the dawn of human intelligence. It is nothing more than an application of the simple principle that nothing can come from or to nothing"; and, in the preface to the second edition, he said (*ibid.*, pp. xl–xli): "But physicists, and especially mathematicians, are puzzled by the circumstance that not only has the law of causality always been applied before any experiential induction was thought of. . . ."

A few remarks by Poincaré on the principle of the conservation of energy on pp. 153–154 and 158–159 of his book *La science et l'hypothèse* (Paris, 6th ed.) are of an epistemological nature.

Cf. also Hans Kleinpeter, "Ueber Ernst Mach's und Heinrich Hertz' principelle Auffassung der Physik," *Archiv für systematische Philos.*, V, 1899, Heft 2; and "J. B. Stallo als Erkenntnisskritiker," *Vierteljahrsschr. für wiss. Philos.*, XXV, 1901, Heft 3.

A short exposition of the view[16] of the "symbolical physicists"—that our thoughts stand to things in the same relation as models to the objects they represent—is given by Ludwig Boltzmann in his article "Models" in the new volumes of the *Encyclopaedia Britannica* (Vol. XXX, 1902, pp. 788–791).

To p. 61.—Mach, in the memoir translated above, used *Erscheinungen* (phenomena) for what he afterwards (*Contributions to the Analysis of the Sensations*, Chicago, 1897, pp. 5, 11, 18) called by the less metaphysical name of "elements," thereby avoiding a verbal trap into which so many philosophers have fallen (see my article, referred to above, in the *Monist*, pp. 218–219, n. 6).

To p. 64.—The principle of the unique determination of natural events by others has been developed by Joseph Petzoldt, starting from Mach's considerations of 1872. Petzoldt's first work was entitled *Maxima, Minima und Ökonomie*, was printed in the *Vierteljahrsschr. für wiss. Philos.*, XIV., 1890, pp. 206–239, 354–366, 417–442, and was also printed separately as a

[16] This view I call the *typonoetic* theory

dissertation (Altenburg, 1891). On p. 12 of the reprint, Petzoldt states that the principles of Euler, Hamilton, and Gauss[17] are merely analytical expressions for the fact of experience that natural events are *uniquely* determined: the essential point is not the minimum but this uniqueness (*Einzigartigkeit*). Petzoldt's view that the thorough determinateness of all occurrences is a presupposition of all science was set forth in his paper: "Das Gesetz der Eindeutigkeit," *Vierteljahrsschr. für wiss. Philos.*, Vol. XIX, 1895, pp. 146–203.

Cf. also Mach's references to Petzoldt in *Mechanics*, pp. 552, 558, 562–563, 571–572, 575–577, 580–581; cf. pp. 10, 502–504, and *Wärmelehre*, pp. 324–327, for Mach's use of the principle of uniqueness, and a note farther on for further details about the principle of economy.

Petzoldt's views of the thoroughgoing uniqueness (*eindeutige Bestimmtheit*) of events were explained in his *Einführung in die Philosophie der reinen Erfahrung*[18]

[17] For German translations of some of the chief memoirs on these principles, very full of historical notes and modern references (by the present writer), see Ostwald's *Klassiker*, Nr 167

[18] Erster Band: *Die Bestimmtheit der Seele*, Leipzig, 1900. A critical notice of this volume was given by W. R. Boyce Gibson in *Mind*, N.S., IX, No. 35 (July, 1900), pp. 389–401. The sentences following, in the text, are quoted from this review, pp. 391–392.

Petzoldt maintained: (1) that the facts upon which the time-worn principle of causation is founded do not justify us in admitting more or less than the unideterminateness of all that happens; (2) that the psychical states being non-unideterminable by each other, the attempt to make them explain one another is scientifically unthinkable; (3) that the only way out of the difficulty is to accept the doctrine of psycho-physical parallelism in the sense of Avenarius. In the sequences of the mental life, there is neither continuity, singleness of direction, nor uniqueness.

—in the first part an interpretation of the philosophy of Avenarius:

Whenever there are a number of possible ways in which, say, the movement of a body would be directed, Petzoldt showed, by a number of examples, that that path is selected, as a matter of fact, which possesses the following three elements of unideterminateness: (1) singleness of direction, (2) uniqueness, (3) continuity; for in satisfying these three conditions all indeterminateness is taken from its changes. The meaning of the first determining element is simply this, that as a matter of fact there is no *actual* ambiguity as to the sense in which any change takes place. Warm bodies left to themselves always grow cooler; heavy bodies left to themselves always fall downwards, not upwards. A first conceivable ambiguity is thus put to rest by Nature herself. In the second place Nature takes care that bodies shall move in such a way relatively to their *Bestimmungsmittel* or media of determination that the actual direction of motion differentiates itself from all the others by its uniqueness. It is only this uniqueness that gives to the actual change its right to be actualized, its right to be chosen in preference to any other possible change. Thus a ball moving freely on a horizontal plane passes from *A* in a rectilinear direction to *B* and on to *C*. It might conceivably have passed from *B* to *D*, where *B D* is not collinear with *A B;* but though this course is a thinkable one it is not realized, because its realization would involve an ambiguity, for no reason could then be given why the direction of *B D* was chosen in preference to the symmetrical direction *B E*. The direction *B C* is in this case the only one that is unique and therefore unambiguous. The third element, that of continuity, secures the possibility of exact quantitative determination.

For every occurrence [says Petzoldt[19]] means of determination can be discovered whereby the occurrence is unambiguously determined, in this sense, that for every deviation from it, supposed to be brought about through the same means, at least one

[19] *Op. cit.*, p. 39.

other could be found which being determined in the same way would be its precise equivalent, and have as it were precisely the same right to be actualized.

By "means of determination" are meant just those means—e.g., masses, velocities, temperatures, distances—by the help of which we are able to grasp an occurrence as singled out by its uniqueness from a number of equally thinkable occurrences The unideterminateness of things is both a fact of Nature and the *a priori* logical condition of there being a cosmos at all instead of a chaos. Our thought demands it from Nature, and Nature invariably justifies the demand. In this one supreme fact of the unideterminateness of all things the mind finds its rest. It is an ultimate fact, and one can no longer ask Why? when one comes to ultimate facts.

To p 65.—On Archimedes's deduction of the law of the lever, and on the *uniqueness* of determination of equilibrium, see *Mechanics*, pp. 8–11, 13–14, 18–19.

To p. 66.—On the very similar methods employed by Galileo, Huygens, and Lagrange to demonstrate the law of equilibrium of the lever, see *Mechanics*, pp. 11–18.

To p. 76.—On Neumann's essay of 1870, cf. Mach's *Mechanics*, pp. 567–568, 572; Stallo, *op cit.*, pp. 196–200; Russell, *op. cit.*, pp. 490–491; the following note to p. 80; and C. Neumann, "Ueber die sogenannte absolute Bewegung" (*Festschrift, Ludwig Boltzmann gewidmet* . . . . , Leipzig, 1904, pp. 252–259).

To p. 80.—On relativity of position and motion, see Stallo, *op. cit.*, pp. 133–138, 183–206; Mach, *Mechanics* pp. 222–238, 542–547, 567–573, and *Mechanik* (5. Aufl., 1904), pp. 257–263; James Ward, *op. cit.*, Vol. I, pp. 70–80; Russell, *op. cit.*, pp. 489–493; and my article in the *Monist*, quoted above, p. 221.

Planck has determined the form of the fundamental equations of mechanics which must take the place of the ordinary Newtonian equations of motion of a free mass-point if the principle of relativity is to be generally valid, in his paper: "Das Prinzip der Relativität und die Grundgleichungen der Mechanik" (*Verh. der Deutschen Phys. Ges.*, Vol. VIII, 1906, pp. 136–141).

To p. 80.—As regards Mach's definition of mass, it is interesting to find that Barré de Saint-Venant, in the paper[20] in which he announced and applied his independent discovery of Hermann Grassmann's[21] "outer multiplication," expressly drew attention to the use of "geometrical quantities" in treating mechanics by only letting space and time combinations enter, and not speaking of "forces." In the definition he gives of mass as a constant for each body, so chosen as to satisfy his "second law of mechanics":

$$mF_{mm}+m'F_{m'm}=0,$$

he is exactly of the same view as Mach.

Cf. also H. Padé ("Barré de Saint-Venant et les

[20] "Mémoire sur les sommes et les différences géométriques, et sur leur usage pour simplifier la mécanique," *Compt. Rend.*, T. XXI, 1845, pp. 620–625. Cf. Hermann Hankel, *Vorlesungen über die complexen Zahlen und ihre Functionen* (I. Theil; "Theorie der complexen Zahlensysteme"), Leipzig, 1867, p. 140.

[21] *Ausdehnungslehre von 1844.* For some account of the use of the methods of Hamilton and Grassmann in questions of mechanics, see *Mechanics*, pp. 527–528, 577–579.

principes de la mécanique," *Rev. générale des sciences*, XV, 1904, pp. 761–767), who points out that, in various points, de Saint-Venant's views coincide with those of Boltzmann.

Mach's definition has been accepted by most modern writers of books on dynamics; for example, Gian Antonio Maggi, *Principii della teoria matematica del movimento dei corpi*, Milano, 1896, p. 150; A. E. H. Love, *Theoretical Mechanics*, Cambridge, 1897, p. 87; Ludwig Boltzmann, *Vorlesungen über die Principien der Mechanik*, I. Theil, Leipzig, 1897, p. 22, and Poincaré—who, however, makes no mention of Mach's name—*La science et l'hypothèse*, 6th thousand, Paris, p. 123.

On criticisms of Mach's definition of mass, see *Mechanics*, pp. 539–540, 558–560.

To pp. 85–86, 93–94.—This analogy between heat and work done by gravity is known as "Zeuner's analogy," after Zeuner's remark in the second edition (1866) of his *Grundzüge der mechanischen Wärmetheorie*. See Georg Helm, *Die Energetik nach ihrer geschichtlichen Entwickelung*, Leipzig, 1898, pp. 254–266.

On the subject of a "comparative physics"—that is to say, a concise expression of extensive groups of physical facts, which is based on the analogies observed between the conceptions in different branches of physics —see Mach, *Mechanics*, pp. 496–498, 583; *Wärmelehre*, pp. 117–119; and *Pop. Sci. Lect.* (1898), p. 250;[22] L.

[22] Cf. also Mach, "Die Ähnlichkeit und die Analogie als Leitmotiv der Forschung" (*Annalen der Naturphilosophie*, Bd I, and *Erkenntnis und Irrtum*, 1906, pp. 220–231).

Boltzmann, in notes to his translation of Maxwell's paper of 1855 and 1856 "On Faraday's Lines of Force" (Ostwald's *Klassiker*, Nr. 69, pp. 100–102); M. Pétrovitch, *La mécanique des phénomènes fondée sur les analogies*, Paris, 1906; and Helm, *op. cit.*, pp. 253–266, 322–366.

It seems to me that the methods of a comparative physics, especially when aided by a calculus so well adapted to dealing with physical conceptions as that of Grassmann, Hamilton, and others, would afford a powerful means of discovering the ultimate principles of physics. Cf. the paper by de Saint-Venant referred to in the preceding note; M. O'Brien's paper "On Symbolic Forms Derived from the Conception of the Translation of a Directed Magnitude," in *Phil. Trans.*, Vol. CXLII, 1851, pp. 161–206; papers by Grassmann on mechanics[23] in his *Ges. Werke*, Bd. II, 2. Teil; and Maxwell, "On the Mathematical Classification of Physical Quantities," *Scientific Papers*, Vol. II, pp. 257–266. Cf. also Maxwell, *A Treatise on Electricity and Magnetism*, Oxford, 1873, Vol. I, pp. 8–29 (on the application of Lagrange's dynamical equations to electrical phenomena, see Vol. II, pp. 184–194); W. K. Clifford, *Elements of Dynamic*, Part I, "Kinematic," London, 1878; Hankel, *op. cit.*, pp. 114, 118, 126, 129, 132, 133, 134, 135, 137, 140; and Grassmann's *Ausdehnungslehre von 1844, passim.*

In this connexion, we may also give the following references: On the principle of energy, cf. Voss *op.*

[23] Especially important is Grassmann's paper: "Die Mechanik und die Principien der Ausdehnungslehre," in *Math. Ann.*, XII 1877.

cit., pp. 104–107; on the Virial and the second law of thermodynamics, *ibid.*, pp. 107–109, and *Wärmelehre*, p. 364; on the localization of energy, Voss, *op. cit.*, pp. 109–115; on the treatment of mechanics by energetics, *ibid.*, 115–116, Mach, *Mechanics*, p. 585, and *Mechanik* (5. Aufl., 1904), pp. 405–406, Max Planck, *Das Prinzip der Erhaltung der Energie*, 2 Aufl., Leipzig and Berlin, pp. 166–213, Helm, *op. cit.*, pp. 205–252.

To p. 88.—A very slight indication of the principle of the economy of thought was, as Boltzmann[24] has remarked, contained in Maxwell's (1855) observation that, in order further to develop the theory of electricity, we must first of all simplify the results of earlier investigations and bring them into a form readily accessible to our understanding.

On the principle of the economy of thought in various branches of science, see Mach, *Mechanics*, pp. x–xi, 6, 481–494, 549, 579–583; *Wärmelehre*, pp. 391–395; *Pop. Sci. Lect.* (1898), pp. 186–213; A. N. Whitehead, *A Treatise on Universal Algebra*, Vol. I, Cambridge, 1898, p. 4; and my above-mentioned article in *Nature*, p. 383.

On Mach's formal principles of economy, simplicity, continuity, and analogy, see Voss, *op. cit.*, p. 20.

[24] Ostwald's *Klassiker*, Nr. 69, p. 100. The whole of the introduction to this paper of Maxwell's is of the greatest epistemological interest, as it states much more clearly than in any other of his writings what has been called the "symbolic" point of view in physics (see *ibid.*, pp. 3–9, 99–102.)

INDEX

Printed in the United States
By Bookmasters